TELECOMMUNICATION SYSTEMS

The Artech House Telecom Library

Telecommunications: An Interdisciplinary Text, Leonard Lewin, ed.
Telecommunications in the U.S.: Trends and Policies, Leonard Lewin, ed.
The ITU in a Changing World by George A. Codding, Jr. and Anthony M. Rutkowski
Integrated Services Digital Networks by Anthony M. Rutkowski
The ISDN Workshop: INTUG Proceedings, G. Russell Pipe, ed.
The Deregulation of International Telecommunications by Ronald S. Eward
The Competition for Markets in International Telecommunications by Ronald S. Eward
Machine Cryptography and Modern Cryptanalysis by Cipher A. Deavours and Louis Kruh
Teleconferencing Technology and Applications by Christine H. Olgren and Lorne A. Parker
The Executive Guide to Video Teleconferencing by Ronald J. Bohm and Lee B. Templeton
World-Traded Services: The Challenge for the Eighties by Raymond J. Krommenacker
Microcomputer Tools for Communication Engineering by S.T. Li, J.C. Logan, J.W. Rockway, and D.W.S. Tam
World Atlas of Satellites, Donald M. Jansky, ed.
Communication Satellites in the Geostationary Orbit by Donald M. Jansky and Michael C. Jeruchim
Communication Satellites: Power Politics in Space by Larry Martinez
New Directions in Satellite Communications: Challenges for North and South by Heather E. Hudson
Television Programming Across National Boundaries: The EBU and OIRT Experience by Ernest Eugster
Evaluating Telecommunication Technology in Medicine by David Conrath, Earl Dunn, and Christopher Higgins
Measurements in Optical Fibers and Devices: Theory and Experiments by G. Cancellieri and U. Ravaioli
Fiber Optics Communications, Henry F. Taylor, ed.
Mathematical Methods of Information Transmission by Kurt Arbenz and Jean-Claude Martin
Principles of Secure Communication Systems by Don J. Torrieri
Telecommunication Systems by Pierre-Gerard Fontolliet
Signal Theory and Processing by Frederic de Coulon
Techniques in Data Communications by Ralph Glasgal
Advances in Computer Communications and Networking, Wesley Chu, ed.
The Public Manager's Telephone Book by Paul Daubitz
The Cable Television Technical Handbook by Bobby Harrell
Proceedings, Conference on Advanced Research in VLSI, 1982, MIT

TELECOMMUNICATION SYSTEMS

Pierre-Girard Fontolliet

Copyright © 1986

ARTECH HOUSE, INC.
610 Washington Street
Dedham, MA 02026

All rights reserved. Printed and bound in United States of America. No part of this book may be reproduced or utilized in any form or by any means, electronic or mechanical, including photocopying, recording, or by any information storage and retrieval system, without permission in writing from the publisher.

International Standard Book Number: 0-89006-184-x
Library of Congress Catalog Card Number: 85-071870

Translation of *Systèmes de Télécommunications*, originally published in French as volume XVIII of the *Traité d'Électricité* by the Presses Polytechniques Romandes, Lausanne, Switzerland. © 1983.

Printed and manufactured in the United States by Bookcrafters, Inc., Chelsea, MI.

INTRODUCTION

"The engineer is a guy who knows what to leave out."
Prof. E. Juillard, 1886-1982

The ability to leave things out is as important a requirement in engineering as the ability to calculate. Knowing how to omit things is a difficult and subtle art which requires in-depth knowledge of nature and techniques, and at the same time sure judgement to evaluate the degree of approximation necessary, which can still be compatible with the prescribed goals. The engineer must know that simple models are false, but also that those which are not simple are unusable. Between a perfectionism which is fatal because of its excessive cost and an empiricism which is dangerous because of its unpredictable nature, an appropriate compromise must be found which satisfies quality requirements and economic imperatives. There are things which can be calculated, and there are others which must be estimated or perceived. The engineer must be capable of the two approaches and, further, have the judgement necessary to distinguish between the two.

Reality is always more complex than the theory which attempts to describe it. In effect, to understand this reality, we are obliged to simplify and schematize it. Reality sometimes has its revenge with malice or violence, reminding the imprudent engineer that he is only human. Nonetheless, it also often lends itself well to this game and gives the engineer the joy of having participated in a creation.

Purpose

Faced with the triple reality of human needs in communication, the imperfect engineering means available to satisfy them, and economic constraints imposed on the cost of these means, telecommunication represents the typical example of a complex technical system, involving theoretical and practical aspects, continually confronted with the hard necessity of a compromise between cost and quality. The interface with human beings predominates, and has definite societal implications. The high degree of quality and reliability expected from telecommunications services imposes strict requirements for planning, operation, and maintenance of systems and networks. The worldwide scope of telecommunications networks requires very strict coordination on the international level.

In this spirit, this book attempts to take a global and macroscopic approach [1] to telecommunications systems and networks by showing the complexity and overlapping of problems from the start. It represents a synthesis of principles, theories, and methods developed in other volumes of the *Traité d'Électricité*, while orienting them toward specific applications.

To this end, the purpose of this book is essentially tutorial. Its goal is to give future electrical engineers some idea of the diversity of problems which a large technical system entails and the richness of the solutions which have been proposed.

Presenting this panorama of telecommunications in one volume inevitably involves choices, short cuts, and distortions. The principles and methods covered should serve as the basis for specific development, of which only a few examples are given in the second half of the book, starting with chapter 9. Systems are divided into functional groups, without going into the details of their technical implementation. However, problems of planning and operation are discussed each time.

Outline of the book

The study of systems as complex as those used in telecommunications cannot be approached by a linear sequence of successive chapters. It has been necessary to adopt a *spiral approach*, in which the subject is clarified little by little by a cumulative integration in the course of a progression, which is necessarily somewhat arbitrary and in which frequent leaps forward or backward are necessary. Figure I-1 below explains the structure of the volume and suggests ways of reading it.

It is noted that digital systems are deliberately presented before analog systems. This order corresponds, on the one hand, to the historical evolution of telecommunications whereby the telegraph preceded the telephone and, on the other hand, to present developments which are definitely oriented toward digital forms of communication. However, knowledge and mastery of digital communication is ultimately achieved only through an in-depth understanding of analog phenomena, because, in fact, there are no digital signals (despite the misuse of language which names them so!) but only analog ones.

Teaching objectives

Designed for a two-semester course, this volume is only a skeleton which needs to be animated by case studies and exercises based on concrete exam-

INTRODUCTION

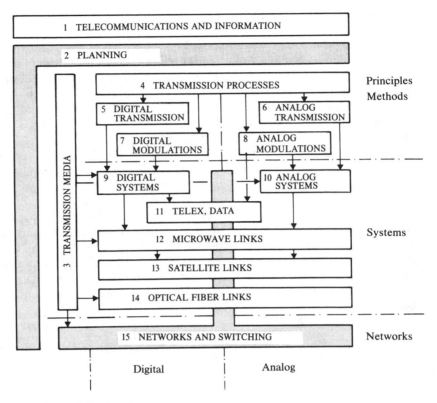

Fig. I-1 Outline of the book.

ples. With this indispensable complement, we can hope to accomplish the following objectives:

To be capable of

- qualitatively and quantitatively defining the problem of information transmission in technical and human contexts;
- evaluating and comparing modulation techniques and real transmission media in order to make intelligent choices;
- applying general theories with common sense to particular concrete cases of transmission;
- planning and designing roughly a digital telecommunications system (by means of its error probability) or an analog system (by means of the noise budget);
- evaluating and comparing complete systems with a view toward justifiable choices.

Conventions

The *Traité d'Électricité* is composed of volumes (vol.) indicated by a roman numeral (vol. V). Each volume is divided into chapters (chap.) indicated by by an arabic numeral (chap. 2). Each chapter is divided into sections (sect.) indicated by by two arabic numerals separated by a dot (sect. 2.3). Each section is divided into sub-sections indicated by two dots (sect. 2.3.11). The internal references stipulate the volume, chapter, and section of the work from the series which is being cited. See the Select Bibliography for the complete list, including English translations by Artech House. In the case of reference to part of this book, the number of this volume is, of course, omitted.

Bibliographic references are continuously numbered throughout this book and indicated by a single arabic numeral between brackets.

The equations outside the text are numbered continuously by chapter and indicated by two arabic numerals placed between parentheses and separated by a dot (3.14); a particularly important equation is referred to by its number printed in boldface. The figures and tables are numbered continuously by chapter and indicated by two arabic numerals preceded by Fig. (Fig. 4.12) or Table (Table 4.13). The figures and tables are numbered in a combined sequence.

A term appears in *italics* the first time it is defined in the text. An important passage is emphasized by being printed in ***bold italics***.

A list of acronyms and abbreviations and a list of graphic symbols can be found at the end of this book.

CONTENTS

INTRODUCTION v

CHAPTER 1 TELECOMMUNICATIONS AND INFORMATION ... 1

1.1 Objectives of telecommunications. 1
1.2 Telecommunications, man, and society 3
1.3 Concept of system 7
1.4 Evolution of telecommunications 8
1.5 Information 11
1.6 Communication 14
1.7 Characteristics of information sources and sinks and their implications for transmission systems 18

CHAPTER 2 PLANNING: OBJECTIVES, CONSTRAINTS, AND METHODS 27

2.1 Conception and design of a telecommunication system 27
2.2 National and international standards 32
2.3 Attenuation and level. 35
2.4 Transmission quality 42
2.5 Switching quality 56
2.6 Reliability. 60
2.7 Economic aspects. 71

CHAPTER 3 TRANSMISSION MEDIA. 77

3.1 Introduction. 77
3.2 General line characteristics in the frequency domain 78
3.3 General line characteristics in the time domain 88
3.4 Crosstalk 92
3.5 Balanced pair lines and cables. 98
3.6 Coaxial pair lines and cables 109
3.7 Optical fibers. 114
3.8 Radio waves 124
3.9 Antennas 130
3.10 Radio transmission. 133
3.11 Critical comparison of transmission media 136

CHAPTER 4 TRANSMISSION PROCESSES. 139

4.1 Limitations due to the channel 139
4.2 Principles of modulation 147

4.3	Sampling	150
4.4	Multiplexing	160
4.5	Mixed modulations	163
4.6	"Two-wire" and "four-wire" links	163

CHAPTER 5   DIGITAL TRANSMISSION ... 173

5.1	Characteristics of digital transmission	173
5.2	Regeneration	178
5.3	Effects of distortion: intersymbol interference	180
5.4	Effects of noise: regeneration errors	189
5.5	Planning digital transmission	195
5.6	Combined effects of interference and noise	201
5.7	Cumulative regeneration	203
5.8	Applications of digital transmission	206

CHAPTER 6   ANALOG TRANSMISSION ... 209

6.1	Characteristics of analog transmission	209
6.2	Amplification	211
6.3	Noise budget	216
6.4	Calculation of the signal-to-noise ratio	219
6.5	Applications of analog transmission	224
6.6	Companding	226

CHAPTER 7   DIGITAL MODULATION ... 233

7.1	Principles and types of digital modulation	233
7.2	Quantizing	235
7.3	Uniform quantizing	239
7.4	Non-uniform quantizing	245
7.5	Pulse code modulation	248
7.6	Measurement of quantizing noise	261
7.7	Effects of transmission errors in PCM	262
7.8	Differential digital modulations	265
7.9	Adaptive digital modulations	275
7.10	Comparison and applications of digital modulations	277

CHAPTER 8   ANALOG MODULATION ... 281

8.1	Sinusoidal carrier modulations	281
8.2	Amplitude modulation (AM)	282
8.3	Suppressed carrier amplitude modulation (DSBSC)	290
8.4	Single sideband modulation (SSB)	294
8.5	Vestigial sideband modulation (VSB)	300
8.6	Frequency modulation (FM)	302
8.7	Phase modulation (ΦM, PM)	319

8.8	Effect of noise	322
8.9	Summary and comparison of sinusoidal carrier modulations	332
8.10	Analog pulse modulations	337
8.11	Discrete analog modulations	352

CHAPTER 9 DIGITAL SYSTEMS 365

9.1	Principles and structure of digital systems	365
9.2	Structure of the digital time-division multiplex	367
9.3	Primary multiplex systems	370
9.4	Higher order systems	376
9.5	Transmultiplexers	383
9.6	Transmission equipment	384
9.7	Planning problems	392
9.8	Digital systems using balanced pairs	394
9.9	Digital systems using coaxial pairs	398
9.10	Operational problems	401

CHAPTER 10 ANALOG SYSTEMS 405

10.1	Carrier systems	405
10.2	Terminal equipment for carrier systems	410
10.3	Line equipment for carrier systems	414
10.4	Multichannel system level	417
10.5	Planning problems	421
10.6	Operational problems	425
10.7	Analog broadcasting systems	426

CHAPTER 11 TELEX AND DATA TRANSMISSION 435

11.1	Introduction	435
11.2	Codes	436
11.3	Telex	440
11.4	Data transmission	443

CHAPTER 12 MICROWAVE LINKS 453

12.1	Principles and structure of microwave links	453
12.2	Propagation conditions	455
12.3	Digital microwave links	459
12.4	Analog microwave links	462
12.5	Comparison between digital and analog microwave links	469

CHAPTER 13 SATELLITE LINKS 473

13.1	Principles and conditions of satellite links	473
13.2	Planning of a link	477
13.3	Equipment on the satellite	480
13.4	Earth stations	483

13.5 Multiple access	487
13.6 Perspectives	492

CHAPTER 14 OPTICAL FIBER LINKS 495

14.1 Principles and structure of optical fiber links	495
14.2 Optical transducers	496
14.3 Transmission modes using optical fibers	501
14.4 Planning of digital links using optical fibers	505
14.5 Transmission systems using optical fibers	512

CHAPTER 15 NETWORKS AND SWITCHING 515

15.1 Types and structure of telecommunications networks	515
15.2 Telephone network	519
15.3 Echoes and stability	527
15.4 Transmission plan	531
15.5 Switching principles	532
15.6 Data networks	537
15.7 Integrated digital network (IDN)	539
15.8 Integrated services digital network (ISDN)	543
15.9 Epilogue	549

CHAPTER 16 APPENDICES 553

16.1 Convention for the representation of spectra	553
16.2 Gaussian function	554
16.3 Absolute power and voltage levels	561
BIBLIOGRAPHY	562
SELECT BIBLIOGRAPHY	565
LIST OF PRINCIPAL GRAPHIC SYMBOLS	566
LIST OF ABBREVIATIONS AND ACRONYMS	569
GLOSSARY	572
INDEX	577

Chapter 1

Telecommunications and Information

1.1 OBJECTIVES OF TELECOMMUNICATIONS

1.1.1 Definition and commentary

Telecommunications (communication engineering), in the broad sense, comprises the whole collection of technical means necessary to convey information between any two points, at any distance, as faithfully and reliably as possible, at reasonable cost.

This overall definition raises the following comments:

- Telecommunications is a ***technology***, and therefore a human endeavor. The need for communication, inherent in all human beings, exists without it. Telecommunications does nothing but enhance and extend the possibilities of satisfying this need. The principal technical means used are of an electromagnetic type.
- In contrast to the postal services, telecommunications concerns only the ***information*** to be transmitted and not its physical support (paper, disk, magnetic tape, *et cetera*). This information can take many different forms: speech, music, still or moving pictures, text, data, *et cetera*.
- The user who trusts his information to telecommunication systems hopes that it will be restored without losses and without alterations. One of the principal problems to be solved is to guarantee a high degree of *fidelity*, i.e., transparency, despite the inevitable imperfections and disturbances that the available means present.
- The user expects a permanent service, available under any circumstances. Assuring this reliability despite unforeseeable and inevitable partial breakdowns is also a primary concern.
- The connection of any two users according to their commands in order to permit them to transmit information is a problem of ***switching***, an important branch of telecommunications.
- The transport of information over a certain distance, which may be very great (space communications) is a problem of *transmission*, another equally important branch of telecommunications.
- The art of the engineer consists of finding the most economical solution to a communication problem, in an eternal compromise between *cost* and *quality*. Cost must be taken in the broad sense: hardware, development

costs, operation, *et cetera*. Telecommunications is not exempt from the laws of the marketplace, and it is ultimately the user who judges whether the costs are "reasonable."

1.1.2 Types of services

The services offered by telecommunications are distinguished by:
- the type of information transmitted
- the number of partners involved
- the respective roles that these partners play (mode of communication, figure 1.1)

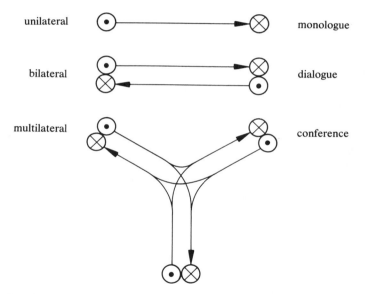

Fig. 1.1 Modes of communication: ⊙ source; ⊗ sink.

When a certain number of users benefit from the same service, all of the physical links between them constitute a *network*. The network may be used to *broadcast* information transmitted unilaterally from one source to several sinks, or, inversely, to *collect* information coming from many sources to the same sink.

If the links between partners are not permanent, but are set up from case to case according to their orders, the network is said to be *switched*. Therefore, in addition to transmission means, it comprises equipment capable of interpreting and executing these orders (exchanges, switches).

In its most common form, the switching network is *shared*, i.e., its transmission and switching devices are made available in common to a great number of users who have access to this network by an individual means of transmission (figure 1.2).

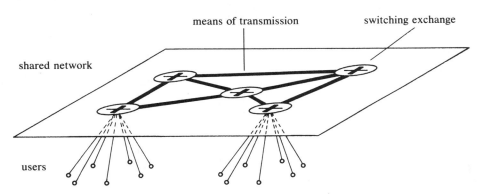

Fig. 1.2 Shared switched network.

Table 1.3 points out some services with respect to the above criteria.

1.2 TELECOMMUNICATIONS, MAN, AND SOCIETY

1.2.1 Concept of service

Telecommunications offers a *service* in the sense that its product is nonphysical and responds to an expressed or latent need. The goal is not primarily profit, but rather the satisfaction of the user. Aside from indispensable economic considerations to assure that the service will be profitable, or at least not operating at a deficit, and that it can be offered at a reasonable price, utility and quality of the service, social equity, and permanence of service are essential aspects which proceed more from an ideal than from cold calculations.

Three groups of partners are implied in the functioning of the service:
- the manufacturer who conceives, develops, produces, and sells the equipment necessary for the implementation of the service
- the manager who plans, specifies, and installs the network, then sees to its proper functioning and to good management of the service

Table 1.3 Examples of telecommunications services.

Type of information	Mode of communication			Network				Service
	unilateral	bilateral	multilateral	collection	fixed point-to-point	switched	shared	
Speech		×				×	×	TELEPHONY
			×			×	×	— teleconference
	×					×	×	— speaking clock
	×			×				RADIO BROADCASTING
			×		×			Intercom
Music	×			×				RADIO BROADCASTING
	×			×				Wire broadcasting
Text	×					×	(×)	(×) TELEGRAPHY
		×	(×)			×	×	TELEX, TELETEX / Electronic mail
Still pictures	×			(×)	(×)	×	×	Telecopying (facsimile)
		×				×	×	Videotex
Moving pictures	×			×	(×)			TELEVISION
		×				×	×	Visiophony
Data		×				×	(×)	TELEMATICS
	×				×			Telemetry
	×				×			Remote surveillance
	×					×	(×)	Remote control

- the user, who is the consumer of the service, a client who is demanding but unaware of technical difficulties

1.2.2 Man-machine interface

Telecommunications cannot ignore human beings, their physiological characteristics, and their psychological reactions. The properties of the voice and

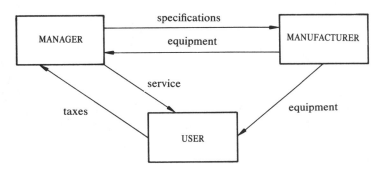

Fig. 1.4 The three partners in telecommunications.

hearing have a profound influence on the conception of a telephone system. The properties of sight dictate the parameters of television transmission. Switching systems must take into account the logical behavior of subscribers, their reaction time, and their unpredictable or disruptive actions.

Telecommunications interposes a machine between the users designed to extend and enhance their senses. The man-machine interface demands a mutual adaptation: the human being must learn a language to communicate with the machine (telegraphic code, dialing a telephone number at a telephone set, the significance of tones, the dialogue procedure in videotex, *et cetera*), and the machine must accept the human being as such (cope with manipulation errors, admit a great variation in sound level, take into account statistical fluctuations of human parameters, *et cetera*).

Also, for the manager, the man-machine interface determines the efficiency of maintenance, (surveillance, fault location, modifications, manual interventions) and of management (levying taxes, traffic monitoring, statistics, *et cetera*). The management personnel must be able to enter into dialogue with the machine, at a more or less high level according to the degree of automation of its functions.

Man intervenes in telecommunication systems in two ways:

- as a *source and receiver* of the transmitted information. In this role, he imposes his type of information (speech, writing, images, *et cetera*), his output, and his demands for quality. The dialogue takes place from person to person, via the machine;
- as a *command element* dictating the orders to the system, and receiving its reactions in return. The dialogue thus takes place between man and machine, according to an appropriate code whose technical level must be adapted to the capabilities of the users.

1.2.3 Public aspect of the service

Most of the telecommunications services involve a vast public which is interested in the service itself and not at all in the technical means used to establish it. This implies a responsibility to the users on the part of manufacturers and managers in response to the trust which the users place in them. In concrete terms, this responsibility is reflected in the following requirements, which are particularly relevant for a shared switched network:

- a sufficient quality of service must be guaranteed
- quantitative long-term predictions must allow for coverage of future needs
- a minimum revenue price must be sought
- for social equity, the service must be able to be offered to those who require it, without geographic or social discrimination, and taxed according to actual use
- the service must be permanent, due to reliability of its equipment
- the equipment made available to users must be simple to use and sturdy. It must withstand wrong manipulations and require only a minimum of previous training for use.
- the secrecy of private communications must be absolutely respected.

As a consequence of these requirements, in the majority of countries, the responsibility for public telecommunications services has been entrusted exclusively to the state in the form of a *monopoly*. This situation limits the freedom of the marketplace by eliminating all competition at the level of service, but, as long as the state takes this task seriously, it guarantees the interests of the users with regard to the quality of services offered.

1.2.4 Social impact

As with many human inventions, telecommunications can be used for good or ill. It is necessary to distinguish, however, between

- *mass* communications (such as radio and television, i.e., unilateral) which are intended to inform and entertain a large public
- *individual* communications, bilateral and switched, of which the telephone is the principal representative. These fill a role of contact and exchange, utility (particularly in business), security, and effectiveness.

While leaving aside the task of analyzing the effect that these two types of communication have had and could have on society, it is nevertheless important to bring up the following points:

- historically the *offering* of services in telecommunications has preceded the demand and the public has either accepted, modified, or refused the offer;
- at present, a *demand* (notably in telematics) appears more pressing and constitutes a challenge to the monopoly of the state;
- telecommunications associated with computing is an important factor in professional life, both public and private, at present and in the future. It can be oriented to influence in one sense or another the model of tomorrow's society (wired city, information society?);
- whatever is technically (and economically) feasible is not necessarily desirable nor expected;
- the individual and collective responsibility of engineers is involved in this evolution, along with that of political authorities and users.

1.3 CONCEPT OF SYSTEM

1.3.1 The need for a restrictive definition

The term *system* is used in a number of contexts and in very different senses, for which the definitions are fluid. If, in the field of telecommunications, we speak of the system, it is understood to mean a *coherent* arrangement of *interdependent* elements.

The coherence of the arrangement implies a *common end*, even that which telecommunications pursues (sect. 1.1.1). The interdependence of the elements expresses the dynamic interactions between them. Seen from the outside, the system appears as a whole which is greater than the sum of its parts.

1.3.2 Systemic approach

Because of their mutual interactions, the parts of the system are necessarily quite incomplete if studied in isolation. A system must be approached from a more global viewpoint through a *systemic approach* which consists of:
- considering the system in its totality and in all its complexity
- studying in particular the *interactions* between the parts
- paying particular attention to the functional organization of the system as a means of coping with increasing technical complexity.
- integrating the *economic, human, social, and legal* aspects with purely technical considerations.

The systemic approach cannot be linear or sequential. Unable to be truly simultaneous and parallel, it adopts a *spiral route* which, through successive steps, treats of the problem by integrating different points of view at different levels. Thus, it is multidisciplinary by nature.

While the classical analytical approach proceeds through systematic decomposition and logical deduction, the systemic approach, which is complementary to it, makes use of induction and intuition, and takes into account the environment (in the broadest sense) and its interactions with the system.

These two approaches are necessary to conceive and understand large technical systems such as telecommunication systems. The two approaches will thus be used interchangeably in this book.

1.4 EVOLUTION OF TELECOMMUNICATIONS

1.4.1 Historical development

The advent of telecommunications goes back to the middle of the 19th century in the context of the industrial revolution and the acceleration of physical means of communication and transport (railways). Digital coded transmission (telegraph) systematically preceded analog transmission, which is much more delicate, particularly at great distances. Very rapidly, and to the surprise of their inventors, the means of telecommunications exhibited an extraordinary growth: shortly after transmission, switching was developed (manual, then automatic); oceans and continents were crossed; radio waves permitted things which were impossible by wire; and the network became worldwide in scope.

The history of telecommunications is the reflection of a great human adventure. Here are some of its particularly important dates:

1837. Samuel Morse invents a system of coded transmission for the letters of the alphabet, which became the ***telegraph***. His code (which is still used) took into account the relative frequency of letters in the English language to optimize the time to transmit a message. In this aspect, it was an intuitive precursor of the theory of information and coding.

1858. At the price of considerable technical and financial efforts, a cable (single-wire, isolated with gutta-percha) was placed across the Atlantic and allowed intercontinental telegraphic transmission (very slowly!). After a month, an insulation defect rendered it unusable. Another cable was installed in 1866.

1870. Telegraphic link by aerial and submarine wire between London and Calcutta (11,000 km).

1876. Alexander Graham Bell files a patent (a few hours before Elisha Gray) concerning a method of transmitting sounds electrically by means of a variable resistance. Following the ideas of Philipp Reis (1860), Bell's invention, in fact motivated by research into deafness, marked the beginning of the *telephone*.

1891. Irritated by the partiality of manual switching operators, Almon Strowger invents an automatic selector remotely controlled by the telephone subscriber's unit.

1901. Guglielmo Marconi transmits a telegram *by radio waves* from England to Newfoundland (wireless telegraphy).

1907. The invention of the *triode* by Lee de Forest permits the analog amplification of signals and opens the way to long-distance telephone transmission.

1927. First transatlantic telephone link by short wave.

1938. Pulse code modulation (PCM), invented by Alec Reeves, permits the digital representation of analog information. Nevertheless, the technology is still poorly adapted to its development.

1948. The invention of the *transistor* unleashes the potential of electronics and its numerous and diverse applications to telecommunications.

1956. One century after its telegraphic precursor, the first transatlantic telephone cable (with 51 submarine repeaters) is placed into service.

1962. A third type of transatlantic link is added to short wave and cable: the active *satellite* "Telstar I" orbiting at low altitude, allows the first transatlantic television transmission.

1965. First geostationary satellite "Early Bird" (Intelsat I).

1969. Direct transmission of the first steps of man on the moon.

1980. A space probe transmits photos of Jupiter and Saturn.

1.4.2 The growth of the telephone

The telephone is the principal telecommunications service and it has had a sustained exponential growth: the number of telephones tied to the world network currently increases by 6% per year (thus, doubling in 12 years) and will pass one billion before the end of the century (figure 1.5).

1.4.3 Development perspectives

The combined effect of three factors has given new technological dimensions to telecommunications:

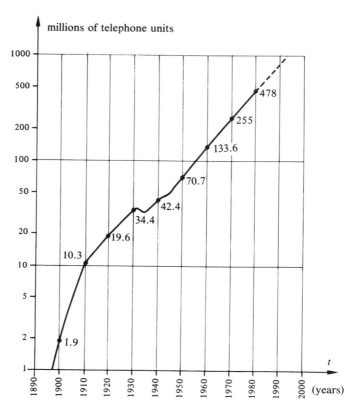

Fig. 1.5 Growth of the worldwide telephone network.

- the possibilities offered by electronics, particularly *microelectronics* and large-scale integrated (LSI) circuits;
- the development of *digital techniques* for transmission and switching;
- the influence of *informatics* which, on the one hand, imposes new demands on telecommunications (telematics) and, on the other hand, brings to it a very powerful tool for the control of switching operations and network management.

Along with the telephone, which will remain the predominant service for a long time, other switched services appear and will be developed. These concern, for the most part, nonverbal information (text, images, data) and, mainly business communications. It is attractive to envision, for the future, an *integrated services digital network, ISDN*, in which all switched services (telephone and others), and perhaps even broadcast services (musical and television programs), would be offered in digital form (sect. 15.8.3).

1.5 INFORMATION

1.5.1 Concept of information

Information, like energy, is a basic concept, which is a part of our daily lives, but for which it is difficult to give a rigorous and satisfactory definition. Without attempting to define it, we can illustrate its different aspects by the following considerations:
- to inform means literally "to give form to";
- information is an abstract concept, originally of a psychological and philosophical nature (ideas), which has been taken up by science and technology to be made into a measurable quantity;
- information is a factor of order and structure, diminishing the unknown and uncertainty. In this aspect it is linked, by antithesis, to the thermodynamic idea of entropy. This is why it is sometimes designated "negentropy";
- the value of information resides in its effect of surprise; the more it is unpredictable, the more it is interesting;
- information is not conserved: it must be generated, and it can be lost;
- information can be carried (transmission), stored (memory), and transformed (processing).

1.5.2 Sources of information: definitions

A source is said to be *discrete* if it produces information from a *finite number of characters* (letters, numbers, symbols, *et cetera*) which, in different combinations, constitute *messages*. The information generated in this way is *digital* in nature.

The collection of n available characters constitutes, by analogy with written language, which is a typically discrete source, an *alphabet*.

If the source is *continuous* in the sense that it expresses an infinite number of subtle nuances, the information produced is *analog*.

It must be noted that the distinction between digital and analog proceeds from a progression, and it depends on the scale of observation, i.e., on the actual sense given to the abstract notion of infinity!

1.5.3 Information content: definition

In the case of a discrete source, each message produced appears with a certain probability. To express the *information content H_i* carried by a message

i, it seems intuitively logical to relate it to the probability of appearance Prob(i) of this message and to fulfill the following conditions:
- if, at a given moment, the message is certain (Prob(i) = 1), its appearance carries no information, because it was perfectly predictable;
- the more a message is improbable, the more information it carries if it appears (surprise effect);
- if two messages, which are statistically independent, with probability Prob (i) and Prob(j), respectively, appear together, the information content H_{ij} that they carry must logically be the sum of their respective information contents H_i and H_j.

These conditions lead to a logarithmic function. For convenience, we have chosen base 2 to express the logarithm (binary logarithm lb). Therefore, the definition of the information content of the message is:

$$H_i = \text{lb}[1/\text{Prob}(i)] = \text{lb}[\text{Prob}(i)] \quad \text{Sh} \qquad (1.1)$$

Therefore, we have:
- $H_i = 0$ if Prob (i) = 1
- H_i increases if Prob (i) decreases
- $H_{ij} = H_i + H_j$ if the messages i and j are independent, because Prob(i, j) = Prob (i) · Prob (j)

If Prob (i) = 0.5, then $H_i = 1$ and the unit thus defined is termed the *shannon* (Sh), often erroneously called the bit (this term is reserved for the unit of decision content, section 1.5.5). It is, in fact, a pseudo-unit, without dimension, because the quantity it expresses is logarithmic.

1.5.4 Source entropy: definition

The *mean* information content H produced by a discrete source is the mathematical expression of the information content H_i carried by each of the messages it generates:

$$H = \text{E}(H_i) = \sum_i \text{Prob}(i)\, H_i = -\sum_i \text{Prob}(i)\, \text{lb}[\text{Prob}(i)] \quad \text{Sh} \qquad (1.2)$$

Each message generally can be constituted from a number of characters, each chosen from the n available. It is on the whole collection of these messages that the mean must be calculated, according to (1.2), with the following condition:

$$\sum_i \text{Prob}(i) = 1 \qquad (1.3)$$

H is called *source entropy*. It expresses the same reality as thermodynamic entropy: the more the source produces messages in an unpredictable manner, the higher the entropy. In particular, source entropy is maximum when the n characters are used by the source with the same probability (equally probable characters) and without any sequential condition. We then have

Prob(i) = $1/n$ with $i = 1 \ldots n$ (1.4)

and (1.2) becomes

$H = H_{\max} = \text{lb } n$ Sh   (1.5)

On the other hand, the term entropy can be confusing when it is applied to the information content transmitted to a sink. In effect, the entropy (in the thermodynamic sense) of the sink diminishes as it receives information.

1.5.5 Decision content: definition

The choice that a discrete source must make from the n available characters corresponds to a certain *decision content D* defined by

$D = \text{lb } n$ bit   (1.6)

If $n = 2$ (binary choice), then $D = 1$ bit, a pseudo-unit representing the minimum elementary choice between two characters.

From this alphabet of n characters, a discrete source can produce an information content H (entropy) which, according to (1.5), is at a maximum equal to D.

For example, the Roman alphabet arranges $n = 27$ characters (26 letters + 1 space). The corresponding decision content is thus $D = \text{lb } 27 = 4{,}755$ bit per character.

The frequency of utilization of characters and groups of characters varies from one language to another. The entropy of the source constituted by a written language depends somewhat on the language. It has been estimated (by statistical tests) to be $H \cong 1.5$ Sh/letter for English, for example.

1.5.6 Redundancy: definition

The *redundancy R* is the difference between the decision content D of an alphabet and the mean information content (entropy) H of the discrete source which utilizes this alphabet.

$R = D - H$ bit   (1.7)

1.5.7 Information and decision rates

A discrete source generates its characters fairly rapidly at fairly regular time intervals.

The *decision rate*, also often termed the *bit rate*, is the product of the average number of characters generated per unit time multiplied by the decision content D of the alphabet from which they are issued. This is a very important value for digital transmission systems. It expresses in bit/s the real transmission rate of the system. It will be written $\dot{D}$ here; the dot above the symbol designates a derivation of the decision content D with respect to time. In the same way, the information rate $\dot{H}$ represents the average information content produced by the source over a unit of time (in Sh/s).

1.5.8 Coding

Coding is a translation operation which assigns a list of symbols to another list without changing the information contained therein.

The inverse operation, decoding, must also be as perfectly reciprocal as possible. It therefore necessitates a prearranged convention between the encoder and the decoder.

For example, the spoken language is a code which translates ideas into words. The written language is another code which represents these words by letters. Morse code in turn transcribes these letters into dots, dashes, and spaces.

1.6 COMMUNICATION

1.6.1 Principles and conditions

Communication is, in the broad sense, a transfer of information from a source to a sink across a medium called a *channel* (fig. 1.6). For information transmission to occur, the following conditions must be fulfilled:
- source and sink must agree on the symbolic representation of the information to be transmitted, for example, on the code being used;
- the channel must be perfectly "transparent," i.e., must be a neutral carrier, which does not interfere with the information transmitted;
- the channel must be adapted (technically and economically) to the type of source and receiver being used;
- likewise, the information to be transmitted must be put into a form compatible with the channel.

1.6.2 Signals: definition

In the case of telecommunications, the channel is always a physical medium, which requires that the information, an abstract concept, must first be put into physical form by electromagnetic signals.

Signal denotes a variable physical quantity, which carries information.

The translation of information (generally already coded) into signals, and reciprocally, is accomplished by electrical *transducers* such as microphones, earphones, loudspeakers, cameras, screens, sensors, actuators, *et cetera*.

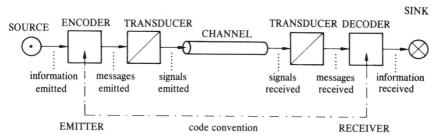

Fig. 1.6 Ideal transmission system.

The definition of a signal given above brings to mind three comments:
- The *signal* is distinguished from *noise* by the fact that it carries information;
- this distinction has meaning only if the relationship between the information and the signal is perfectly defined and identically for both partners (fig. 1.6). A receiver interprets the signals it receives as noise, if it cannot make sense of them, while the same information effectively carried by these signals may be perfectly intelligible to other receivers;
- signals are, by nature, *always analog*, i.e., their parameters can be the object of continuous variations. However, the information that they carry can be either analog or digital. The difference resides in the interpretation that the receiver gives to them (chapters 5 and 6). Through a regrettable abuse of language, "digital signal" is sometimes used to refer to (analog!) signals carrying digital information.

1.6.3 Real channels

Real channels used in telecommunication systems present the following principal flaws:

- they *deform* the signals that they transmit (distorsion, sect. 2.4.7 and 2.4.10)
- they introduce undesirable *interference* (noise, crosstalk, sect. 2.4.11)
- they are *expensive* and must be used as economically as possible.

By using a prearranged code, we attempt to adapt better the information to be transmitted to the known characteristics of the available channel (fig. 1.7). This coding has two purposes, which are partly contradictory:

- *Source coding* to reduce the redundancy of the source and transmit a minimum bit rate as close as possible to the real information rate of the source. Morse code, for example, is an attempt to do this, which optimizes the transmission time (sect. 11.2.2). In image transmission, we attempt to exploit the statistical dependence of a point of the image on its neighbors or with the same point of the preceding image.
- *Channel coding* to protect against the (inevitable) effects of channel imperfections. This coding requires the introduction of a desired redundance, e.g., to permit the detection, and indeed the correction, of transmission errors or to eliminate the continuous component of a signal (e.g., AMI mode, sect. 9.6.4).

These two operations are expensive, the first in equipment, the second in decision rate in the channel. They are only justified by a global view of system optimization.

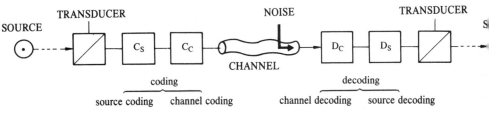

Fig. 1.7 Real transmission system.

1.6.4 Digital representation of analog information

We can, under certain conditions, leave out subtle details of analog information and substitute digital information which is more or less equivalent. The inverse operation is impossible because the details of analog information are definitively lost.

The modes of digital representation, imperfect by nature, are treated in chapter 7. Digital representation rests on a simple principle, that of *quantizing*, or analog-digital conversion, which consists of replacing an exact analog value with the closest from among a collection of finite (digital) discrete values. Despite the name, digital-analog converters do not restore analog values, but rather the discrete value assumed to be its representative after quantizing.

Our interest in digital representation stems from the relative insensitivity of this form of information to channel imperfections (chapter 5), compared to the severe demands imposed by analog transmission (chapter 6).

The price paid for this advantage is in the complexity of the converters and in the degradation of transmission quality due to the loss of analog details.

1.6.5 Symbols: definitions

The signals used to carry digital information are composed of a series of elementary time signals, called *symbols*. The characteristic parameters (amplitude, frequency, phase) of each of these remain constant over the entire time T_M of the symbol and represent the digital information carried by this symbol. Generally, this parameter can take m discrete values. The symbols are thus termed *m-ary*.

According to the number m, we have

- $m = 2$ binary symbols
- $m = 3$ ternary symbols
- $m = 4$ quaternary symbols
- $m = 5$ quinary symbols

By virtue of (1.6), the decision content corresponding to an m-ary moment is

$$D_M = \text{lb } m \qquad \text{bit} \qquad (1.8)$$

1.6.6 Symbol rate

The average number of symbols transmitted per unit time is called the symbol rate $\dot{M}$ and is expressed in baud (Bd) to distinguish it from the decision rate $\dot{D}$ expressed in bit/s.

$$\dot{M} = 1/T_M \qquad \text{Bd} \qquad (1.9)$$

The decision rate $\dot{D}$ corresponding to an m-ary symbol rate $\dot{M}$ is calculated by multiplying $\dot{M}$ by the decision content D_M of each symbol, according to (1.8):

$$\dot{D} = \dot{M} D_M = \dot{M} \text{ lb } m \qquad \text{bit/s} \qquad (1.10)$$

In digital transmission, the symbol rate $\dot{M}$ takes on particular importance because it expresses the speed of physical variation of the signal parameters.

In this respect, it is directly linked to the bandwidth of the channel (sect. 4.1.4).

For the same decision rate $\dot{D}$, it is possible to diminish $\dot{M}$ by increasing m. The symbol rate is maximum for a binary transmission ($m = 2$). In this case, and in this case only, it is equal to the decision rate $\dot{D}$.

1.7 CHARACTERISTICS OF INFORMATION SOURCES AND SINKS AND THEIR IMPLICATIONS FOR TRANSMISSION SYSTEMS

1.7.1 The human voice and speech

The voice constitutes a primary analog information vector in direct or indirect human communications over a telecommunication system. Its study is made difficult by the physiological aspects of sound reproduction, on the one hand, and by psychological elements (temperament, mood, *et cetera*) which influence it, on the other. Only a statistical approach is possible. This leads to distributions of continuous unpredictable variables, affected by strong variances in time and from one individual to another.

For telecommunications needs, it is useful to outline roughly the following principal characteristics concerning the voice and the sounds it produces, on the one hand, and, on the other, the words which structure these sounds in a spoken language:

- The voice is a non-stationary and random phenomenon which is discontinuous in time. It can nevertheless be decomposed into a series of elementary sounds called *phonemes*, which are considered to be quasi-stationary.
- The *frequency distribution* (spectrum) of the energy produced is very different depending on the sound being made: *voiced* sounds (vowels, consonants such as j,l,m,n,v) have a range of harmonic tones (men: 100 — 200 Hz, women: 200 — 400 Hz); their spectral envelope, and particularly their maxima (termed formants), are specific to the sound emitted (fig. 1.8(a)); *unvoiced* sounds (consonants such as f,s,p,ch) are characterized by a continuous spectrum, either stationary or transitory, but non-uniform (fig. 1.8(b)).
- The spectral range of the entire collection of vocal sounds extends from 80 Hz to 12 kHz, with a strongly decreasing energy as we approach the higher frequencies.
- The statistical distribution of instantaneous amplitudes is particularly complex and subject to variation. It can be roughly approximated by an

TELECOMMUNICATIONS AND INFORMATION 19

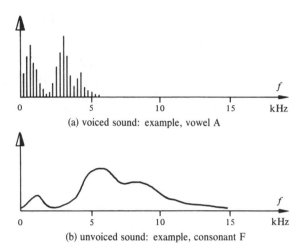

Fig. 1.8 Examples of the spectra of vocal sounds.

exponential distribution of absolute values, of which the rms value (U_{eff}) varies greatly from one individual to the next (fig. 1.9).
- The *time structure* is irregular: words and phrases are separated by pauses (>100 ms) which represent approximately 50 percent of the monologue time and 75 percent of the time for each speaker in a dialogue.

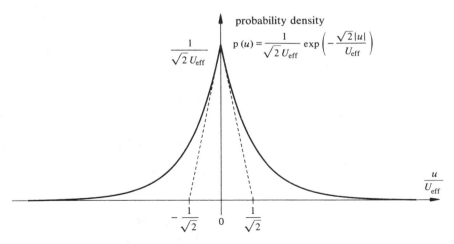

Fig. 1.9 Approximation of the statistical distribution of instantaneous amplitudes in speech.

- The mean *rate* of speech can be evaluated at 80 — 200 words/minute. At an average of 5 letters per word and $D = 4.7$ bits/letter, this represents an average decision rate of $\dot{D} = 30 - 80$ bit/s. In assuming that $H = 1.5$ Sh/letter, the actual corresponding average information rate is only $\dot{H} = 10 - 25$ Sh/s. These two rates express only the semantic output of speech (i.e., the literal significance of the words used). In reality, speech transmits an entire subjective envelope (identity, timber, humor, nuances, *et cetera*) which is difficult to evaluate.

1.7.2 Hearing

The following few properties of the human ear and sense of hearing are particularly relevant to telecommunications:
- the sensitivity of the ear depends on the frequency and intensity of the sound. The audible range extends roughly from 20 Hz to approximately 16 kHz (well adapted to speech);
- the ear is practically *insensitive to the relative phase* between two sound components;
- the effect of nonlinear distortions is more noticeable when the signal occupies a large frequency band.

1.7.3 Implications for telephony

The properties mentioned in sections 1.7.1 and 1.7.2 are utilized by telephone systems in the following way:
- restriction of the frequency band to a range of 300 — 3,400 Hz, judged sufficient for intelligibility (sect. 2.4.9), to the detriment of the ability to distinguish between certain unvoiced sounds such as "s" and "f";
- the preponderance of low-frequency components in the spectrum justifies the use of certain differential modulations such as DPCM or ΔM (sect. 7.8);
- small amplitudes, relatively more frequent, can be enhanced by a companding process (sect. 6.6);
- the insensitivity of the ear to phase shifting allows the use of modulations which do not conserve phase (SSB modulation in carrier systems, sects. 8.44 and 10.1);
- the measurement of noise takes into account the frequency response of the ear by way of a psophometric weighting filter, which simulates the effect that this noise would have on a user (sect. 2.4.13);

- pauses in a telephone conversation, suitably detected, can be used to interpose parts of other conversations (time assignment speech interpolation, TASI); for example, in very expensive channels such as transoceanic cables, or for simultaneously transmitting data and speech.

1.7.4 Music

Although the receiver (the ear) is the same as in the case of speech, the source of continuous information which music represents is distinguished by:
- an average long-term range which extends upward to the higher frequencies
- low frequencies down to 15 Hz
- a different statistical distribution which is less structured in time
- less frequent pauses

As a result, the demands placed on the channel are more severe, particularly regarding the upper limit of the frequency band (10 — 15 kHz), the signal-to-noise ratio, and the distortion rate.

1.7.5 Text

A text is a document composed of characters (letters, numbers, signs, *et cetera*) taken from a finite alphabet. It contains digital information, produced by a discrete source, according to semantic rules (significance of words) and syntax (grammar) specific to a language.

The minimum Roman alphabet is comprised of:
- 26 letters
- 10 digits
- the space between words
- a few punctuation signs

This is the alphabet used by ***telegraph*** and ***telex***.

Furthermore, ***teletex***, a new teletype service related to text processing, uses an alphabet of 128 characters and commands, permitting the distinction between upper and lower case, accenting of lower-case letters (as in French, for example), and several special signs.

Alphanumeric characters of a text can still be augmented by graphic elements, which provide for the possibility of composing diagrams or signs. ***Videotex*** is based on such an alphabet used to represent written and graphic information on home television screens (sect. 15.6.3).

1.7.6 Still black-and-white images

An image is a two-dimensional document (surface). In order to be transmitted, it must be transformed by sequential analysis, line by line (scanning), spirally, or point by point (framing).

The fineness of this analysis, with respect to the dimensions of the image, is determined by the resolution power of the receiver (human eye: approximately 1.5' of angle).

The image, originally analog in nature, can be represented in digital form, after point-by-point analysis, by discrete points whose luminous intensity is quantized into a finite number g of grey levels. The digital equivalent (decision content D) of an image of surface A, analyzed with a resolution of r points per unit length is thus:

$$D = Ar^2 \text{lb } g \qquad \text{bit} \qquad (1.11)$$

In order to transmit this image across a channel of decision rate $\dot{D}$, requires a time T:

$$T = D/\dot{D} \qquad (1.12)$$

The linear (lines) or two-dimensional (surfaces) dependence between neighboring points of a coherent image renders the corresponding information content H (which is difficult to evaluate!) much lower than D (redundancy effect). By appropriate source coding, it is possible to decrease D, and thus T.

Facsimile allows the transmission of still images with the following characteristics:

- $g = 2$ black-and-white only, no grey tones
- maximum format A4: $A = 1/16 \text{ m}^2$
- resolution: $r = 3.85$ points/mm, a total of approximately 10^6 points and $D \cong 1$ Mbit per page.

The transmission time of an A4 page across a normal telephone channel ranges from 6 minutes (without reduction of redundancy and $\dot{D} = 2,400$ bits/s) to approximately 1 minute (with source coding and $\dot{D} = 4,800$ bit/s).

1.7.7 Moving black-and-white images

The illusion of movement is given by a succession of still images sufficiently rapid to fool the eye (cinematic effect, retinal persistence). Given i images per unit time, the decision rate $\dot{D}$, corresponding to images put into digital form according to (1.11), is expressed by

$$\dot{D} = iAr^2 \text{ lb } g \qquad \text{bit/s} \qquad (1.13)$$

Besides the intrinsic redundancy of each still image, successive images are manifestly not independent, but, on the contrary, they are generally very slightly different (dynamic redundance). This double redundance appears as a preponderance of low frequencies in the signal spectrum resulting from the line-by-line analysis of a moving image, which is termed a *video signal*. Nevertheless, the high-frequency components, although rare, are very important for sudden changes (transition to the background, very wide movements) and render redundancy reduction difficult.

The mean statistical distribution of the amplitude of a video signal (luminance) can be assumed to be uniform (equally probable grey levels).

Commercial *television* decomposes each image into 625 lines (in Europe) or 525 lines (USA, Canada, Japan), along the length of which the luminous intensity is transmitted in an analog manner. The transmission takes place at $i = 25$ images per second. In the American system, the video signal thus obtained occupies a frequency band ranging from 0 to approximately 4.2 MHz.

For eventual *visiophony* (a switched low-quality video service accompanying the telephone), we try to reduce this enormous bandwidth by reducing the image format, its luminous resolution and, if possible, its redundance.

1.7.8 Color images

In addition to luminous intensity, or *luminance*, of a black-and-white image, the color information, or *chrominance*, breaks down into two supplementary parameters:
- the hue, i.e., the spectral composition to the sense of sight;
- the degree of saturation of this color (density, pale or darkened).

These three parameters can be derived from three judiciously chosen monochrome components, for example, red, green and blue. Addition of these can practically allow the reconstitution of any color nuance.

Nevertheless, to avoid tripling the decision content of a color image with respect to the same image in black-and-white (and consequently tripling the bandwidth necessary to transmit it), we can make use of three physiological properties of the eye:
- visual acuity for color nuances is weaker than for an image in black-and-white. The result is that the chrominance parameters can be transmitted with fewer details (i.e., a narrower bandwidth) than the luminance;
- visual acuity for chrominance depends on the color; it is maximum for contrasts between complementary colors — orange/sky blue;
- the sensitivity of the eye depends on the color, and is maximum for yellow ($\lambda \cong 550$ nm).

As a consequence, for transmission requirements, television systems effect a transformation of the image by linear combination of the three monochrome components, R (red), G (green), and B (blue) captured by the camera and restored by the screen. This transformation gives the following signals:
- luminance Y by addition of the three components, weighted by the sensitivity curve of the eye

$$Y = 0.30R + 0.59G + 0.11B \qquad (1.14)$$

This signal offers compatibility with black and white receivers;
- The first chrominance signal I, along the orange/sky-blue axis, derived from R, G, and B by linear combination

$$I = 0.60R - 0.28G - 0.32B \qquad (1.15)$$

- second chrominance signal Q, along the magenta-green axis, orthogonal to the first

$$Q = 0.21R - 0.52G + 0.31B \qquad (1.16)$$

The various systems used (NTSC, PAL, SECAM) are distinguished by different technical solutions to the problem of transmitting these three kinds of signals (sect. 10.7.7).

1.7.9 Data

The term *data* is generally used to designate digital information, conventionally represented in a coded form, for the purpose of processing, generally by automatic means.

The sources and sinks of data are, thus, in the first place, **machines** (terminals, computers, processors) and indirectly, according to the particular case, human beings.

The term data is often used in a general sense to include all **non-verbal information**, which can also be transmitted in digital form over a telephone network. In this respect, the term covers, for example, the following types of information:
- text
- still pictures (facsimile)
- handwritten messages
- measurements (telemetry)
- commands (remote control)
- alarms, *et cetera*

The decision rate $\dot{D}$ produced by a source or accepted by a data sink constitutes the principal parameter of a link, as seen by the user. It must be arranged with respect to the possible decision rate across the available channel, for example:

- telegraph channel: $\dot{D}$ = 50 (100 or 200) bits/s
- analog telephone channel: $\dot{M}_{max} \cong$ 1,600 baud, $\dot{D}$ = 1,200, 2,400, (4,800, 9,600) bits/s
- digital telephone channel: $\dot{D}$ = 64 kbit/s

The probability of appearance of the different characters that the source makes use of can be anything. Characters are generally produced in an appar-

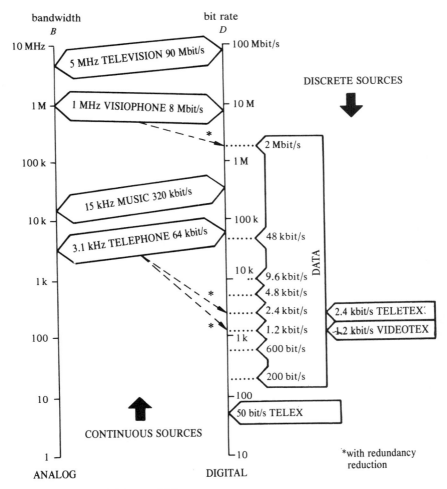

Table 1.10 Bandwidths and bit rates.

ently random manner, but repetitive motifs are not excluded (uniform areas in images, chains of identical signs, *et cetera*).

1.7.10 Comparison of the requirements for different services

Table 1.10 gives some idea of the comparative importance of different telecommunications services with respect to
- the bandwidth B allowable for analog transmission
- the bit rate $\dot{D}$ necessary for digital transmission, taking into account the possibility for digital representation of analog information.

Chapter 2

Planning: Objectives, constraints, and methods

2.1 CONCEPTION AND DESIGN OF A TELECOMMUNICATION SYSTEM

2.1.1 Quantitative and qualitative planning

Planning is understood to mean the determination of all the parameters of a system as a function of present demands and their presumed evolution in time. In the case of telecommunications, this operation has two distinct aspects:
- qualitative planning, which consists of conceiving and designing the system in its entirety, and each of its organs in particular, as a function of objectives expressed in terms of service quality, particularly quality of transmission. The telephone user expects, for example, that a certain intelligibility of conversations will be guaranteed in all circumstances;
- quantitative planning, which concerns the evaluation of a number of organs (lines, channels, routes, equipment, *et cetera*) to be provided in a network to cope with the traffic (of random nature) produced by the users. The criteria of quantitative planning also refer to the idea of the service quality required by users, but what is meant in this case is the quality of switching. For example, the number of public telephone channels available between two cities must be sufficient so that not too many calls will be refused at peak hours nor will users be made to wait too long.

2.1.2 Service quality

In telecommunications, the notion of service quality serves as a fundamental reference for all phases of planning. Three of the principal aspects of the service quality presented below will be treated in sections 2.4, 2.5, and 2.6.

The *transmission quality* concerns the fidelity of the transmitted information. The information entrusted by the source to the system must arrive at the sink reasonably free of losses, additions, and alterations. The overall criteria of quality depend on the type of service, for example:

- intelligibility in telephony
- fidelity and clarity in musical transmission
- conformity in transmission of images
- rate and probability of errors in telegraphy and data transmission

On the technical level, particular criteria concerning the parameters of the system must be considered. These are principally:

- the general attenuation of the link
- the propagation time
- the bandwidth
- the behavior in the presence of noise
- the influence of interference (noise, crosstalk)

The *switching quality* concerns the routing of communications across a switched network. The commands from the source (the calling subscriber) specifying the identity of the information sink (the called subscriber) must be executed exactly and rapidly. The principal criteria of switching quality are:

- the probability of congestion in the network
- the behavior of the system in the case of congestion (losses, delays)
- the mean set-up time to establish connections
- the probability of false connections
- the accuracy of charging

Reliability is a very important aspect of service quality. It expresses the ability of the system to satisfy the operating requirements over a given period of time. It is specified by, and evaluated with the aid of, the following criteria:

- the probability of partial and total breakdowns
- the reparability of the system (mean duration of breakdowns)
- the consequences of a breakdown
- the behavior of the system in case of breakdown (detection, location, isolation)

2.1.3 Compromise between cost and quality

It is always possible to improve the quality of a system, on the condition that you pay the price. In any case, perfectionism is a destructive attitude for an engineer. The art of planning consists, therefore, of guaranteeing the minimum quality required at a reasonable, competitive price, taking into account the capabilities of present technology.

Defining the minimum quality required is a delicate matter, even more so in telecommunications because it is often tied to human appreciation factors, which are subjective by nature. Therefore, we try to set a limit of quality, based on extensive statistical tests, such that a substantial percentage (e.g., 95%) of users will declare themselves to be satisfied. The criteria of quality thus defined have been, for the most part, the object of international conventions (section 2.2), which are themselves already the result of a compromise between cost and quality.

It is clear, for example, that an increase in the upper limit of the telephone frequency band of 3.4 kHz (present convention) to 5.5 kHz would noticeably improve the intelligibility of communications, but the capacity of long-distance transmission systems would be reduced by one-third and their cost per channel increased in similar proportion.

2.1.4 Way of designing a system

Initially, three factors determine the design of a system:

- the *objectives* to be attained, specified in terms of qualitative and quantitative needs. These are, for example: servicing of a region, introduction of a new service, operating rationale, increasing the transmission capacity, exploration of a new technology, covering a potential market, *et cetera*;
- the *constraints* to be respected, principally economic in nature (profitability). To this is added conformity to national and international standards and recommendations, compatibility with other existing systems, climatic conditions, and the particular demands of operation, transport, or setting up. Time schedules, often imperative and always felt to be too tight, also constitute a significant constraint among all those which limit the freedom of the engineer;
- the *means* available, i.e., on the one hand, the hardware (components, circuits, equipment practice, *et cetera*) and software (methods of computation, programming, *et cetera*); on the other hand, developmental personnel resources and experience.

The precise statement of objectives and constraints constitutes the *specification* of the system, which serves as the general instruction and as a reference for the entire design, conception, and evaluation of the system. On the basis of this specification and the means available, the design of the system is elaborated and continuously submitted to an ongoing process of evaluation by:

- confrontation with the specification
- analysis of means (feasibility)
- comparison of alternatives
- choice of basic options
- cost estimates, notably the extent and duration of development.

This work develops a *system concept* on the basis of which the system can be divided into subsystems or units, each defined in relation to the others by a set of specifications which serves as the basis of development.

The exploratory phase (conception, evaluation) should not be underestimated either in time or difficulty. Well-founded decisions at this level allow the design engineer to escape costly disappointments in the course of development.

Figure 2.1 summarizes the genesis of a technical system in schematic form.

2.1.5 Specifications

In the general category of system definition, the *specification* is the technical document which is the basis and reference for the development or acquisition of a subsystem or unit. In this respect, it must unambiguously deal with the following points:
- location of the unit in the overall system
- exactly define its *functions*
- carefully specify the *interfaces*, i.e., the contact points with other units of the system, at the functional level (exchange of information, codes, sequences, *et cetera*) as well as electrical (polarity, level, impedance, *et cetera*)
- prescribe the *tolerances* of all the essential parameters
- establish the principles of monitoring (alarms, displays, controls, *et cetera*), maintenance, and service (man-machine interface, manual interventions, *et cetera*)
- state the requirements of *reliability*, the acceptable consequences of breakdown (availability, time to repair, *et cetera*)
- give the construction parameters (modular construction, dimensions, connectors, type of wiring, mechanical interfaces, *et cetera*)
- indicate the sources of energy available (power supply by mains, by battery, *et cetera*)
- specify the ambient conditions (climate, shaking, *et cetera*)

When the unit must perform logical functions (sequences of operations, conditional execution of operations), as is the case with control units and alarm or test units, a document generally is not sufficient to specify unambiguously the exact and complete operation of its functions, i.e., the *program* in the broad sense (independently of its mode of operation and whether it is wired or not). It must, therefore, use symbolic representations (Boolean algebra, Petri network, flow charts, structure diagrams, *et cetera*).

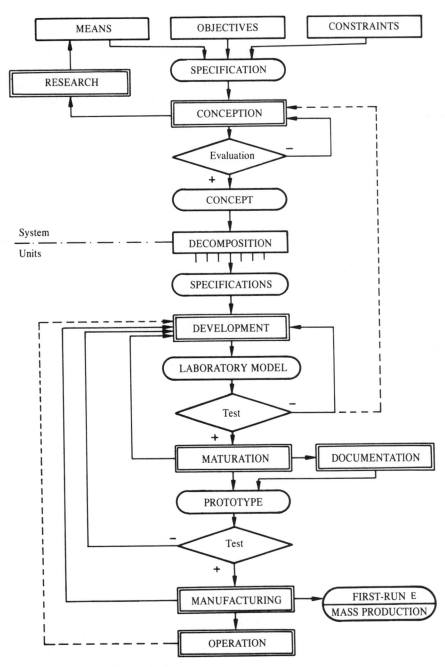

Fig. 2.1 Genesis of a technical system.

2.1.6 Development and manufacturing

The process of *development*, like that of design, is also repetitive. The results of tests, technical and economic evaluations, and contending with the specifications allow the design engineer to modify the options selected in the course of development and to approach an optimal solution (provided that the time schedule allows for it and that the process converges).

The development phase is followed by a phase of *maturation*, i.e., of exact description of all the details, and the parallel development of complete *technical documentation* (diagrams, list of parts, programs, assembling prescriptions, adjustment, testing, operating instructions, *et cetera*). This phase is as extensive as that of development, strictly speaking, and should directly lead to the manufacturing of the unit.

The importance of documentation is tremendous. It constitutes the connection between the research department or the laboratory and the production departments (manufacturing plant, workshop, trial premises), and ultimately the client (installation, start-up, operation, repairing, *et cetera*). Because the development engineer generally does not himself construct the things that he designs, he must be able to communicate his ideas to others intelligibly, without ambiguity, in a conventional language (symbols, drawings, plans, *et cetera*).

For a complex system, the stages of development are generally the following:

- laboratory models (first edition of documentation)
- prototype in experimental use (second edition of documentation)
- first-run units in actual service (third edition of documentation)
- mass production

A considerable amount of work is necessary to specify the system and put it into service (several hundred man-years in the case of new transmission or switching systems of large scale).

2.2 NATIONAL AND INTERNATIONAL STANDARDS

2.2.1 Need for coordination

The necessity for telecommunications to transmit information at great, indeed very great, distances has been recognized since the beginning of its history. For this reason, early on, it was necessary to come to an agreement on both sides of national frontiers on a certain number of fundamental points so as to guarantee the quality of service. This concerns primarily the following:

- technical questions: definition of service quality and determination of the parameters which influence it; detailed and very precise specification of the interfaces (electrical and logical characteristics), notably the nature of the signals used for transmission and the conventions defining the exchange of information (signaling) for automatic switching;
- general network planning: structure of the international network, mode of routing communications, distribution of attenuation (transmission plan), distribution of call numbers (numbering plan);
- problems of operation and management: in particular, charge accounting of international communications and monitoring of traffic.

On the *national* level, strict standardization of the network and its equipment is also absolutely necessary for

- guaranteeing the *compatibility* of the elements of systems from different producers;
- assuring the same minimum service quality to all users;
- adherence to international conventions.

2.2.2 Role of the International Telecommunication Union

In 1865, during the early development of the transcontinental and intercontinental telegraph lines and eleven years before the invention of the telephone, the first conference marking the debut of the International Telegraph Union was held in Paris. The name was changed in 1947 to the *International Telecommunication Union (ITU)*, which is a Specialized Agency of the United Nations.

The ITU is comprised of four permanent bodies:

- General Secretariat;
- the International Telegraphy and Telephone Consultative Committee (CCITT, *Comité consultatif international télégraphique et téléphonique*), which deals with line transmissions modulation and multiplexing equipment, switching, and telephone, telegraph, and data networks;
- the International Radiocommunications Consultative Committee (CCIR, *Comité consultatif international des radiocommunications*), concerned with radio broadcasting (audio and video), microwave links, satellite links, radio astronomy, and the use of the radio waves;
- the International Frequency Registration Board (IFRB), which is in charge of the assignment of frequency bands to different services and the rational use of the limited available spectrum.

The CCITT and the CCIR are composed of numerous commissions in which the delegates of member countries and observers or advisors from the industry participate. Their work results in "recommendations" ratified by the Plenary Assemblies held every four years.

These recommendations are published in "Books" designated by the color of their covers (e.g., CCITT Green Book, CCITT Yellow Book) typical of each respective Plenary Assembly. While the ITU has no means of enforcing these recommendations, in practice they have the effect of standards to which countries adhere in their own interest and which serve as the basis for the specifications by national administrations.

These recommendations thus take on great importance for telecommunications. They define references for the quality of service offered to users and guarantee a certain amount of compatibility between systems at a worldwide level. In this sense, they certainly limit the freedom of the engineer who designs systems, but assure rational inclusion of his developments in the general context of public telecommunications service. Frequent reference will be made to these recommendations in the following chapters.

Table 2.2 locates the areas of the various commissions and the recommendations they have produced.

Table 2.2
Commissions and recommendations of the CCITT.

Service	Function					
	Tariff	Operation	Equipment	Switching	Transmission	Maintenance
Telephony	(III)	(II) / E	(XVI) World Links — (XII) / P	(XI) / Q	(XV) / G,H,J	(IV) / M,N,O
Telegraphy		(I) / F	(VIII) / S	(X) / U	(IX) / R	
Data	D		(XIV) Facsimile / T — (VII) New Data Networks / X		(XVII) / V	
Digital Networks			(XVIII) PCM Integrated Networks / G			

○ Commission Number   □ Letter designation of its recommendations

2.3 ATTENUATION AND LEVEL

2.3.1 Attenuation and gain: general definitions

Generally, the *attenuation A* or the *gain G* is defined by the logarithm of the ratio of two apparent powers P_1 and P_2 characterizing one of the conditions at the input of a quadripole (line, circuit, filter, *et cetera*), the other one at the output:

$$A = 10 \lg \frac{P_1}{P_2} \quad \text{dB} \qquad (2.1)$$

$$A = \frac{1}{2} \ln \frac{P_1}{P_2} \quad \text{Np} \qquad (2.2)$$

$$G = -A \qquad (2.3)$$

It must be noted that:
- we speak of gain when $P_2 > P_1$ ($G > 0$, $A < 0$);
- according to the base chosen for the logarithm, the attenuation and the gain are expressed in *decibels* dB, which is more practical than the *bel* (base 10), or in *neper* Np (base e), which are two similar pseudo-units (dimensionless);
- the neper, mathematically more "natural" because it is directly tied to the exponential function, appears logically in the theory of quadripoles and lines (chapter 3). Nevertheless, the decibel is much more widely used. One can be converted to the other by the equations:

$$1 \text{ Np} = 20 \lg e \text{ dB} \cong 8.68 \text{ dB} \qquad (2.4)$$

$$1 \text{ dB} = (1/20) \ln 10 \text{ Np} \cong 0.115 \text{ Np} \qquad (2.5)$$

- the usefulness of a logarithmic representation stems from the possibility of adding the attenuation and partial gain instead of multiplying the power ratios.

Table 2.3 gives some important values of ratios in logarithmic form.

2.3.2 Composite attenuation, gain, and phase change: definitions

In practice, it is rarely possible to terminate exactly a line, circuit, or quadripole (e.g., a filter) on its characteristic impedance, and less so because it is generally not purely real and depends on the frequency.

In the case of the termination of a quadripole on two impedances $\underline{Z}_1$ at the

Table 2.3 Logarithmic expressions of power ratios.

$\dfrac{P_1}{P_2}$	$10 \lg \dfrac{P_1}{P_2}$	$\dfrac{1}{2} \ln \dfrac{P_1}{P_2}$
	dB	Np
1	0	0
2	~3	~0.35
3	~4.7	~0.55
10	10	~1.15
10^n	10n	~1.15n

input and $\underline{Z}_2$ at the output, the *composite attenuation* A_{cp} is defined according to the general relation (2.1) by:

- P_1 = maximum apparent power that the source of internal impedance $\underline{Z}_1$ can provide. It is obtained by loading the source directly with an impedance equal to $\underline{Z}_1$ (fig. 2.4(b));
- P_2 = apparent power supplied by the source across the quadripole on the impedance $\underline{Z}_2$ (fig. 2.4(a)).

The composite attenuation is, thus,

$$A_{cp} = 10 \lg \frac{P_1}{P_2} = 10 \lg \frac{U_0^2}{4Z_1} \cdot \frac{Z_2}{U_2^2} = 20 \lg \frac{U_0}{2 U_2} - 10 \lg \frac{Z_1}{Z_2} \quad \text{dB} \quad (2.6)$$

If $P_1 < P_2$, we speak rather of *composite gain*:

$$G_{cp} = -A_{cp} \quad \text{dB} \quad (2.7)$$

In the same way, the *composite phase change* b_{cp} is defined by the difference in phase between the source voltage $\underline{U}_0$ and the output voltage $\underline{U}_0$.

$$b_{cp} = \arg(\underline{U}_0) - \arg(\underline{U}_2) \quad (2.8)$$

It must be noted that:

- if and only if $Z_1 = Z_2$ (in modulus), then

$$A_{cp} = 20 \lg \frac{U_0}{2 U_2} \quad \text{dB} \quad (2.9)$$

- the composite attenuation does *not* express the ratio of the voltages U_1 at the input and U_2 at the output of the quadripole, **unless** $\underline{Z}_1 = \underline{Z}_2 = \underline{Z}_c$ = the characteristic impedance of the quadripole. In this case, we obtain the *image attenuation*.

PLANNING: OBJECTIVES, CONSTRAINTS, AND METHODS 37

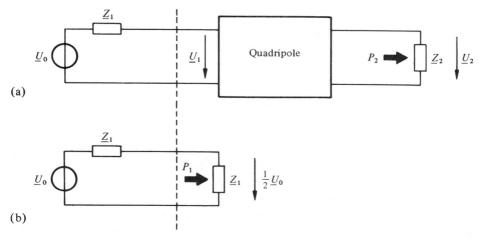

Fig. 2.4 Definition of the composite attenuation.

2.3.3 Concept of level: definition

The *level* is the logarithmic expression of the ratio of a quantity, generally a power P_x, and a reference value of the same type, for example, a power P_{ref}:

$$L_x = 10 \lg \frac{P_x}{P_{ref}} \qquad \text{dB} \qquad (2.10)$$

$$L_x = \frac{1}{2} \ln \frac{P_x}{P_{ref}} \qquad \text{Np} \qquad (2.11)$$

For a given impedance, the power is proportional to the square of the voltage. We can, therefore, also write:

$$L_x = 10 \lg \frac{U_x^2}{U_{ref}^2} = 20 \lg \frac{U_x}{U_{ref}} \qquad \text{dB} \qquad (2.12)$$

$$L_x = \frac{1}{2} \ln \frac{U_x^2}{U_{ref}^2} = \ln \frac{U_x}{U_{ref}} \qquad \text{Np} \qquad (2.13)$$

The concept of level is directly related to that of attenuation or gain. In fact, according to (2.1):

$$A = 10 \lg \frac{P_1}{P_2} = 10 \lg \frac{P_1}{P_{ref}} - 10 \lg \frac{P_2}{P_{ref}} = L_1 - L_2 \qquad \text{dB} \qquad (2.14)$$

Attenuation is thus synonymous with a decrease in level between two points; likewise, a gain expresses an increase in level.

2.3.4 Absolute level: definition

If the reference value P_{ref} is defined independently of the system, the level referred to this value is called the *absolute power level*. It is common practice in telecommunications to choose a power of 1 mW as a reference:

$$P_{ref} = 1 \text{ mW}: \qquad L_x(\text{re 1 mW}) = 10 \lg \frac{P_x}{1 \text{ mW}} \qquad \text{dBm} \qquad (2.15)$$

The suffix m added to the abbreviation dB indicates the reference to 1 mW. This notation, extremely widespread in designating absolute levels, is nonetheless not in accordance with the rules of the IEC according to which we should write dB (1 mW).

The absolute level is a way of expressing the signal power. The correspondence between absolute level and power is graphically presented in the appendix (fig. 16.4). For example:

$L_x(\text{re 1 mW}) = 0$ dBm corresponds to $P_x = 1$ mW

$L_x(\text{re 1 mW}) = +20$ dBm corresponds to $P_x = 100$ mW

$L_x(\text{re 1 mW}) = -13$ dBm corresponds to $P_x = 50$ μW

Across a resistance of $R_{ref} = 600 \,\Omega$, almost universally adopted for the terminal load of circuits, and especially for low-frequency measuring instrument interfaces, the reference power $P_{ref} = 1$ mW produces a voltage:

$$U_{ref} = \sqrt{P_{ref} R_{ref}} = 774.6 \ldots \text{ mV} \cong 775 \text{ mV} \qquad (2.16)$$

which can be taken as a reference voltage in order to define an absolute voltage level:

$$L_x(\text{re 775 mV}) = 20 \lg \frac{U_x}{775 \text{ mV}} \qquad \text{dB} \qquad (2.17)$$

The choice of this reference voltage has the result that, across a resistance of 600 Ω (and with this value alone), the absolute voltage level is expressed by the same value, in decibels or nepers, as the absolute power level:

$$L_x(\text{re 775 mV}) = L_x(\text{re 1 mW}) \quad (\text{across 600 }\Omega) \qquad (2.18)$$

To each absolute voltage level, independently of the impedance at this point, there corresponds a certain voltage, as shown in the diagram of figure 16.5 given in the appendix. In particular, for example:

PLANNING: OBJECTIVES, CONSTRAINTS, AND METHODS

$L_x(\text{re } 775 \text{ mV}) = 0 \text{ dB}$ corresponds to $U_x = 775 \text{ mV}$

$L_x(\text{re } 775 \text{ mV}) = +20 \text{ dB}$ corresponds to $U_x = 7.75 \text{ V}$

$L_x(\text{re } 775 \text{ mV}) = -65 \text{ dB}$ corresponds to $U_x = 435.8 \text{ μV}$

In telecommunications, the measurement of voltage is often carried out with special voltmeters called *level meters*, which are graduated in dB and which directly indicate the absolute voltage level with respect to 775 mV, independently of the impedance at the point of measurement.

Other reference voltages are sometimes used, for example:

$U_{\text{ref}} = 1 \text{ V}$: $L(\text{re } 1 \text{ V})$ in dBV

$U_{\text{ref}} = 1 \text{ μV}$: $L(\text{re } 1 \text{ μV})$ in dBμV

At high frequencies, the reference resistance is normally 50 Ω or 75 Ω (coaxial transmission cables). The reference power of 1 mW produces a voltage of 223.6 mV and 273.9 mV, respectively. We, therefore, have

$$L_x(\text{re } 1 \text{ mW}) = L_x(\text{re } 223.6 \text{ mV})$$
$$= L_x(\text{re } 775 \text{ mV}) + 10.8 \text{ dB} \quad (\text{across } 50 \text{ Ω}) \quad (2.19)$$

$$L_x(\text{re } 1 \text{ mW}) = L_x(\text{re } 273.9 \text{ mV})$$
$$= L_x(\text{re } 775 \text{ mV}) + 9.0 \text{ dB} \quad (\text{across } 75 \text{ Ω}) \quad (2.20)$$

2.3.5 Relative level: definition

The term *relative level* is applied to the **difference** in absolute level at point X of a system and the absolute level at a reference point of the system (generally the source), where the power of the signal is P_1.

$$L_x(\text{re } P_1) = 10 \lg \frac{P_x}{P_1} = L_x(\text{re } 1 \text{ mW}) - L_1(\text{re } 1 \text{ mW}) \quad \text{dB} \quad (2.21)$$

The relative level is therefore equal (by definition) to the attenuation A, with the sign changed, between the reference point and the point X being considered:

$$L_x(\text{re } P_1) = -A \quad (2.22)$$

It is sometimes useful, but against the rules of the IEC, to express the relative level in dBr to distinguish it from the absolute level expressed in dBm.

2.3.6 Level diagram: example

The *level diagram* gives the absolute or relative level at all points of a system. It shows all the decreases (attenuations) and increases (gains) in level intervening in the course of transmission over the lines, radio links, and in amplifiers. The advantage of logarithmic representation is made apparent by the ease of combination of these elements by simple addition or substraction of decibels.

Figure 2.5 gives an example of a level diagram relative to a transmission line of attenuation 6 dB per kilometer and supported by two amplifiers of identical gain equal to 35 dB located at 5 and 13 km from the start of the line. The absolute level at the input is -12 dBm.

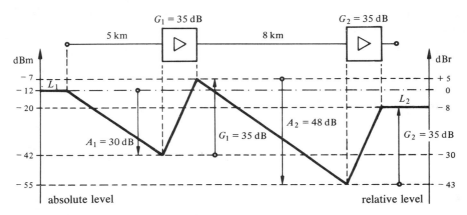

Fig. 2.5 Example of a level diagram.

The output level L_2 is calculated immediately by the algebraic summation of the gains and attenuations:
- absolute level:
$$L_2(\text{re } 1 \text{ mW}) = L_1(\text{re } 1 \text{ mW}) - A_1 + G_1 - A_2 + G_2$$
$$= -12\text{-dBm} - 30 \text{ dB} + 35 \text{ dB} - 48 \text{ dB} + 35 \text{ dB}$$
$$= -20 \text{ dBm}$$
- relative level:
$$L_2(\text{re } P_1) = A_1 + G_1 - A_2 + G_2 = -30 \text{ dB} + 35 \text{ dB} - 48 \text{ dB}$$
$$+ 35 \text{ dB} = -8 \text{ dBr}$$

This relative level of -8 dBr means that the output level is located 8 dB lower than the source level, thus an absolute level of L_2 (re 1 mW) = L_2(re P_1) $-$ L_1 (re 1 mW) = -8 dBr $-$ 12 dBm = -20 dBm.

2.3.7 Addition of signals

When *n* **independent** signals (in the statistical sense, i.e., not correlated) are combined, the *power* of the resulting signal is equal to the *sum of the powers* P_i of the *n* signals. If these signals have the same rms value U_{eff} (thus the same power *P*), we obtain:

$$P_{\text{res}} = \sum_{i=1}^{n} P_i = nP \tag{2.23}$$

given

$$U_{\text{res eff}} = \sqrt{n}\, U_{\text{eff}} \tag{2.24}$$

and for the resulting level:

$$L_{\text{res}} = L + 10 \lg n \quad\quad \text{dB} \tag{2.25}$$

In the very particular case of *n* **isochronous signals**, i.e., signals of the same form which are perfectly synchronized in time, and therefore completely correlated, it is the *rms values* of the signals which are added to give the rms value of the resulting signal. If these signals not only have the same form, but the same power (identical signals), we obtain

$$U_{\text{eff res}} = \sum_{i=1}^{n} U_{\text{eff i}} = nU_{\text{eff}} \tag{2.26}$$

given

$$P_{\text{res}} = n^2 P \tag{2.27}$$

and the resulting level:

$$L_{\text{res}} = L + 20 \lg n \quad\quad \text{dB} \tag{2.28}$$

In the practice of telecommunications, one often deals with the sum of random signals, which are manifestly independent, such as, for example, in trying to evaluate the resulting level of *n* simultaneous telephone conversations. We, therefore, have recourse to *power addition*, according to (2.25). Nonetheless, in certain cases the correlation between the *n* signals is not exactly zero, without their being isochronous (e.g., interference of a digital transmission by a small number of other, neighboring, or even equal frequencies). In this case, the *voltage addition*, according to (2.28) gives an indication which, although pessimistic, may be useful.

2.4 TRANSMISSION QUALITY

2.4.1 Overall loss: definition

The impedance conditions at the terminal ends of a telecommunications network (local lines, terminal units, translators, *et cetera*) are not always very well defined.

In order to *dimension* and *measure* the overall attenuation of a telephone network from one end to the other, it is necessary to terminate it with a *pure resistance* of 600 Ω. The composite attenuation thus defined is called *overall loss* A_{eq}.

We have, according to (2.9), with $\underline{Z}_1 = \underline{Z}_2 = 600\ \Omega$ (real):

$$A_{eq} = 20\ \lg \frac{U_0}{2U_2} \qquad \text{dB} \qquad (2.29)$$

Our interest in the idea of overall loss is that it allows a practical measurement and clear comparison of the different systems. The value of 600 Ω, chosen for historical reasons, (sect. 3.5.2), is currently the conventional nominal value of the impedance at the low-frequency interfaces of equipment and measuring instruments.

2.4.2 Reference equivalent: definition

The quality (intelligibility) of a telephone link depends essentially on the intensity of the sound reaching the ear of the subscriber. This intensity depends, in turn, on
- the output of the sending transducer (microphone)
- the output of the receiving transducer (earphone or loudspeaker)
- the attenuation (overall loss A_{eq}) of the entire transmission system between the two transducers

In order to permit a comparative evaluation of different systems, a *standard system* has been defined on an international basis (CCITT), principally comprised of a specified microphone and earphone, and designated NOSFER (Nouveau Système Fondamental pour la détermination des Équivalents de Référence) (fig. 2.6).

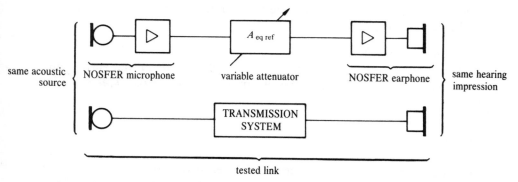

Fig. 2.6 Measurement of the reference equivalent.

The *reference equivalent* $A_{eq\ ref}$ is the attenuation that must be used in the NOSFER system to obtain the same sound intensity (evaluated by subjective comparison) as across the tested link.

$A_{eq\ ref} < 0$ if the link is better
$A_{eq\ ref} > 0$ if the link is worse } than NOSFER

2.4.3 Worldwide reference equivalent

To guarantee sufficient intelligibility in telephone communication between two terminal units connected to the worldwide network, it is necessary to assure that the reference equivalent does not exceed a critical value, whatever the types of telephone unit and interconnected lines, and to plan the worldwide network accordingly. This critical value has been fixed by international convention (CCITT, Recommendation G.111) as:

$$A_{eq\ ref} \leq 36\ \text{dB} \tag{2.30}$$

Taking into account the properties of the NOSFER system, this value corresponds to the sound intensity received when a direct acoustical link is made at a distance of approximately 3 m.

To allow the planning of national networks, the worldwide reference equivalent is distributed in the following fashion:

- national network on the sending side (including the electroacoustic efficiency of the microphone):

    $A_{eq\ ref} \leq 21\ \text{dB}$

- international network:

    $A_{eq} \leq 3\ \text{dB}$

- international network on the receiver side (including the electroacoustic efficiency of the earphone):

$$A_{eq\ ref} \leq 12\ dB$$

Countries distribute the overall loss within their national network (transducers, local lines, long distance lines) according to a *transmission plan* (section 15.4) specific to each country.

2.4.4 Elements of telephonometry

Electroacoustic transducers (microphone or earphone) transform a variation in acoustical pressure into a variation in electrical voltage and *vice versa*. Their efficiency is directly applicable to evaluation of the national reference equivalent. A transducer of good quality, i.e., with a small reference equivalent, allows a larger attenuation tolerance in the national network. A delicate compromise must be found between the efficiency of the transducers (very critically tied to their cost, given the great number of telephone sets) and the agreed upon investment in the network (gauge of local and rural lines, amplifiers) required to guarantee the overall loss. The sum of the transducer reference equivalent and the overall loss of the network must not exceed the value assigned by the worldwide transmission plan, whatever subscriber being considered.

Figure 2.7 shows that the quality (reference equivalent for emission) of a microphone is interchangeable with that of the network. This can lead to an economically optimal result that is technically feasibile.

The measurement of the transducer reference equivalent by direct comparison with the NOSFER reference system is very unpractical. It is made rather with the aid of artificial mouths, which produce a standardized acoustical pressure, or with the aid of artificial ears with a standard microphone, which allow the measurement of an acoustical pressure under carefully specified conditions. We can find the reference equivalent by the comparison of the electrical voltage delivered to 600 Ω by the microphone being tested, excited by the artificial mouth, with the known value of the voltage produced by the NOSFER microphone under the same conditions. In the same way, the reference equivalent for the earphone can be found with the artificial ear.

2.4.5 Transmission delay

The time that it takes for information to travel from the source to the sink can be a significant factor in transmission quality. It is essentially the result of the propagation time by radio waves or by wire, but in certain cases of data

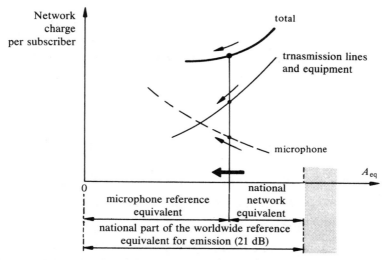

Fig. 2.7 Effect of microphone improvement on network costs.

transmission, there is additional delay due to switching (messsage switching, sect. 15.5.3).

The delay is not critical in one-way communications (radio broadcasting, television, facsimile, *et cetera*). It becomes so when a response is expected in the other direction (conversation, interactive mode).

In a telephone conversation, the delay is noticeable from a certain time of propagation (a single route) of approximately 150 ms. For psychological reasons, conversation becomes difficult, even impossible without particular training, when this delay exceeds 400 ms.

Delays on the order of 250 ms are found in geostationary satellite links. Consequently, it is necessary to avoid making telephone communications via two satellite links connected in series.

2.4.6 Conformal transmission

In analog transmission, the information is contained in the form of a signal. It is, therefore, a question of safeguarding it. For a conformal transmission, i.e., without deformation, the signal received $u_R(t)$ cannot differ from the emitted signal $u_E(t)$, except by

- a constant factor a
- a constant delay τ.

These two signals are associated by the relation:

$$u_R(t) = a\, u_E(t - \tau) \tag{2.31}$$

This identity in the time domain corresponds, by Fourier transform (delay theorem, VI. 4.1.9), to the following relation in the complex frequency domain:

$$\underline{U}_R(f) = a\, \underline{U}_E(f)\, \exp(-j2\pi f\tau) \tag{2.32}$$

The result is that

- the *attenuation* A of the transmission must be *constant*, and independent of frequency:

$$A = 20\, \lg[U_E(f)/U_R(f)] = 20\, \lg(1/a) = \text{const.}\ \forall f \quad \text{dB} \tag{2.33}$$

- the *phase change* b must be a *linear function* of frequency:

$$b = \arg[\underline{U}_E(f)/\underline{U}_R(f)] = 2\pi f\tau \sim f \tag{2.34}$$

These are the two conditions required for an ideal conformal transmission.

In digital transmission, the conformity is unnecessary. Because the received signal is sampled before the digital information is extracted, it is sufficient that *the interference between symbols* is zero, i.e., that the value of the signal at an instant of sampling depends only on the digital information carried by this symbol and not on the values at neighboring sampling instants (section 5.3).

2.4.7 Linear distortion: definition

The conditions (2.33) and (2.34) cannot be perfectly satisfied in practice. On the contrary, the transmission channels present imperfection in this regard, called *linear distortions*. In particular,

- *attenuation distortions* if the attenuation A varies with frequency
- *phase distortions* if the phase change b does not increase linearly with frequency, i.e., if the propagation delay $\tau = b/\omega$ is not constant.

As a consequence of linear distortion, a purely sinusoidal signal at emission remains sinusoidal at reception, but the form and spectrum of any other signal are modified. Nonetheless, no new spectral component appears and the principle of superposition remains valid, from which we obtain the name "linear" distortion.

2.4.7 Concept of bandwidth

The *bandwidth B* of a real channel is a vague concept, but one with widespread use and obvious practical interest.

PLANNING: OBJECTIVES, CONSTRAINTS, AND METHODS

We can attempt to define the bandwidth by the range of frequencies in which linear distortions, and particularly attenuation distortion, remain below a certain limit. This limit is often arbitrarily set at an increase in attenuation ΔA of 3 dB with respect to a reference frequency (fig. 2.8).

The precariousness of the bandwidth concept results from this description of the frequency response curve $A(f)$ by only two points. The ambiguity always arises except in the case of an ideal filter, which is of little practical interest.

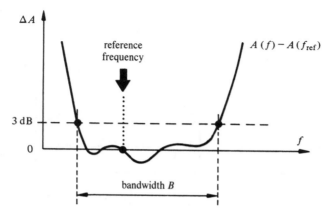

Fig. 2.8 Example of bandwidth.

The bandwidth concept also applies to a signal as an indicator of the frequency range occupied by its spectrum. Here, also, a conventional approximation is necessary to pass from one band of frequencies, often theoretically infinite, to a bandwidth of practical significance.

By selecting a channel for a particular transmission, we endeavor to adapt the bandwidth to that of the signals to be transmitted. The transmission quality depends on this adaptation, with a degree of perfection limited on the other hand by economic considerations.

2.4.9 Conventional telephone band

Such a compromise between cost and quality had to be made in order to define by international convention the bandwidth necessary for transmitting a telephone conversation.

A number of subjective intelligibility tests with logatoms (nonsensical monosyllables) have shown that the transmission quality is generally judged to be satisfactory with the following band limitations:

- telephone bandwidth: f = 300 Hz to 3,400 Hz giving B = 3,100 Hz.

These values are universally accepted and serve as a basis for the planning of all public telephone systems.

By international convention, a scheme (fig. 2.9) has also been established to limit in-band attenuation distortions with respect to 800 Hz (CCITT, Recommendation G.132).

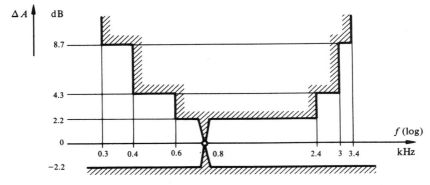

Fig. 2.9 Worldwide telephone circuit frequency response standard.

2.4.10 Nonlinear distortions

If the attenuation factor a of relation (2.31) varies with the amplitude of the signal, the system displays *nonlinear distortion*. The transfer characteristic $u_R = f(u_E)$ is no longer linear. It can be broken down into a series of powers:

$$u_R = a_1 u_E + a_2 u_E^2 + a_3 u_E^3 + \ldots \qquad (2.35)$$

If the transmitted signal is purely sinusoidal with frequency f, the received signal contains in addition to the frequency f its multiples or *harmonics* of frequency $2f$, $3f$, *et cetera*. The term in u_E^2 causes a *second-order harmonic* of frequency $2f$, by decomposition of

$$\cos^2 \omega t = \frac{1}{2} + \frac{1}{2} \cos 2\omega t \qquad (2.36)$$

The term in u_E^3 gives rise to the appearance of a *third-order harmonic* of frequency $2f$ by

$$\cos^3 \omega t = \frac{3}{4} \cos \omega t + \frac{1}{4} \cos 3\omega t \qquad (2.37)$$

If the signal emitted is formed from several sinusoidal components of frequency $f_1, f_2, \ldots, f_i$, and the principle of superposition is no longer valid due to the nonlinearity of the system, then *intermodulation products* will be found in the received signal, in addition to these frequencies and their harmonics. The frequencies of the intermodulation products are linear combinations of $f_1, f_2, \ldots, f_i$.

For example, for three sinusoidal components in f_1, f_2, and f_3, the term in u_E^2 through the decomposition of

$$(\cos\omega_1 t + \cos\omega_2 t + \cos\omega_3 t)^2 = \sum_i \cos^2\omega_i t + 2\sum_{i\neq j} \cos\omega_i t \cos\omega_j t$$

$$= \frac{3}{2} + \frac{1}{2}\sum_i \cos 2\omega_i t + \sum_{i\neq j} \cos(\omega_i + \omega_j)t + \sum_{i\neq j} \cos(\omega_i - \omega_j)t$$

with $i, j = 1, 2, 3$ \qquad (2.38)

gives rise to **second-order intermodulation products** at frequencies $f_1 \pm f_2$, in addition to the harmonics at $2f_1$ and $2f_2$.

The term in u_E^3 causes **third-order intermodulation products** at frequencies $2f_i \pm f_j$ and $\pm f_1 \pm f_2 \pm f_3$, as well as harmonics at $3f_1$, $3f_2$ and $3f_3$ by

$$(\cos\omega_1 t + \cos\omega_2 t + \cos\omega_3 t)^3 = \sum_i \cos^3\omega_i t + 3\sum_{i\neq j} \cos^2\omega_i t \cos\omega_j t$$

$$+ 6\cos\omega_1 t \cos\omega_2 t \cos\omega_3 t = \frac{15}{4}\sum_i \cos\omega_i t + \frac{1}{4}\sum_i \cos 3\omega_i t$$

$$+ \frac{3}{4}\sum_{i\neq j} \cos(2\omega_i \pm \omega_j)t + \frac{3}{2}\sum_{i\neq j\neq k} \cos(\omega_i \pm \omega_j \pm \omega_k)t \qquad (2.39)$$

with $i, j, k = 1, 2, 3$.

Table 2.10 summarizes the frequencies at which second- and third-order intermodulation products are found.

Table 2.10
Harmonics and intermodulation products.

Frequencies	1st order	2nd order	3rd order
Original signals	f_i		
Harmonics		$2f_i$	$3f_i$
Intermodulation products		$f_i \pm f_j$ $i \neq j$	$2f_i \pm f_j$ $f_i \pm f_j \pm f_k$ $i \neq j \neq k$

In the case of a signal formed by two sinusoidal components f_1 and f_2, figure 2.11 gives a qualitative idea of the emission and reception spectra in the presence of second- and third-order nonlinear distortions.

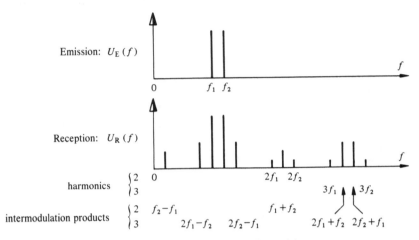

Fig. 2.11 Spectra in the presence of nonlinearities.

If the signal emitted is contained in a frequency band B located between f_1 and f_2, the harmonics and intermodulation products are located in the bands shown in figure 2.12.

Second-order products can be eliminated by filtering as long as $f_2 < 2f_1$.

On the other hand, *third-order products always appear in the original signal band*. This is why third-order nonlinearities (term in u_E^3) of the transfer characteristic are the most critical. They cause new spectral components in the original band (nonharmonic distortions) which finally manifest themselves as noise, called *intermodulation noise*, superimposed over the signal.

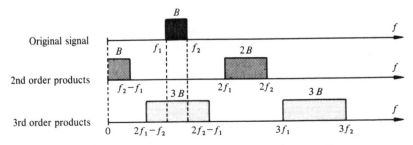

Fig. 2.12 Spectra of harmonics and intermodulation products.

2.4.11 Interference, noise, and crosstalk: definitions

In the course of transmission across a real channel, ***interference*** is added to the usable signal and alters its characteristics. The information extracted from the received signal is degraded, whether it is analog or digital.

Noise is generally used to mean any received signal, which, when interpreted by the receiver for information content, delivers incoherent information of no interest to the receiver.

The *background noise* exists even in the absence of any useful signal carrying information.

The presence of distortion in a transmission channel (nonlinearities, sect. 2.4.10; aliasing in sampling, sect. 4.3.5; quantizing, sect. 7.2.3) gives rise to a supplementary noise, which only appears when the channel is occupied by a signal. This is called *self-interference*.

Crosstalk, generally speaking, is an undesirable influence between useful signals transmitted on neighboring channels (in space, frequency, or time). The receiver in one channel receives parts of signals destined for other receivers and transmitted on other channels. If the information extracted in this way makes sense to the receiver, we have what is called *intelligible* crosstalk, a particularly serious condition because the secrecy of communication is breached. The opposite case, *unintelligible* crosstalk, causes *noise* whose importance depends on the signals carried by other channels.

There are many causes of interference in a system (table 2.13). Their analysis, and that of their effects, is one of the principal concerns in telecommunication systems.

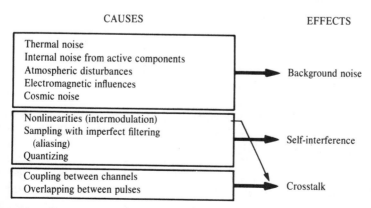

Table 2.13 Principal causes of interference.

2.4.12 Thermal noise

Thermal noise constitutes the most widespread, if not the most important, source of noise in telecommunication systems. Without going into the details of the physical phenomenon (chap. VI.6), we will use a simplified but realistic model of thermal noise in the form of noise which is simultaneously

- *white*, i.e., with a uniform power spectral density in the frequency range being considered (below 1000 GHz)
- *Gaussian*, i.e., the statistical distribution of instantaneous amplitudes is normal, with a zero mean, and a variance equal to the power P_{Nth}.

By virtue of the laws of thermodynamics, the maximum power of the thermal noise that a thermal resistance R can deliver in an infinitesimal frequency band df is independent of R and equal to

$$dP_{Nth} = kT df \qquad (2.40)$$

where k is Boltzmann's constant (1.38×10^{-23} Ws/K) and T is the absolute temperature.

By extension, the thermal noise power in a frequency bandwidth B is

$$P_{Nth} = kTB \qquad (2.41)$$

For the usual value of ambient absolute temperature of $T_0 = 290$K, we have

$$kT_0 \cong 4 \times 10^{-21} \qquad \text{Ws} \qquad (2.42)$$

which leads to the following values for the absolute noise level:

- telephone channel ($B = 3.1$ kHz):

$$L_{Nth\ tph} \text{ (re 1 mW)} \cong -139 \text{ dBm} \qquad (2.43)$$

- television ($B = 5$ MHz):

$$L_{Nth\ TV} \text{ (re 1 mW)} \cong -107 \text{ dBm} \qquad (2.44)$$

2.4.13 Psophometric weighting

In order to take into account the subjective effect that noise has on the user of an acoustical link, the background noise is measured across a filter which simulates the sensitivity of the ear at different frequencies in the conventional band for telephony or radio (music). This filter is incorporated into special voltmeters, called *psophometers*, which measure the true rms value of the noise. The attenuation of this filter, which is equivalent to a frequency dependent weighting according to the curves of figure 2.14, has been fixed by international convention (CCITT, Recommendation P.53).

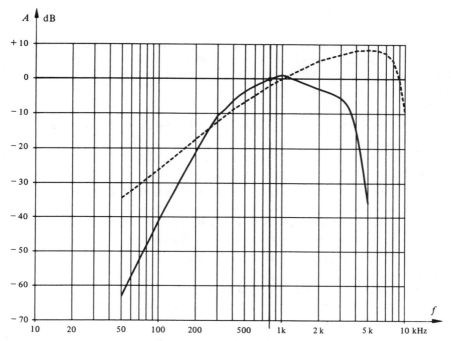

Fig. 2.14 Psophometric weighting curve: —for telephony; ---for broadcasting.

The noise levels measured across the psophometric filter are often expressed in dBmp to indicate the weighting that has been applied.

When white noise is measured in the frequency band ranging from 300 Hz to 3.4 kHz (e.g., thermal noise), across the psophometric filter used in telephony, the noise level is attenuated by 2.5 dB with respect to the unweighted value. Thus, after weighting, the thermal noise level in a telephone channel, according to relation (2.43), becomes

$$L_{\text{Nth tph p}} \text{ (re 1 mW)} = -141.5 \text{ dBmp} \tag{2.45}$$

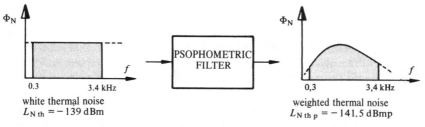

Fig. 2.15 Effect of psophometric weighting on thermal noise (telephony).

2.4.14 Qualitative planning: nominal signal

In analog systems, the signals representing the information to be transmitted have quite unpredictable characteristics (amplitude, phase, frequency). Under these conditions, it is difficult to define or measure the real signal level at a given point in the system.

For

- qualitative planning (*calculation* of levels, transmission plans, evaluation of signal-to-noise ratio, *et cetera*)
- testing (*measurement* of levels, of signal-to-noise ratio, of crosstalk, *et cetera*)

of the system, a *nominal signal*, or test signal, has been defined. This serves as a reference for all the characteristic values of the system, (emission power, modulation rate, modulation index, level, signal-to-noise ratio, *et cetera*).

The nominal signal has the following characteristics:

- it is purely sinusoidal (thus deterministic);
- its baseband frequency is conventionally chosen in telephony to be 800 Hz (820 Hz in systems sampled at 8 kHz);
- its level at any point X of the system defines the *nominal level* $L_{x\,nom}$ of the system.

In this way, level diagrams, such as that in figure 2.5, always represent the development of the *nominal* level along the length of the system.

2.4.15 Signal-to-noise ratio: definition

At a given point in a telecommunication system, the power P_N of the noise and its level L_N are meaningful only with respect to the power P_{Snom} and to the level L_{Snom} of the *nominal signal* at this point.

The severity of the interference depends on the ratio of these powers, called the *signal-to-noise ratio SNR*, and designated by

$$\xi = \frac{P_{Snom}}{P_N} \qquad (2.46)$$

or, more commonly, in logarithmic form:

$$10 \lg \xi = 10 \lg \frac{P_{Snom}}{P_N} = L_{Snom} - L_N \quad \text{dB} \qquad (2.47)$$

This ratio plays an essential role in the understanding of transmission quality. It is important to note that it is **always calculated with respect to the nominal signal level**.

2.4.16 Signal-to-noise ratio: requirements

The planning of analog transmission systems requires the following values, established after numerous subjective tests.

- telephony ($B = 3.1$ kHz):

$$10 \lg \xi \geq 50 \text{ dB} \tag{2.48}$$

musical transmission ($B = 5$ MHz):

$$10 \lg \xi \geq 47 \text{ dB} \tag{2.49}$$

television transmission ($B = 5$ MHz):

$$10 \lg \xi \geq 52 \text{ dB} \tag{2.50}$$

These values raise the following points:

- the value of 50 dB in telephony corresponds to a faintly perceptible noise. On the other hand, at 30 dB the noise would be quite pronounced and intelligibility would be affected;
- taking bandwidth ratios into account, the requirements for music and television are much more severe than with telephony.

2.4.17 Level referred to zero: definition

In telecommunications, the *level referred to zero* at a point X of a system is the difference between the **real** level (mean statistical level) of a random signal (telephone signal, video signal, *et cetera*), or of the noise at this point, and the **nominal** level (conventional test signal) at the *same* point.

$$L_{x0} = L_{x\text{ real}} - L_{x\text{ nom}} \qquad \text{dB} \tag{2.51}$$

L_{x0} can be interpreted as a **virtual** level; it is the real level which would be established at the point X if the nominal level were not $L_{x\text{ nom}}$ at this point, but 0 dBm. This justifies the expression, which is at first surprising, of level referred to zero. It would be more correct to say it is referred to the nominal level.

In the technical literature concerning telecommunications, this level is expressed in dBm0 to distinguish it from the absolute level expressed in dBm.

2.4.18 Noise level referred to zero

At a point X of a system, if the real noise level is L_{Nx}, the signal-to-noise ratio ξ_x at this point is given by (2.47). Further, the noise level referred to zero at this point is, from (2.51):

$$L_{Nx0} = L_{Nx} - L_{Sx\ nom} = -10 \lg \xi_x \tag{2.52}$$

The noise level referred to zero is thus equal to the signal-to-noise ratio (in dB), with the sign changed. In other words, it is nothing but the *"noise-to-signal" ratio* at this point.

2.4.19 Real level of telephone communication

The real signal resulting from the occupation of the system by a telephone communication is a widely variable random signal. The statistical distribution (probability density) of its instantaneous values has zero mean, but is not Gaussian (fig 1.9).

By international convention (CCITT, Recommendation G.223) a ratio

$$\frac{P_{real}}{P_{nom}} = 32 \cdot 10^{-3} \tag{2.53}$$

has been established between the real and nominal powers at the same point of a telephone system, which again means that the *average real level* of a telephone communication in time, including pauses, and over a large number of communications, is 15 dB lower than the nominal level of the system:

$$L_{tph\ real} = L_{nom} - 15 \text{ dB} \tag{2.54}$$

The mean level referred to zero of a telephone communication, from (2.51), is thus:

$$L_{tph\ 0} = L_{tph\ real} - L_{nom} = -15 \text{ dBm0} \tag{2.55}$$

This value has been chosen so that there is low probability of the random signal corresponding to a telephone conversation exceeding the nominal level.

This value is important, for example, for the evaluation of the overloading probability of an amplifier in real service (sect. 10.4.3) or the effective bandwidth in use in a microwave link (sect. 12.6).

2.5 SWITCHING QUALITY

2.5.1 Concept of traffic

In a public network, the means of transmission are at the disposal of a large number of users (e.g., telephone subscribers) who have access to it through the intermediary of a switching center (exchange). These users constitute sources whose random activity is generally small. They generate *calls*, of which the duration τ and number c per unit time are continuous random variables and which constitute the *traffic*.

The *(intensity of) traffic* Y routed by a collection of n equivalent units (for example, transmission channels), which constitute a *trunk group*, is the product of the mean duration τ_m of the occupations and their mean frequency c_m:

$$Y = \tau_m\, c_m \qquad (2.56)$$

Y is a dimensionless quantity, expressed in *erlangs* (E). A unit routes a traffic of 1 E if it is occupied for 100% of the observation period. Y also represents the average number of units occupied simultaneously in the trunk group. Figure 2.16 uses an example to illustrate the occupation of units in a six-line group routing a traffic of 2.6 E. It is shown that an average of 2.6 of the six lines are occupied.

The traffic Y_i routed by a unit i is the product of the mean time of occupation τ_m by the mean frequency c_{im} of the occupations of that unit:

$$Y_i = \tau_m\, c_{im} \leq 1 \qquad \text{E} \qquad (2.57)$$

Y_i expresses the rate of occupation of the unit i, i.e., the *probability* of finding the unit i occupied at a given moment.

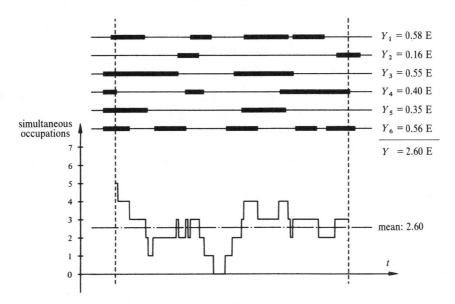

Fig. 2.16 Random occupation of a trunk group.

The traffic routed by the n trunk is the sum of the traffic Y_i:

$$Y = \sum_{i=1}^{n} Y_i \leqq n \qquad \text{E} \qquad (2.58)$$

It must be noted that Y is not a probability, while Y_i is.

In the same way, we can define the *offered traffic* A, by a group of N sources, by the product of the mean frequency of *call attempts* c_{Am} of the system by the sources and the average duration τ_m of *occupations* (completed calls):

$$A = \tau_m \, c_{Am} \qquad \text{E} \qquad (2.59)$$

2.5.2 Statistical speculations, losses, delays

When a large number N of sources of low traffic have access to a smaller number n of lines, the probability that all n lines will be occupied simultaneously is not zero. There is, then, *congestion*.

In a *loss system*, any new call attempt will then be refused (for example: a busy signal is generated). The traffic routed by the system is thus lower than the traffic offered by the sources.

In a *delay system*, the call attempt is delayed until a line is freed by the end of a communication. All the traffic offered is eventually routed, as long as $A < n$.

The evaluation of the probability of losses or waiting, as well as the delays which accompany them, are the subjects of teletraffic theory, which will not be elaborated here.

These probabilities are quantitative measures of the network's switching quality. Here, too, a compromise must be made between the size of the trunk group (cost) and the probability of congestion (quality).

2.5.3 Quantitative planning: traffic estimation

All the quantitative planning of the public switched network rests on the maximum value of daily traffic, which has been established on the basis of traffic volume during the *busy hour*. The rest of the time, the network is oversized.

For a prescribed switching quality (loss probability or characteristics of delays), determining the number of units necessary, requires knowledge of the actual or estimated traffic offered.

Statistical observation (counting call attempts, occupations, and their duration), permits evaluation of the traffic offered, which must then be extrapolated on the basis of experience in analogous situations.

PLANNING: OBJECTIVES, CONSTRAINTS, AND METHODS 59

Such a value based on experience is, for example, the mean value (determined for a great number of subscribers) of the traffic offered A_1 by a subscriber:

$$A_1 \cong 0.05 \text{ E/subscriber} \tag{2.60}$$

This means that on the average a subscriber is occupied for three minutes during the busy hour. This value permits an estimation of the traffic offered by a large group of subscribers. In certain groups of particular subscribers, much higher traffic can be established (e.g., $A_1 \cong 0.2$ E/subscriber in the business district of a large city).

2.5.4 Prediction of user density

An essential element of the long-term planning of a telcommunications network is the estimation of the future development of the number of users in relation to the population statistics (user density D).

A mathematical model of *exponential growth* is often assumed (fig. 2.17), starting with a density D_0 at time $t = 0$.

$$D = D_0 \exp(t/\tau) \tag{2.61}$$

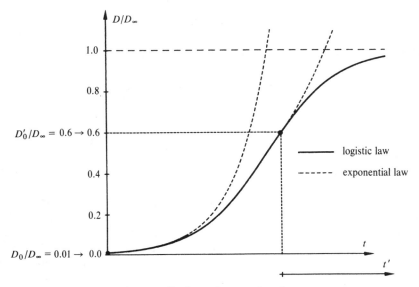

Fig. 2.17 Models for the prediction of user density.

Its rate of growth is

$$\frac{dD}{D\,dt} = \frac{1}{\tau} = \text{const.} \tag{2.62}$$

Nevertheless, this model is unrealistic over an extended period because the density D cannot increase infinitely, but must rather approach an asymptotic value, corresponding to a saturation point. One way of accounting for this phenomenon is by the *logistic curve*. This proceeds from the hypothesis that the rate of increase is not constant, but proportional to the difference between D and its asymptotic value D_∞:

$$\frac{dD}{D\,dt} = \frac{1}{\tau}\frac{D_\infty - D}{D_\infty} \tag{2.63}$$

The result is the *logistic law* (fig. 2.17):

$$D = \frac{D_\infty}{1 + \left(\dfrac{D_\infty}{D_0} - 1\right)\exp\left(-\dfrac{t}{\tau}\right)} \tag{2.64}$$

where D_0 is the density at time $t = 0$.

2.6 RELIABILITY

2.6.1 Definition of reliability and its implications

Reliability is an equipment characteristic expressed by:
- the probability that it will operate,
- carrying out a required function,
- under given conditions,
- during a given time.

The *equipment* can be a component, a unit, a system, a network, or even a program.

The *function* required must be specified clearly (specifications), in such a way that the limits (tolerances) between them can be considered satisfactory.

The *conditions* are related to the climatic environment (temperature, humidity), mechanical environment (shocks, vibrations), chemical environment (corrosion) or electrical environment (static discharges, over-voltages, under-voltages). These conditions must be carefully specified.

Reliability is related to the idea of *duration* of good operation and is thus related to time.

Finally, reliability is quantifiable, but always in the form of a *probability*.

Reliability is a general concern in technology. Nevertheless, it takes on particular importance in telecommunication systems, from which we naturally expect a permanent service of high and consistent quality. This is why the strict demands of reliability (maximum probability and duration of total or partial breakdowns) are systematically included in the specifications. The concern for reliability must be included from the beginning of the design of a system.

2.6.2 Improvement of reliability and cost

There are, in principle, two methods of obtaining a highly reliable system:
- utilizing components of high intrinsic reliability and avoiding any overload;
- choosing redundant structures which allow good functioning of the system, even in certain cases of failures at the component level.

The two possibilities, often combined, are expensive. Reliability is costly, but the higher cost of the more reliable system is compensated for by reduced operating costs (fewer breakdowns, thus fewer repairs). An optimum must be found by considering the sum of system costs and its costs of operation (fig. 2.18).

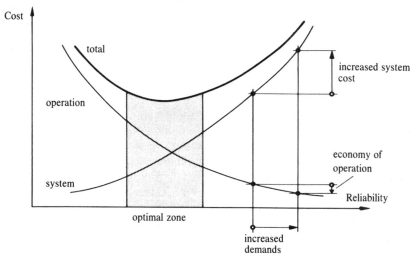

Fig. 2.18 Reliability and overall cost.

2.6.3 Reliability requirements for a telecommunications network

All breakdowns in a system do not have the same importance. In the case of a telecommunications network serving a large number of users, breakdowns can be evaluated according to two criteria:
- *severity*, i.e., the relative loss of quality of service experienced by users involved (100% is equivalent to a total interruption of service). The probability of congestion in a network as the result of a breakdown is, for example, a good measure of the severity of the breakdown;
- *extent*, i.e., the proportion of users affected (0% represents the limit case of a single user affected, while 100% represents a loss of quality experienced by all users).

The tolerable probability of appearance of a breakdown depends on its severity and extent. In particular, *total breakdown* of a system or network, which deprives all users (100% extent) of all service (100% severity) must be a *highly improbable* event.

A realistic order of magnitude demanded of public telecommunications network reliability is the following:
- the cumulative duration of total breakdowns of the system must remain *less than two hours* during a 30-year period of operation.

In order to conform to such a severe demand, it is imperative to have recourse to redundant structures, which permit reduction of the severity and the probability of appearance of an isolated defect. On the other hand, the segmentation of the system into units, independent of each other from the point of view of their reliability, allows us to limit the extent of the breakdowns.

2.6.4 Behavior in the event of a breakdown

The majority of telecommunication systems are *reparable systems*, which means that in the event of a breakdown, the defective unit is repaired or replaced and the integrity of the service is re-established after a more or less long time, called *time to repair* or *down time*.

This delay, which varies from case to case according to the accessibility of the unit to be repaired, the availability of maintenance personnel, and the severity of the defect, includes, as well as the time to repair *per se*, the time necessary to detect the breakdown, locate it, arrive at the location, and put the system back into service.

Evidently, telecommunications satellites are, for the moment at least, *non-reparable systems*, i.e., only the further development of their reliability char-

PLANNING: OBJECTIVES, CONSTRAINTS, AND METHODS

acteristics between their placement into service and the first breakdown need be considered.

2.6.5 Hypotheses

The theory of reliability elaborates on mathematical models, which, using certain simplifying hypotheses, allows us to describe the probable behavior of the system's reliability parameters. Only a very elementary approach will be made here, with the following hypotheses:

- total failures: the complete and sudden disappearance of a function. The units and the system will, therefore, have a binary behavior: either functioning well or broken;
- random failures: the defects are assumed to be statistically independent. The breakdown of one unit does not involve that of another. The exact moments of breakdown are strictly unpredictable;
- reparable system;
- independent repairs: in the case of simultaneous failures of units, the repairs must also take place simultaneously (a personnel problem!);
- perfect repairs: the repaired system is in all respects identical with the intact system as far as reliability is concerned.

These hypotheses are convenient, but debatable. In particular, operating experience shows that the majority of system breakdowns are not due to random component defects, but rather to *the clumsy intervention of man* (repairs, modifications of wiring or programming, maintenance, cleaning, *et cetera*), which escapes all mathematical prediction! On the other hand, *catastrophic breakdowns*, originating outside the system (disaster, sabotage, accidental cutting of a cable, *et cetera*) are also very difficult to express as a mathematical probability. Their *possibility* (rather than their probability) of appearance can, nevertheless, have a very important influence on the design of a system or network.

2.6.6 Parameters of reliability: definition

The behavior, binary by hypothesis, of a reparable unit or system can be described in time by a random series of the conditions "functioning well" and "broken" (fig. 2.19).

The duration of a breakdown τ_{0i} (down time) and the interval between two breakdowns τ_{1i} are continuous random variables. Their mean value, or mathematical expectation, is designated by:

- *mean time to repair* MTTR:

$$\tau_0 = E(\tau_{0i}) \qquad (2.65)$$

- *mean time between failures* MTBF:

$$\tau_1 = E(\tau_{1i}) \qquad (2.66)$$

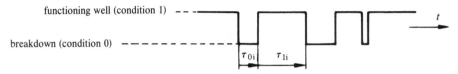

Fig. 2.19 Behavior in time.

The *availability* A is defined by

$$A = \frac{\tau_1}{\tau_1 + \tau_0} \qquad (2.67)$$

Thus, A expresses the stationary probability (long-term) of finding the equipment functioning well at a given moment.

In the fortunately frequent case of highly reliable equipment, i.e., for which

$$\tau_1 \gg \tau_0 \qquad (2.68)$$

the reliability A is very close to 1. It is thus more practical to define the *unavailability* $1 = A$

$$1 - A = \frac{\tau_0}{\tau_1 + \tau_0} \cong \frac{\tau_0}{\tau_1} \qquad (2.69)$$

The value $1 - A$ is also a stationary probability, that of finding the equipment broken at a given moment. The unavailability gives a better idea of the degree of reliability of a unit or a system than the MTBF. For example, the requirement stated in section 2.6.3 corresponds to an unavailability $1 - A$ = 2h/30 years = $7.6 \cdot 10^{-6}$.

2.6.7 Structure of the system. Reliability diagram

The dependence of the reliability of a system on that of the different units which compose it can be schematized by a reliability diagram in which the units are represented by blocks arranged in the following manner:

- in *series*: units or groups of units whose functioning is *necessary* to the overall functioning of the system (fig. 2.20(a));

PLANNING: OBJECTIVES, CONSTRAINTS, AND METHODS 65

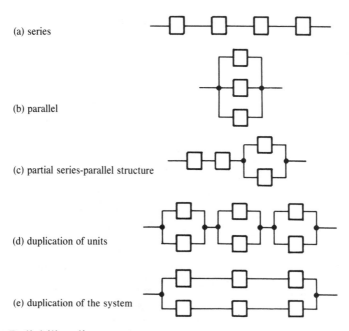

(a) series

(b) parallel

(c) partial series-parallel structure

(d) duplication of units

(e) duplication of the system

Fig. 2.20 Reliability diagrams.

- in *parallel*: units equivalent in function, such that the functioning of one of them is *sufficient* for the whole group of units to function (fig. 2.20(b)).

These blocks are symbolically equivalent to open interrupters (broken unit) or closed interrupters (functioning unit). The system is considered to be broken if there is no possible route across the reliability diagram.

In this way, the majority of systems can be represented by series arrangement of parallel structures or parallel placement of series chains (fig. 2.20(c), (d), (e)).

It is, therefore, sufficient to analyze these two arrangements and to calculate the availability of the partial system from that of the units which compose it in the case of series or parallel structures.

2.6.8 Series system (without redundance)

The series system functions if, and only if, all the units function. Consequently, and taking into account the statistical independence of the breakdowns, the probability of proper functioning (availability A_S) of the system is equal to the product of the availabilities A_{Uj} of the units.

The availability of the series system is, therefore,

$$A_S = \prod_j A_{Uj} \tag{2.70}$$

and its inavailability is

$$(1 - A_S) = 1 - \prod_j [1 - (1 - A_{Uj})] \tag{2.71}$$

If $1 - A_{Uj} \ll 1$, then

$$\prod_j [1 - (1 - A_{Uj})] \cong 1 - \sum_j (1 - A_{Uj}) \tag{2.72}$$

and

$$(1 - A_S) \cong \sum_j (1 - A_{Uj}) \tag{2.73}$$

During the time T, there is an average of T/τ_{1Uj} breakdowns of unit j, given N breakdowns of the series system, with

$$N = \sum_j \frac{T}{\tau_{1Uj}} \tag{2.74}$$

The mean time between failures (MTBF) of the system τ_{1S} is, thus,

$$\tau_{1S} = \frac{T}{N} = \left(\sum_j \frac{1}{\tau_{1Uj}}\right)^{-1} \tag{2.75}$$

2.6.9 Parallel system (redundant)

The parallel system is down if, and only if, all the units are down.

The system is composed of several units which are equivalent in their function, but not necessarily identical in their reliability characteristics. The units are all in service ("hot" reserve), and only one is needed for the system to function.

Assuming that unit breakdowns are independent, the unavailability of the parallel system is, thus,

$$1 - A_S = \prod_j (1 - A_{Uj}) \tag{2.76}$$

and the mean time to repair (MTTR)

$$\tau_{0S} = \left(\sum_j \frac{1}{\tau_{0Uj}}\right)^{-1} \tag{2.77}$$

PLANNING: OBJECTIVES, CONSTRAINTS, AND METHODS

It must be noted that, in the case of a parallel system, the *total* breakdowns (of the system) are much less frequent than with the series system. The unavailability is considerably reduced. On the other hand, *partial* breakdowns, and hence repairs, are more frequent the higher the redundance of the system (increased number of units, thus increased risk of breakdown).

Table 2.21 summarizes the characteristic properties of series and parallel structures, and emphasizes the duality of relations (2.70) and (2.75) with respect to (2.76) and (2.77).

Table 2.21 Reliability characteristics of series and parallel structures.

	Series	Parallel
Reliability diagram	—[1]–[2]–[3]⋯[m]—	[1],[2],[3],…,[m] in parallel
Availability of the system	$A_S = \prod_j A_{Uj}$	$A_S = 1 - \prod_j (1 - A_{Uj})$
Inavailability of the system	$1 - A_S \cong \sum_j (1 - A_{Uj})$	$1 - A_S = \prod_j (1 - A_{Uj})$
System MTBF (mean time between failures)	$\tau_{1S} = \left(\sum_j \frac{1}{\tau_{1Uj}}\right)^{-1}$	$\tau_{1S} \cong \dfrac{\tau_{0S}}{1 - A_S}$
System MTTR (mean time to repair)	$\tau_{0S} \cong (1 - A_S) \cdot \tau_{1S}$	$\tau_{0S} = \left(\sum_j \frac{1}{\tau_{0Uj}}\right)^{-1}$
Specific case: m identical units (with respect to reliability)	$1 - A_S \cong m(1 - A_U)$ $\tau_{1S} = \dfrac{1}{m}\tau_{1U}$	$1 - A_S = (1 - A_U)^m$ $\tau_{0S} = \dfrac{1}{m}\tau_{0U}$
Subscripts: S = system, Uj = unit j	0 = malfunctioning 1 = functioning well	

2.6.10 Example of a reparable system with variations

A system is composed of four units, all necessary to its functioning, and the following reliability characteristics (table 2.22). Several possible structures, with or without redundancy, are compared in table 2.23.

It is noted that the duplication of each unit, apparently without supplemen-

Table 2.22 Reliability characteristics (example).

Unit	MTBF τ_{1Uj}	Repair delay (time to repair) τ_{0U}	Inavailability $1 - A_{Uj}$
A	5 000 h		10^{-3}
B	500 000 h	5 h	10^{-5}
C	10 000 h		$5 \cdot 10^{-4}$
D	10 000 h		$5 \cdot 10^{-4}$

tary hardware, provides better reliability than duplication of the entire system. Nevertheless, if the independence of breakdowns is to be maintained, this parallel arrangement poses decoupling problems in practice.

2.6.11 Application to telecommunications

The telecommunication systems and networks often make systematic use of redundant structures to increase the reliability of the service.
Some examples:

- duplication of radio links on other frequencies or other antennas (diversity, sect. 3.8.5);
- parallel use of microwave links and cables in the lines between cities, on one hand, and satellites and submarine cables used in parallel for intercontinental links on the other;
- supply from two sources: mains with batteries, and diesel motor generators (in "cold" reserve, i.e., put into use only in the case of a mains breakdown);
- placement of cables along different routes (protection against catastrophic breakdowns);
- duplication (even triplication) of control processors in exchanges;
- principle of twin contact (two points of virtually independent mechanical contact in relays and switching units);
- redoubling of important nodes of the network (intercity and international exchanges) in geographically distinct sites;
- meshed network with the possibility of using indirect routes.

2.6.12 Software failures

It may appear surprising to speak of failures or reliability of programs. However, in systems controlled by computers, notably in modern switching exchanges, certain breakdowns are due to software failures.

Table 2.23 Variations of structures for a system (example).

Structure	Reliability diagram	Inavailability of the system	System MTBF	Mean duration of failure	Cumulative duration of failures in 30 years	Mean time between repairs
Series (without redundance)	—A—B—C—D—	$2.01 \cdot 10^{-3}$	103.7 days	5 hr.	22 days	103.7 days
System duplication	A-B-C-D / A'-B'-C'-D'	$4.04 \cdot 10^{-6}$	70 yr.	2.5 hr.	1.1 hr.	51.8 days
Unit duplication	parallel pairs A/A', B/B', C/C', D/D'	$1.5 \cdot 10^{-6}$	190 yr.	2.5 hr.	24 min.	51.8 days
Partial duplication	A/A' — B—C—D	$1.01 \cdot 10^{-3}$	205.8 days	4.98 hr.	11.1 days	69.2 days

This is actually a matter of imperfect reasoning or programming during the development of the system. Therefore, they do not, unlike physical failures, have a random character, but exist potentially in the system. Nevertheless, they are revealed in a quasi-random manner, sometimes after months of service for the most underhanded of them, as the result of an unforeseen conjunction of rare events. In the worst of cases, they can lead to paralysis of the system.

As with physical breakdowns, in repairing the system there is the highest risk of introducing new errors. The correction of a programming error can lead to a series of unforeseen consequences. We are, therefore, never sure that a complex system has been completely debugged of all its programming errors.

2.6.13 Maintenance

Maintenance operations are intended to assure, and, according to the case, to re-establish proper functioning of the system. Maintenance is thus a factor that contributes to the reliability of a system when it is in service.

Among the measures of preventive maintenance (prevention of breakdowns) we can cite:
- *upkeep* (cleaning, lubrication, replacement of worn components, *et cetera*)
- *monitoring* through routine tests, traffic measurement, state statistics, *et cetera*

Upon appearance of a failure, maintenance requires corrective measures in the following order:
- immediate *detection* of the breakdown by the negative results of routine tests and triggering of alarms;
- *diagnosis*, which leads to the location, as precisely as possible, of the defective unit or program;
- emergency *repair*, which consists of either repairing the defective unit, or, more often, replacing it with a reserve unit in order to repair the defective unit later in the workshop. This phase includes its placement back into service with all the necessary tests and reinitializations.

The concern for maintenance must be given primary consideration from the beginning stages of the design (hardware and software) of a system. Maintenance plays an important role in the cost of a system (test possibilities, diagnostic programs, man-machine interfaces), especially the operating costs (maintenance and monitoring personnel, inspection, repair shops). The current trend is in the direction of automated maintenance, intended to reduce operating costs, at the price, however, of an increase in the complexity, and thus the cost, of the initial system (fig. 2.24).

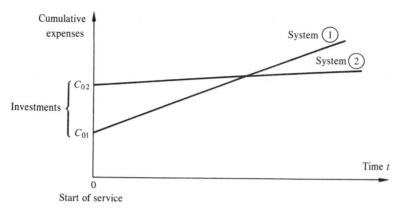

Fig. 2.24 Economic comparison of two systems, the first (1) at low initial cost, but increased operating costs, the other (2) more expensive at acquisition, but less expensive to operate.

2.7 ECONOMIC ASPECTS

2.7.1 Nature of costs

The cost of telecommunications service consists of two categories of expenses (fig. 2.24):

- *investments*, or initial costs (capital C_0), necessary for the acquisition of goods (land, buildings) and materials (equipment, their transport, construction and placement into service). They also include costs of research and development, as well as the initial training of personnel;
- *operating costs* (total amount e per year) incurred by the existence and use of the system. These cover the operation, maintenance, and management of the service. Repetitive in nature and directly proportional to time, in the majority of cases they correspond to manpower costs (salaries of operating personnel), but also take into account energy consumption, wear and tear of materials, *et cetera*

Operating costs are not necessarily in direct relation to the investment, as the two examples in figure 2.24 show. In any case, for a given type of material and under specified operating conditions, the experience acquired allows us to express annual operating costs as an empirically derived percentage of the capital invested (table 2.26).

2.7.2 Amortization of investments or back-calculation of annual costs

In order to be able to combine and compare the initial (investment) and recurrent (operating costs) system cost factors, it is necessary to combine these two cost categories into one, taking into account the compound interest relative to the sum, as a one-time cost (single outlay) or distributed over time.

Two methods of calculation are applicable, by choice:

- *amortizing*, i.e., to convert the capital C_0 into a series of n identical sums, called *annuities a*, to be deposited at the end of each year during the n years corresponding to the estimated time of operation (expected lifetime) of the system. The sum of these n deposits with their compounded interest at the end of n years gives the same value C_n as that which the capital C_0 gives after n years of interest compounded at the same rate i. This equivalence is illustrated in figure 2.25. The value of the corresponding amortization annuity a is given by the relation:

$$a = C_0 \frac{i(1+i)^n}{(1+i)^n - 1} \tag{2.78}$$

The *annual effective cost* of the system and its operations is, therefore, given by the sum of the amortization annuity a and annual operating costs e

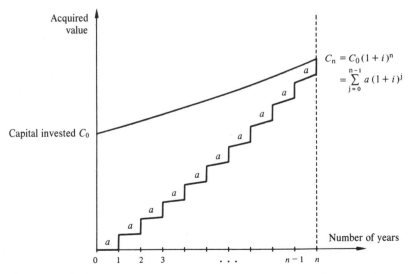

Fig. 2.25 Equivalence between acquired value C_n of capital C_0 after n years and by n step annuities a.

PLANNING: OBJECTIVES, CONSTRAINTS, AND METHODS

Table 2.26 Estimated time of operation and operating costs.

Materials	Time of operation n	Annual operating costs related to capital e/C_0
Cables	25 ... 40 yr.	0.3%
Digital systems Carrier systems	15 yr.	1.5%
Microwave links	15 yr.	2.5%
Satellites	5 yr.	2%

- *Back-calculating* the operating costs, i.e., to calculate the capital E_0 which it would have required at the beginning of the n year period of operation to cover the annual costs of operation e during this period, taking into account the compound interest at the rate i. The corresponding capital is equal to:

$$E_0 = e \, \frac{(1 + i)^n - 1}{i \, (1 + i)^n} \qquad (2.79)$$

The *overall effective investment* for the system and its operation is thus given by the capital C_0, necessary for the acquisition of the system, and the back-calculated value E_0 of its annual operating costs.

The two methods of calculation (amortization or back-calculation) are exactly equivalent. They allow economic comparison of systems on the basis of annual effective costs or the overall effective investment. The estimated time of operation (n years) is an essential element of comparison. The interest rate i, which depends on economic factors, can also exert a determining influence.

Table 2.26 provides only an indication of some of the usual values of telecommunications equipment.

2.7.3 Example

Two equivalent technical systems, designed to function for $n = 30$ years, have the following economic characteristics, (corresponding qualitatively to the case of fig. 2.24):

- System 1: initial costs
  $C_{01} = 10,000$ units; operating costs $e_1 = 200$ units/year

- System 2: inital costs $C_{02} = 8,000$ units, operating costs $e_2 = 400$ units/year

Which is the most economical, if the interest rate is 4%?

Table 2.27 compares the effective annual costs (amortization method), on one hand, and the overall effective investment (back-calculation method) of the two systems, on the other. The comparison is advantageous to system 1.

Let us note that the result would have been reversed if the time of operation had been reduced to 13 years or if the interest rate changed to 8%.

Table 2.27 Example of economic comparison of two systems.

	System 1	System 2
Amortization annuity $a =$	736	589
Annual operating costs $e =$	200	400
Effective annual costs $a + e$	936	989
Initial costs $C_0 =$	10,000	8,000
Present value of annual operating costs $E_0 =$	2,720	5,436
Effective overall investment $C_0 + E_0 =$	12,720	13,436

2.7.4 Effect of size on costs

The overall costs C (inital overall investment or effective annual costs) of a system can be broken down into a fixed part (base cost) and a variable part (incremental cost) that increases as a function of a quantity x, which can be, for example:

- the capacity (number of channels) of the system
- the transmission distance

Empirically, it has been established that the overall cost C increases with x, according to the law (fig. 2.28):

$$C = K_1 x^m + K_2 \qquad (2.80)$$

with $m = 0 \ldots 1$ and where K_1 and K_2 are system-dependent constants.

If $m = 1$, the incremental cost is constant and there is less interest in going toward large values of x (small effect of size) when K_2 is smaller. Conversely, $m = 0$ indicates a maximum effect of size: the greater x, the more economical the system. This is the case particularly with satellite links, where the cost is independent of distance, while with a cable link, it increases proportionally to distance ($m = 1$).

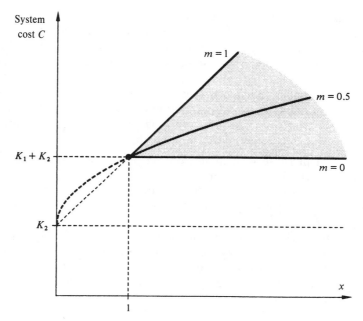

Fig. 2.28 Effect of size on system costs.

If we compare the cost of a great number of multi-channel systems as a function of their capacity $x = z$ (number of channels), we can establish a remarkable concurrence towards a value of $m \cong 0.6$ that explains the tendency to design systems with larger and larger capacity, which are more economical than a multiplicity of small-channel systems with the same overall capacity.

Often, the function $C(x)$ experiences leaps, for example, due to the necessity of installing a supplementary system (new investment, but unchanged incremental cost) in order to absorb increased capacity, or because of the need to amplify, or to regenerate signals after a certain transmission distance.

Comparison of the functions $C(x)$ for different systems defines the optimal levels (total number of channels, transmission distance) of utilization for these systems (sect. 15.2.8).

2.7.5 Techno-economic inertia

Three factors bring an enormous inertia to the telecommunications network:
- the *size of investments* devoted to the development and expansion of the network. In the case of the worldwide telephone network, the size of

investments can be evaluated very roughly at some $1000 US per subscriber link, approximately 30% of which is manpower costs.
- The *longevity* of the equipment: the investments are made on the basis of a time of operation on the order of 10 to 30 years, often exceeded in reality;
- the indispensable *compatibility* of equipments among themselves; local, regional, and international compatibility (section 2.2) of contemporary systems, but also compatibility in time between systems from different generations whose operation times overlap.

The constant increase in demand, on the one hand, and the stagnation of investment possibilities and techno-economic inertia, on the other hand, put telecommunications in a conflict whose solution is found in the development of new systems and technology, which are more economical and require fewer personnel.

The worldwide telecommunications network is in constant quantitative and qualitative evolution. Nevertheless, this evolution, while very prominent, must be slow. A system, conceived today to be operated tomorrow, but which would not be compatible with yesterday's systems, will remain a utopia in the future.

Chapter 3

Transmission Media

3.1 INTRODUCTION

3.1.1 Lines and radio waves: definitions

Since its beginnings, telecommunications technology has used metallic wires to guide information-carrying signals, then it freed itself from this physical link between emitter and receiver to make direct use of omnidirectional radiation (radio broadcasting) or more or less directed radiation (links) of electromagnetic waves.

Line is the term used to designate a finite transmission material.
It can consist of for example,

- two bare parallel metallic wires: open-wire line
- two insulated and twisted metallic wires: balanced pair
- two concentric conductors: coaxial pair
- a metallic tube: waveguide
- a translucent dielectric filament guide: optical fiber.

The name *cable* is given to a construction unit composed of one or several lines and protected from exterior physical, chemical, or electromagnetic influences (in the air, water, or earth).

Although, in the case of line transmission, the energy (and thus the information) is also in fact transported by a guided electromagnetic wave, the term *radio wave transmission* is reserved for the case in which no specific physical element guides the signals along the length of the path from the emitter to the receiver.

3.1.2 Objective

The aim of this chapter is to show the essential characteristics of these media from the point of view of telecommunication systems which use them. The results of the line and wave propagation theory (vol. III) will serve as a basis for these considerations.

3.2 GENERAL LINE CHARACTERISTICS IN THE FREQUENCY DOMAIN

3.2.1 Hypothesis

The line is assumed to be *uniform*, i.e., its local electrical characteristics are the same along the length of the line. This is generally true in telecommunications lines, at least by section, and on a macroscopic scale with respect to the wavelengths used.

The case of lines which are nonuniform, but loaded with periodic obstacles, is discussed in vol. III, sect. 8.8.

3.2.2 Primary parameters

An elementary section of line of length dx can be modeled by the equivalent scheme of figure 3.1.

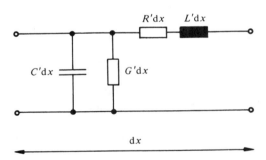

Fig. 3.1 Primary parameters.

The elements of this equivalent scheme are called *primary parameters* of the line. They are:
- the *resistance per unit length R'*, which expresses the series resistance of the two conductors per unit length (in practice, generally 1 km) of the line, taking into account the skin effect.
- the *inductance per unit length L'*, one part of which is due to the magnetic field inside the conductors (internal conductance, dependent on the skin effect), and the other is due to the magnetic field between the conductors (external conductance, practically independent of frequency);

- the *capacitance per unit length C'* related to the permittivity of the dielectric located between the conductors;
- the *leakage conductance per unit length G'* whose origin is partly due to insulation defects (generally negligible) and, on the other hand, dielectric losses whose importance increases with frequency according to relation:

$$G' = \omega C' \tan \delta \tag{3.1}$$

where δ is the angle of loss of the dielectric.

Table 3.2 gives some useful values of the principal materials used for telecommunications lines.

Table 3.2 Materials for lines.

Conductors	Resistivity ρ at 20°C	Temperature coefficient $\dfrac{\Delta \rho}{\rho \Delta \theta}$
Electrolytic copper	17.5 nΩm	+3.9 0/00 °C^{-1}
Aluminum	27.8 nΩm	+4 0/00 °C^{-1}
Bronze (open wire)	20–55 nΩm	+2 0/00 °C^{-1}
Dielectrics	Relative permittivity ϵ_r	Loss factor $\tan \rho$
Paper (dry)	2–2.5	$\sim 3 \times 10^{-3}$ at 800 Hz
Polyethylene (PE)	2.3	$\sim 3 \times 10^{-3}$ at 1 MHz
Polystyrene (PS)	2.5	$\sim 2 \times 10^{-4}$ Af
		$\sim 2 \times 10^{-4}$ Af

3.2.3 Skin effect in a coaxial pair

When a conductor is traversed by an alternating current, the current density is not uniform on the section of the conductor but tends to be concentrated at the surface. This phenomenon, called the *skin effect*, is the subject of vol. IX, sect. 1.4.

In the case of concentric conductors, of circular cross section (coaxial pair, fig. 3.3), the current is concentrated close to the external surface of the interior conductor and near the internal surface of the exterior conductor. The equivalent section of the two conductors can thus be evaluated with the aid of the *depth of penetration* ϑ (vol. IX, eq. (1.108))

$$\vartheta = \sqrt{\dfrac{\rho}{\pi \mu f}} \tag{3.2}$$

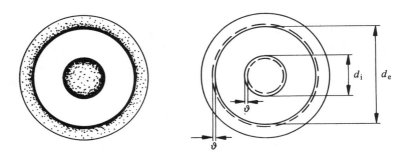

Fig. 3.3 Skin effect in a coaxial pair.

In the case of copper (resistivity $\rho = 17.5$ nΩm, permeability $\mu \cong \mu_0$), the depth of penetration depends on the frequency, as shown in figure 3.4.

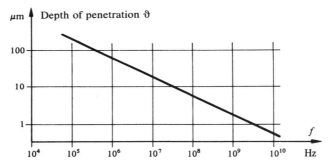

Fig. 3.4 Depth of penetration in copper.

For $\vartheta \ll d$, the equivalent section of each conductor is

$$A \cong \pi d \vartheta = \pi d \sqrt{\frac{\rho}{\pi \mu f}} \tag{3.3}$$

with $d = d_i$ or d_e and $\rho = \rho_i$ or ρ_e (the two conductors are not necessarily of the same metal).

In order to account for the skin effect, the resistance per unit length R' must be replaced by a complex impedance $\underline{Z}'_R$.

Its real part is the resistance R'_P corresponding to the equivalent section A defined by the depth of penetration; its imaginary part, of inductive character, expresses the variation of the internal inductance or the conductors.

We have, therefore, according to [12]

$$R' \rightarrow \underline{Z}'_R = R'_P(1 + j) \tag{3.4}$$

with

$$R'_P = \frac{\rho_i}{A_i} + \frac{\rho_e}{A_e} = \sqrt{\frac{\mu}{\pi}} \left(\frac{\sqrt{\rho_i}}{d_i} + \frac{\sqrt{\rho_i}}{d_e} \right) \sqrt{f} \qquad (3.5)$$

Consequently, we should note that
- the resistance increases proportionally to $\sqrt{f}$
- the resistance no longer depends on ρ but on $\sqrt{\rho}$.

3.2.4 Proximity effect in a balanced pair

In the case of a line composed of two juxtaposed wires (balanced pair) with circular cross section, the physical phenomenon, which is the source of the skin effect, leads to a current density distribution approximating the form sketched qualitatively in figure 3.5.

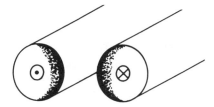

Fig. 3.5 Proximity effect.

The theoretical analysis of this phenomenon is extremely complex [13]. It has been established in practice that this effect only appears by a detectable increase of the resistance starting with relatively high frequencies, i.e., in a range where other inconveniences (crosstalk, sect. 3.5.5) limit practical use of balanced pairs in any case.

3.2.5 Effect of temperature on the primary parameters

The most noticeable effect is that caused by the variation in the resistivity ρ with temperature. With continuous current, R' varies proportionally with ρ, i.e., it increases by approximately 4⁰/₀₀ per degree. At high frequencies, due to the skin effect, R' increases only proportionally to $\sqrt{\rho}$, thus approximately 2⁰/₀₀ per degree. These variations are not negligible in long-distance transmission systems (section 10.3).

The effect of a variation in temperature on other primary parameters appears to be negligible upon initial analysis.

3.2.6 Secondary parameters

The primary parameters are of limited practical interest to the line user. The so-called *secondary parameters*, which are directly measurable, even on a very long line, and can describe the behavior of the line more concretely, are therefore preferred. These are

- the *characteristic impedance* $\underline{Z}_c$

$$\underline{Z}_c = Z_c \exp(j\zeta_c) \qquad (3.6)$$

defined in vol. III, sect. 8.3.3. This is the impedance that would be measured at the input of an infinitely long line or of a finite line terminated with the same $\underline{Z}_c$. It does not depend on the length of the line.
- the *propagation coefficient* γ

$$\gamma = \alpha + j\beta \qquad (3.7)$$

of which the real part is the *attenuation coefficient*, and the imaginary part the *phase change coefficient*. These parameters are related to a unit line length (generally 1 km).

For a uniform line of length l, the propagation coefficient is

$$\underline{\Gamma} = A + jb = \gamma l = \alpha l + j\beta l \qquad (3.8)$$

where $A = \alpha l$ is the attenuation and $b = \beta l$ is the phase change.

The transfer function $\underline{H}(l)$, which relates the complex output voltage $\underline{U}(l)$ to that of the input $\underline{U}(0)$ of a line **terminated by its characteristic impedance at both ends** is given, from vol. III, table 8.15, by

$$\underline{H}(l) = \underline{U}(l)/\underline{U}(0) = \exp(-\underline{\Gamma}) \qquad (3.9)$$

Two comments are necessary on this subject:

- γ, α and β are *image parameters* (vol. XIX, sect. 3.1) in the sense that they are what we would observe in the case (generally unrealistic) of the line terminated with $\underline{Z}_c$;

because of the exponential function in (3.9), A and α are expressed in **neper** (Np) and **neper per kilometer** (Np/km), respectively. To obtain the dB or dB/km equivalent, a multiplicative factor of 20 lg e $\cong$ 8.68 dB/Np is necessary (sect. 2.3.1).

3.2.7 Speed and time of propagation

During a time interval Δt, the phase of a sinusoidal signal at a point x of the line varies with

$$\Delta\varphi = 2\pi\Delta t/T = 2\pi\Delta t\, f = \Delta t\, \omega \tag{3.10}$$

The same phase change is established simultaneously between the point x and a point $x + \Delta x$ such that

$$\Delta x = \Delta\varphi/\beta = \Delta t\, \omega/\beta \tag{3.11}$$

The speed of propagation or *phase velocity* v_φ is the ratio:

$$v_\varphi = \Delta x/\Delta t = \omega/\beta \qquad \text{m} \cdot \text{s}^{-1} \tag{3.12}$$

Its inverse is called the *phase delay per unit length* τ_φ:

$$\tau_\varphi = 1/v_\varphi = \beta/\omega \qquad \text{s} \cdot \text{m}^{-1} \tag{3.13}$$

In the (frequent) cases in which β is not proportional to ω, it is useful to define a *group delay per unit length* τ_g

$$\tau_g = \frac{d\beta}{d\omega} \qquad \text{s} \cdot \text{m}^{-1} \tag{3.14}$$

τ_g takes on particular importance in the case of modulated signals (sect. 8.2.6).

The propagation times are obtained by multiplying these delays by the length of the line.

3.2.8 Relationship between primary and secondary parameters

The line theory (vol. III, chap. 8, in particular sect. 8.3.3 and table 8.11) provides the following relations in the general case of a line with losses:

$$\underline{Z}_c = \sqrt{\frac{R' + j\omega L'}{G' + j\omega C'}} \tag{3.15}$$

$$\underline{\gamma} = \sqrt{(R' + j\omega L')(G' + j\omega C')} \tag{3.16}$$

The principal characteristics of all types of lines proceed from these two fundamental relations.

3.2.9 Asymptotic behavior when the inductive reactance is negligible

The criterion under consideration is

$$\omega L' \ll R' \tag{3.17}$$

It is assumed that the leakage loss is negligible: $G' \cong 0$
Relations (3.15) and (3.16) become

$$\underline{Z}_c \cong \sqrt{\frac{R'}{j\omega C'}} = \sqrt{\frac{R'}{\omega C'}}\, \exp(-j\pi/4) \tag{3.18}$$

$$\underline{\gamma} \cong \sqrt{j\omega R'C'} = \sqrt{\tfrac{1}{2}\omega R'C'} + j\sqrt{\tfrac{1}{2}\omega R'C'} \tag{3.19}$$

that means

$$\alpha \cong \sqrt{\tfrac{1}{2}\omega R'C'} \qquad \text{Np/km} \tag{3.20}$$

and

$$\beta \cong \sqrt{\tfrac{1}{2}\omega R'C'} \qquad \text{rad/km} \tag{3.21}$$

In conclusion, when $\omega L' \ll R'$

- the characteristic impedance is complex and varies proportionally with $1/\sqrt{f}$
- the attenuation coefficient (in Np/km) and the phase change coefficient (in rad/km) become equal and proportional to $\sqrt{f}$, which implies that there are attenuation and phase distortions (sect. 2.4.7).

3.2.10 Asymptotic behavior when the inductive reactance is much greater than the resistance

The criterion under consideration is

$$\omega L' \gg R' \tag{3.22}$$

It is assumed that the leakage loss is negligible: $G' \cong 0$
Relations (3.15) and (3.16) become

$$\underline{Z}_c \cong \sqrt{\frac{L'}{C'}} = Z_c \tag{3.23}$$

$$\underline{\gamma} \cong \sqrt{-\omega^2 L'C' + j\omega R'C'} = j\omega\sqrt{L'C'}\sqrt{1 - j\frac{R'}{\omega L'}} = Z_c$$

$$\cong j\omega\sqrt{L'C'}\left(1 - j\frac{R'}{2\omega L'}\right) \tag{3.24}$$

which means

$$\alpha \cong \frac{R'}{2}\sqrt{\frac{C'}{L'}} = \frac{R'}{2Z_c} \qquad \text{Np/km} \tag{3.25}$$

and

$$\beta = \omega\sqrt{L'C'} \qquad \text{rad/km} \tag{3.26}$$

In conclusion, when $\omega L' \gg R'$

- the characteristic impedance is real and independent of frequency

- the phase change coefficient increases linearly with frequency; there is, therefore, no phase distortion
- the attenuation coefficient is proportional to R' and can only be considered frequency-independent if the skin effect is negligible. Otherwise, according to (3.5), as R'_P increases proportionally to $\sqrt{f}$, α *also becomes proportional to* $\sqrt{f}$ and attenuation distortions appear (this is the case for lines used at high frequencies).

3.2.11 Effect of leakage

At low frequencies, the leakage conductance G', which according to (3.1) is proportional to frequency, can be ignored. This is not the case at high frequencies, notably in the case of section 3.2.10.

With the hypotheses $\omega L' \gg R'$ and $\omega C' \gg G'$, relation (3.16) becomes

$$\gamma \cong \sqrt{R'G' - \omega^2 L'C' + j\omega(R'C' + L'G')} =$$

$$= j\omega \sqrt{L'C'} \sqrt{-\underbrace{\frac{R'G'}{\omega^2 L'C'}}_{\cong 0} + 1 - j\left(\frac{R'}{\omega L'} + \frac{G'}{\omega C'}\right)} \cong \qquad (3.27)$$

$$\cong j\omega\sqrt{L'C'}\left[1 - j\left(\frac{R'}{2\omega L'} + \frac{G'}{2\omega C'}\right)\right]$$

given

$$\alpha \cong \frac{R'}{2}\sqrt{\frac{C'}{L'}} + \frac{G'}{2}\sqrt{\frac{L'}{C'}} = \underbrace{\frac{R'}{2Z_C}}_{\alpha_R} + \underbrace{\frac{G'Z_C}{2}}_{\alpha_G} \qquad (3.28)$$

and

$$\beta \cong \omega\sqrt{L'C'} \qquad (3.29)$$

The attenuation coefficient α_G due to increased losses increases linearly with frequency as G' increases. It thus gains importance with respect to α_R, which increases only proportionally to $\sqrt{f}$.

3.2.12 General behavior of lines as a function of frequency

Figures 3.6, 3.7, 3.8, and 3.9 summarize the results of the preceding sections. These figures raise the following comments:

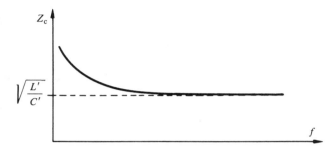

Fig. 3.6 Modulus of the characteristic impedance $\underline{Z}_c$.

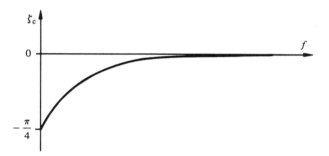

Fig. 3.7 Argument of the characteristic impedance $\underline{Z}_c$.

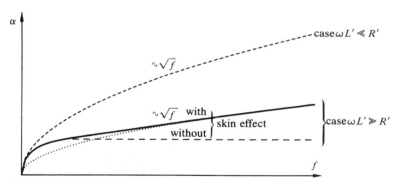

Fig. 3.8 Attenuation coefficient.

- the law for $\sqrt{f}$ regulating the linear attenuation is equally valid at high and low frequencies for completely different reasons: the effect of the capacitance C' at low frequencies and the skin effect at high frequencies;

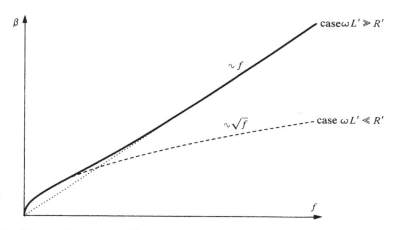

Fig. 3.9 Phase change coefficient.

- the transition between the two regions is scarcely perceptible on certain lines, but gives rise to a noticeable plateau on others;
- a line represents an ideal transmission medium (without linear distortion, sect. 2.4.6) only in the case in which $\omega L' \gg R'$ and when the skin effect can be ignored. This situation is present in a limited range of frequencies on loaded lines (sect. 3.5.6).

3.2.13 Determining the secondary parameters from open-circuited and short-circuited lines

The impedance $\underline{Z}_1$, at the input of a line whose secondary parameters are $\underline{Z}_c$ and $\underline{\gamma}$, terminating at a distance l on an impedance $\underline{Z}_2$ (fig. 3.10), is, according to the theory of lines (vol. III, sect. 8.5.3, eq. (8.40)):

$$\underline{Z}_1 = \underline{Z}_c \frac{\underline{Z}_2 + \underline{Z}_c \tanh \underline{\gamma} l}{\underline{Z}_c + \underline{Z}_2 \tanh \underline{\gamma} l} \tag{3.30}$$

If the output is short-circuited ($\underline{Z}_2 = 0$), we have

$$\underline{Z}_{1sc} = \underline{Z}_c \tanh \underline{\gamma} l \tag{3.31}$$

If it is open-circuited ($\underline{Z}_2 = \infty$), we have

$$\underline{Z}_{1op} = \underline{Z}_c \frac{1}{\tanh \underline{\gamma} l} \tag{3.32}$$

From these two input impedances, we can derive the secondary parameters of the line:

$$\underline{Z}_c = \sqrt{\underline{Z}_{1sc} \underline{Z}_{op}} \tag{3.33}$$

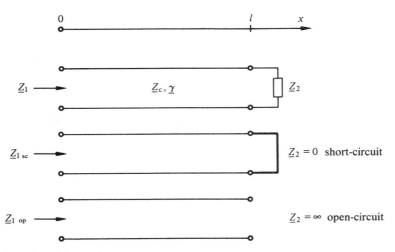

Fig. 3.10 Input impedance.

$$\tanh \underline{\gamma} l = \sqrt{\frac{\underline{Z}_{1sc}}{\underline{Z}_{1op}}} \qquad (3.34)$$

These relations also provide a practical method for measuring the secondary parameters by two measurements of the input impedance after short-circuiting, then opening the line.

3.3 GENERAL LINE CHARACTERISTICS IN THE TIME DOMAIN

3.3.1 Translating from the frequency domain to the time domain

The attenuation $A = \alpha l$ and the phase change $b = \beta l$ allow us to describe what happens if a sinusoidal signal of frequency f is applied at the input of the line. By varying this frequency f, we obtain the *frequency responses* of the line at attenuation $A(f)$ and at phase $b(f)$, which together define the **transfer function** $\underline{H}(f)$ of the two-port equivalent of the line. From (3.9) and (3.8), we have

$$\underline{H}(f) = \exp(-\underline{\gamma} l) = \exp(-A - jb) \qquad (3.35)$$

Because the line is considered to be a linear medium, the transfer function allows us to find the complex Fourier transform of the output signal $\underline{U}_y(f)$, if that of the input $\underline{U}_x(f)$ signal is known (vol. IV, chap. 2):

$$\underline{U}_y(f) = \underline{H}(f) \, \underline{U}_x(f) \qquad (3.36)$$

When the line transmits pulse signals, it is more useful to know its behavior in the time domain, particularly its *impulse response*, i.e., the result $y(t)$ at the output with a unit pulse (Dirac pulse) applied at the input, with a Fourier transform equal to 1. The result from (3.36) is that

$$\underline{Y}(f) = \underline{H}(f) \tag{3.37}$$

and, thus,

$$y(t) = F^{-1}[\underline{H}(f)] = h(t) \tag{3.38}$$

In conclusion, we must remember that

- the impulse response $h(t)$ and the complex transfer function $\underline{H}(f)$ are Fourier transforms of each other. Each is sufficient to describe entirely the behavior of the line.

3.3.2 Step response

The experimental measurement of the impulse response of a line meets practical difficulties due to the generation of Dirac pulses at the input. This is why the *step response* is generally preferred, i.e., the input signal $e(t)$ correspond to a unit step at the output, which is easier to generate (straight side of a square pulse). Because the unit step is the integral of the Dirac pulse, this relation is kept between the step response $e(t)$ and the impulse response $h(t)$:

$$e(t) = \int h(t)\, dt \tag{3.39}$$

which, with (3.38), provides a way to calculate the step response, if the transfer function is known:

$$e(t) = \int F^{-1}[\underline{H}(f)]dt \tag{3.40}$$

By ccommuting the integration and the inverse Fourier transform and substituting $\underline{H}(f)$ from (3.35), we have

$$e(t) = F^{-1}\left[\frac{1}{j2\pi f}\exp(-\underline{\gamma}l)\right]$$

$$= \int_{-\infty}^{+\infty}\frac{1}{j2\pi f}\exp(j2\pi tf)\exp(-\underline{\gamma}l)d f \tag{3.41}$$

3.3.3 Step response of a line when the inductive reactance is low

The criterion considered is: $\omega L' \ll R'$.
It is assumed that the leakage loss is negligible: $G' \cong 0$.
Taking $\underline{\gamma}$ from (3.19), we obtain the step response from (3.41):

$$e(t) = \int_{-\infty}^{+\infty} \frac{1}{j2\pi f} \exp(j2\pi tf)\exp[-\sqrt{\pi R'C'}\ (1+j)l\sqrt{f}]df \qquad (3.42)$$

The calculation of this integral leads to a complementary integral Gaussian function (appendix, sect. 16.2)

$$e(t) = 2G_c(\sqrt{R'C'/2t}\ l) \qquad (3.43)$$

The function $G_c(1/\sqrt{x})$ is represented in figure 3.11. In this case the argument x is directly proportional to the time

$$x = (2/R'C'l^2)t \qquad (3.44)$$

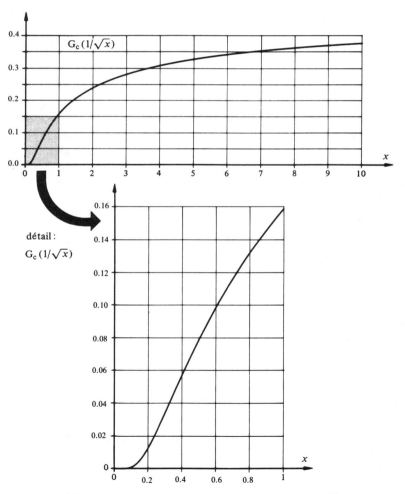

Fig. 3.11 Complementary integral Gaussian function of $1/\sqrt{x}$.

3.3.4 Step response of a line accounting for the skin effect

The criterion considered is: $\omega L' \gg R'$.
It is assumed that leakage loss is negligible: $G' \cong 0$.
The linear transfer coefficient from (3.24), accounting for the skin effect from (3.4), becomes

$$\gamma \cong j2\pi f\sqrt{L'C'} + \frac{R'_P}{2}\sqrt{\frac{C'}{L'}}(1+j) \qquad (3.45)$$

Relation (3.5) can be written

$$R'_P = K\sqrt{f} \qquad (3.46)$$

where $K = \sqrt{\mu/\pi}(\sqrt{\rho_i}/d_i + \sqrt{\rho_e}/d_e)$ is a constant that depends only on the physical and geometric construction of the line.
The step response from (3.41) then becomes

$$e(t) = \int_{-\infty}^{+\infty} \frac{1}{j2\pi f} \exp(j2\pi tf)\exp(-j2\pi\sqrt{L'C'}lf)\exp[-\tfrac{1}{2}\sqrt{C'/L'}K(1+j)l\sqrt{f}]df \qquad (3.47)$$

This relation has a form very similar to (3.42), except for the term $\exp(-2j\pi\sqrt{L'C'}lf)$, which expresses a delay of $\sqrt{L'C'}l$. This delay is the time of phase propagation on the line, calculated by (3.13) and (3.26):

$$t_\varphi = (\beta/\omega)l = \sqrt{L'C'}\, l \qquad (3.48)$$

By making

$$t' = t - \sqrt{L'C'}\, l \qquad (3.49)$$

we obtain, as with (3.43), a step response in the form of a complementary integral Gaussian function:

$$e(t) = 2G_c[\sqrt{C'/(8\pi L't')}\, Kl] \qquad (3.50)$$

Figure 3.11 gives the form of this response in the form of the function $G_c(1/\sqrt{x})$ with the argument x proportional to the delayed time t'

$$x = [8\pi L'/(C'K^2l^2)]t' \qquad (3.51)$$

3.3.5 Remarks on the step response of transmission lines

Equations (3.43) and (3.50), which concern the properties of the function $G_c(1/\sqrt{x})$ presented in figure 3.11, bring up the following remarks:

- as in the case of the frequency response with attenuation $\alpha(f)$ (fig. 3.8), the step response of transmission lines obeys the same law whatever the

frequency range being considered. Here, also, causes as different as the influence of the capacitance in the hypothesis $\omega L' \ll R'$ or the skin effect when $\omega L' \gg R'$ lead to similar results because of a fortunate coincidence;
- the step response starts from zero with a *horizontal tangent*;
- the time taken to attain a given relative value is *proportional to the square of the length of the line*;
- the response to a rectangular pulse of duration τ is obtained by the difference between two step responses shifted by τ. Depending on whether the step response is close to its asymptote or not for t (or t') = τ, the output pulse will be either only slightly deformed, or, otherwise, stretched out considerably (fig. 3.12). It can be assumed that for $x(\tau) \gg 10$, the pulse duration at medium height is noticeably conserved (*lines said to be "short"* with respect to τ), whereas it is considerably increased if $x(\tau) \gg 10$ (*lines said to be "long"* with respect to τ). Thus, the same line can be considered "short" for long pulses and "long" for short pulses.

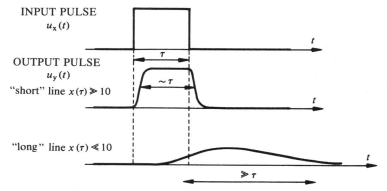

Fig. 3.12 "Short" and "long" line response.

3.4 CROSSTALK

3.4.1 Coupling between lines and crosstalk

When two lines are very close (e.g., two pairs situated on the same cable), they exert a mutual influence on each other caused by three types of coupling:
- *galvanic coupling* due to a common resistance of the two different lines. This phenomenon notably appears in the case of single-wire lines with a common ground;

- *capacitive coupling* due to the presence of capacitances between the conductors of the two lines;
- *inductive coupling* due to the magnetic field of one of the lines crossing the other (mutual inductance).

These couplings depend essentially on the geometric configuration of the conductors, particularly their proximity to each other. They are generally distributed more or less uniformly along the length of the line.

As a result of these couplings, part of the signal transmitted by one of the lines appears on the other line and *vice versa*. The result is *crosstalk* (sect. 2.4.11), for communications carried by these lines. This is highly undesirable, since this type of crosstalk is intelligible.

3.4.2 Near- and far-end crosstalk

While the phenomenon is at first glance reciprocal, we can distinguish (fig. 3.13):

- the *interfering line*, one end (near end) of which is connected to a generator across an impedance $\underline{Z}$ and the other end (far end) is terminated on the same impedance $\underline{Z}$;
- the *interfered line* of which both ends terminate on the same impedance $\underline{Z}$.

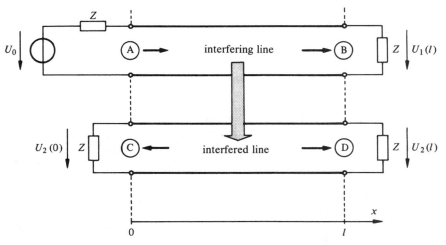

Fig. 3.13 Lines with crosstalk.

Signals coupled during the infered line propagate toward the far end, as well as back to the near end. We designate:

94   TELECOMMUNICATION SYSTEMS

- *near-end crosstalk NEXT*, the crosstalk that appears at the near end of the interfered line, i.e., the voltage $U_2(0)$;
- *far-end crosstalk FEXT*, the crosstalk that appears at the far end of the interfered line, i.e., the voltage $U_2(l)$.

These two phenomena always appear simultaneously, but their effects can be very different according to whether the lines being considered support unidirectional channels in the same direction (in which case only far-end crosstalk is significant), unidirectional channels in the opposite direction (near-end crosstalk), or bidirectional channels (near- and far-end crosstalk).

3.4.3 Attenuation and signal-to-crosstalk ratios: definitions

The ensemble of the two coupled lines of figure 3.13, supposedly identical, constitute a four-port equivalent with an input (A) and three output ports (B,C,D), in which the composite attenuations (sect. 2.4.14) are defined as follows:

Transmission attenuation (A → B):

$$A = 20 \lg \frac{U_0}{2U_1(l)} \qquad (3.52)$$

Near-end crosstalk attenuation (A → C):

$$A_{xp} = 20 \lg \frac{U_0}{2U_2(0)} \qquad (3.53)$$

Far-end crosstalk (A → D):

$$A_{xt} = 20 \lg \frac{U_0}{2U_2(l)} \qquad (3.54)$$

The effect of crosstalk interference must not be judged with respect to the source signal U_0, which is its cause, but rather with respect to the nominal signal (sect. 2.4.14), which would occur if it had been transmitted normally by the line. In this way, the far-end crosstalk voltage $U_2(l)$ acts as an interferer only to the extent that it approaches the nominal signal voltage $U_1(l)$ at the far end of the line. In the same way, the near-end crosstalk voltage should be compared with the voltage, equal to $U_1(l)$, which the usable signal would have at the near end if the same source U_0 were at the far end. In this way, the *signal-to-crosstalk ratios* are defined:

Near-end signal-to-crosstalk ratio:

$$A_{xp0} = 20 \lg \frac{U_1(l)}{U_2(0)} = A_{xp} - A \qquad (3.55)$$

Far-end signal-to-crosstalk ratio:

$$A_{xt0} = 20 \lg \frac{U_1(l)}{U_2(l)} = A_{xt} - A \qquad (3.56)$$

3.4.4 Crosstalk level diagrams

The evolution of the relative level of useful and interfering signals along the length of both lines is presented in the form of a level diagram in figure 3.14.

These level diagrams show qualitatively that near- and far-end crosstalk behave very differently.

- near-end crosstalk: nearby couplings give rise to relatively less attenuated signals, whereas signals coupled at a certain distance x are attenuated first along the path through the interfering line, and a second time by the return trip through the interfered line. The result is that the first few

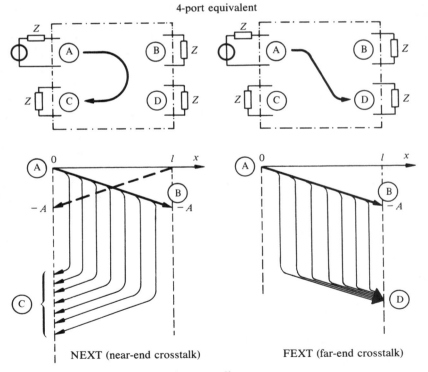

Fig. 3.14 Near-end and far-end crosstalk.

meters of the line are responsible for the bulk of the near-end crosstalk, and thus near-end crosstalk attenuation is not very dependent on the length of the line;
- far-end crosstalk: whatever the location of the coupling, the total attenuation of the coupled signals remains the same—one part x of the path being made on the interfering line and the complementary component on the interfered line. As a result, the effect of these coupled signals relative to the nominal signal increases in proportion to the length of the line.

3.4.5 Estimated calculation

The exact calculation of crosstalk between two lines [13] is difficult and of limited practical interest due to the imprecison with which the geometrical data of the problem are known. Here we will give a rough, simplified method designed to show in a more quantitative manner the different properties of near- and far-end crosstalk discussed in section 3.4.4. The hypotheses of calculation will be the following:
- sinusoidal interfering signal of angular frequency ω (nominal signal);
- capacitive and inductive couplings are assumed to be uniformly distributed along the line, and be expressed per unit length by a resulting mutual capacitance k' and a mutual inductance m';
- the lines are terminated by their characteristic impedance. The composite attenuations defined in (3.52), (3.53), and (3.54) become image attenuations.

A small line segment dx is reduced, with respect to crosstalk, to the schematic equivalent of figure 3.15.

The contribution of this element dx to crosstalk in x depends on the current of the capacitive coupling $d\underline{I}_k$ and on the inductive voltage $d U_m$, according to the relation:

$$d\underline{U}_2(x) = \frac{1}{2} \underline{Z}_c d\underline{I}_k + \frac{1}{2} d\underline{U}_m \tag{3.57}$$

likewise, for the contribution to crosstalk in $x + dx$:

$$d\underline{U}_2(x + dx) = \frac{1}{2}\underline{Z}_c d\underline{I}_k - \frac{1}{2} d\underline{U}_m \tag{3.58}$$

Let us note, on the other hand, that

$$d\underline{I}_k \cong j\omega k' \underline{U}_1(x) dx \qquad \text{if} \quad 1/\omega k' l \gg Z_c \tag{3.59}$$

TRANSMISSION MEDIA

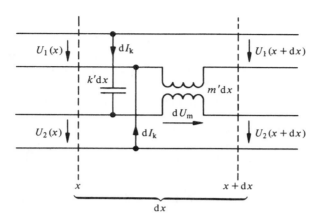

Fig. 3.15 Small segment of coupled lines.

$$d\underline{U}_m \cong j\omega m' \underline{I}_1(x)dx \quad \text{if} \quad \omega m'l \ll Z_c \tag{3.60}$$

and that the voltage $\underline{U}_1(x)$ and the current $\underline{I}_1(x)$ depend on the conditions at the input of the line $\underline{U}_1(0)$ of the propagation coefficient $\underline{\gamma}$, according to the relations:

$$\underline{U}_1(x) = \underline{U}_1(0)\exp(-\underline{\gamma}x) \tag{3.61}$$

$$\underline{I}_1(x) = [\underline{U}_1(0)/\underline{Z}_c]\exp(-\underline{\gamma}x) \tag{3.62}$$

The contribution of dx to the **near-end crosstalk** reaches the near end ($x = 0$) multiplied by $\exp(-\underline{\gamma}x)$, and the integral of all these contributions finally gives

$$\underline{U}_2(0) = \int_{x=0}^{x=l} \exp(-\underline{\gamma}x) d\underline{U}_2(x)$$

$$= \underline{U}_1(0)(1/4\underline{\gamma})j\omega[k'\underline{Z}_c + (m'/\underline{Z}_c)][1 - \exp(-2\underline{\gamma}l)] \tag{3.63}$$

Likewise, for **far-end crosstalk**, the contribution of dx appears at the far end ($x = l$) multiplied by $\exp[-\underline{\gamma}(l - x)]$, which gives, after integration of all these contributions,

$$\underline{U}_2(l) = \int_{x=0}^{x=l} \exp[-\underline{\gamma}(l - x)] d\underline{U}_2(x + dx)$$

$$= \underline{U}_1(0) \exp(-\underline{\gamma}l) \frac{1}{2} j\omega[k'\underline{Z}_c - (m'/\underline{Z}_c)]l \tag{3.64}$$

In conclusion, we can note that:

- inductive and capacitive couplings add in one case and subtract in the other. It is, therefore, not possible to minimize near- and far-end crosstalk by compensating one type of coupling for the other;
- the term $\exp(-2\gamma l)$ of relation (3.63) decreases rapidly as the line length l increases, which renders the near-end crosstalk practically independent of l;
- in (3.64) $\underline{U}_1(0)\exp(-\gamma l) = \underline{U}_1(l)$ represents the voltage of the useful signal at the output of the line. It has been shown that $\underline{U}_2(l)/\underline{U}_1(l)$ is proportional to the length l, which means that the far-end signal-to-crosstalk ratio A_{xt0} from (3.56) decreases as l increases.

These last two remarks confirm the qualitative impressions given in section 3.4.4.

3.5 BALANCED PAIR LINES AND CABLES

3.5.1 Definition

A *balanced pair* is a two-wire line in which the two conductors are identical and, in particular, have the same capacitance and the same leakage with respect to ground.

The balance in question here is, therefore, an electrical balance, which is intended to reduce, indeed to eliminate, the galvanic and capacitive couplings between lines.

3.5.2 Open-wire lines

Open-wire lines are composed of bare conductors supported by insulators. The choice of material and wire diameter is dictated more by mechanical considerations than electrical considerations.

Their principal characteristics are:

- Common materials: bronze, copper, copper-clad steel, and aluminum alloys
- Wire diameter: 2–4 mm
- Wire spacing: generally 30 cm
- Primary parameters: the large diameter gives rise to a low resistance R', but is influenced by the strong temperature variations that the line can undergo. Likewise, the capacitance C' (calculable from vol. III, table 8.12) is low (order of magnitude 5 nF/km). However, the inductance L'

(vol. III, table 8.12) is significant (approximately 2.2 mH/km). The leakage conductance G' depends on the meteorological conditions, but remains negligible, at least at voice frequencies.
- Secondary parameters: the criterion $\omega L' \gg R'$ is already sufficiently satisfied in the voice-frequency range. The result, from section 3.2.10, is a characteristic impedance that is approximately real and independent of frequency. Its value of around 600 Ω for large diameter open-wire lines is historically the basis for the conventional choice of this resistance as a nominal load of all low-frequency circuits (in particular see the definition of the overall loss, sect. 2.4.1). The attenuation coefficient α is low (less than 0.1 dB/km at 800 Hz).
- Crosstalk: the mutual inductance between two parallel lines is significant. We try to reduce the inductive coupling through *transpositions*, which consist of crossing the wires of each line periodically and with different periodicity (order of magnitude: several hundred meters).
- Use: open-wire lines are principally used at voice frequencies, notably for the connection of subscribers in rural or suburban areas. In certain countries, they also serve as a support for low-capacity carrier systems (e.g., 12 channels) for low to medium distance intercity links. The economic consideration of a low initial investment is, however, balanced by the maintenance problems and a sensitivity to external interference sources (storms, lightning bolts, influence of high-voltage power lines, *et cetera*).

A particularly interesting example of open-wire lines is the high-voltage power transmission line, often used as an accessory to carry telephone conversations, telemetry, or data. The huge diameters of the conductors used gives rise to minimal attenuation, which allows long-distance transmission (several hundred km) without intermediate amplification.

3.5.3 Construction of balanced pair cables

The cable consists of an often large number of pairs of ***insulated*** wires, which are ***stranded***, i.e., twisted in such a way as to reduce the inductive coupling between the pairs. The whole arrangement is protected from external influences by an ***insulating jacket***.

These pairs exhibit the following characteristics:
- Insulation: the insulating material most commonly used for telephone cables is a special ***paper*** (pure cellulose, without glue) in a ribbon form, wound obliquely around each conductor with a fine thread of paper slipped into it. The paper, very sensitive to humidity (causes increased

leakage), neverthless, has the advantage of impeding the penetration of moisture inside the cable by way of *hygroscopic swelling*. The insulation defect can thus be located with precision. *Polyethylene*, equally used, does not have this property. The moisture, once it has accidentally infiltrated the cable, can propagate and damage the conductors by corrosion or create capacitive asymmetries. To remedy this, it has been proposed that the cable be filled with appropriate hydrophobic substances, or to combine a polyethylene insulation with the implantation of cellulose fibers in this first layer.

- Conductors: *copper* remains the most favorable material, as much from an electrical point of view as for its machining qualities. *Aluminum* makes for lighter and more economical cables, but it is much more difficult to solder, particularly on the ground.
- Stranding: the four wires, corresponding to two pairs, are stranded together to form a *quad*. This type is called *star quad*. Another type, known as *twin quad* or *DM quad* (from the names of the inventors, Dieselhorst-Martin), is created by stranding two twisted pairs with different twists (fig. 3.16), which gives them better crosstalk properties and advantages for phantom circuits (sect. 3.5.7) but needs more space. The cable is formed either by successive layers of quads arranged spirally with a different step and direction from one layer to the next, or by assembling several separately stranded bundles.
- **Sheath:** the sheath's role is to protect the quads from external mechanical, chemical, or electrical influences. *Lead* (pressed tightly around the cable) is still very desirable despite its weight and poor electrical insulating qualities. *Aluminum* does not present these inconveniences, but its technology is more complex and it is sensitive to alkaline corrosion. If the cable must be pulled through a duct and not simply placed, an *armoring* of spiral (strap-braided) steel bands is necessary. Neither lead nor steel is sufficient to assure good shielding. In critical cases, a supplementary screen of aluminum or copper is necessary.

3.5.4 Properties of balanced pair cables

These cables are used in frequency ranges where the approximation $\omega L' \ll R'$ from section 3.2.9 is generally valid. Consequently,

- α and β are approximately proportional to $\sqrt{f}$
- $\underline{Z}_c$ is not real and its modulus varies with $1/\sqrt{f}$

Whatever the diameter of the conductors, the linear capacitance and inductance have the following approximate values (paper insulation):

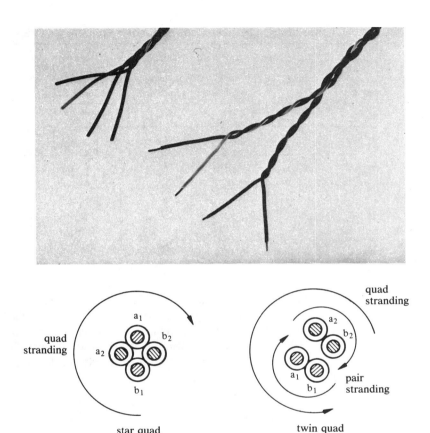

Fig. 3.16 Types of quads.

- $C' \cong 35$ nF/km
- $L' \cong 0.7$ mH/km.

Table 3.17 gives the characteristics of the principal types of balanced pair cables. Because $\underline{Z}_c$ is neither real nor constant, the image attenuation is without great practical interest. This is why table 3.17 contains the *composite attenuation coefficient* α_{cp} on 600 Ω ("overall loss," according to sect. 2.4.1). These are the values which form the basis of network planning. It must be noted, nevertheless, that because the line is not terminated by its characteristic impedance, its composite attenuation is not, strictly speaking, exactly proportional to its length.

Table 3.17 Characteristics of balanced pair cables (planning values).

Nominal conductor diameter (copper)	Stranding star quad	Stranding DM quad	R' at 10°C	Unloaded α_{cp} at 800 Hz for 600 Ω	Loaded α_{cp} at 800 Hz for 600 Ω
mm			Ω/km	dB/km	dB/km
0.4	×		267	2	–
0.6	×		119	0.85	0.43
0.8	×		66.8	0.56	0.25
0.9		×	52.8	0.48	0.22
1.0	×	×	42.8	0.43	0.17
1.2	×		29.7	0.35	0.13
1.4		×	21.8	0.29	0.10
1.5		×	19.0	0.26	0.087

*Loaded: L_p = 88.5 mH across 1830 m.

Figure 3.18 illustrates the typical frequency behavior of some subscriber cables beyond the usual frequency range on these cables, with regard to eventual new services (visiophony, digital transmission). It is shown that the ohmic termination with 600 Ω, customarily applied at low frequencies ($f < 10$ kHz), must be replaced at higher frequencies by a termination closer to the asymptotic value of $\underline{Z}_c$, for example 150 Ω.

3.5.5 Crosstalk in balanced pair cables

The crosstalk attenuation between the balanced pairs of the same cable depends greatly on the relative position of the two pairs and the frequency considered. In the case of near-end crosstalk, the phase variations of different couplings along the line, as a function of distance and frequency, provoke great fluctuations in the frequency behavior of the near-end crosstalk attenuation.

There are six parasitic capacitances the four wires of a quad (fig. 3.19).

The resulting capacitance between the two pairs, $a_1 + b_1$ and $a_2 + b_2$ represent the mutual capacitance k' per unit length.

$$k' = \frac{C'_1 C'_4 - C'_2 C'_3}{C'_1 + C'_2 + C'_3 + C'_4} \tag{3.65}$$

The capacitive coupling is eliminated when the bridge formed by the four capacitances is balanced. In practice, we can artificially tip this balance by crossing the wires of a spliced quad, or by compensating for the differences

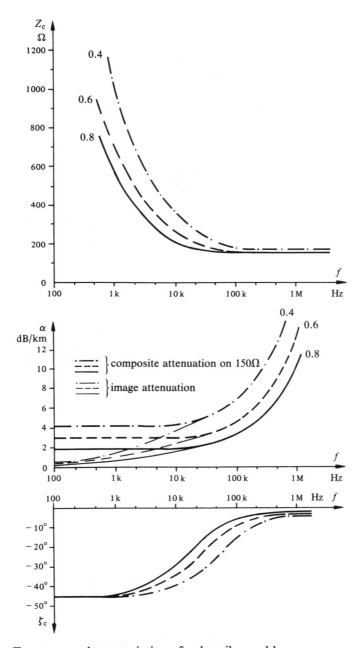

Fig. 3.18 Frequency characteristics of subscriber cables

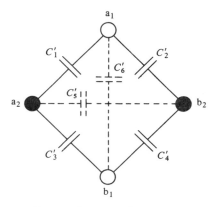

Fig. 3.19 Parasitic capacitances in a quad.

between the measured parasitic capacitances, by the addition of discrete condensers.

The study of crosstalk between balanced pairs is further complicated by the presence of indirect coupling caused by an intermediate third circuit, which can be, for example, the shielding of the cable.

In an approximate and general way, we can establish that:

- the far-end crosstalk signal-to-noise ratio A_{xt0} decreases by approximately 3 dB when the length doubles (and not 6 dB as relation (3.64) would predict because of the statistical independence of distributed coupling). It also decreases by approximately 15 dB when the frequency is multiplied by 10;
- the far-end crosstalk attenuation A_{xp} is only slightly dependent on the length of the line and decreases when the frequency increases to approximately 15 dB per decade. An order of magnitude for these values is presented in figure 3.20.

Some typical values of signal-to-crosstalk ratios in the range of interest for the development of digital systems (sect. 9.8) are given in table 3.21.

3.5.6 Loading of balanced pairs

The characteristics of a line are nearly ideal when (sect. 3.2.12):
- $\omega L' \gg R'$
- the skin effect is negligible

We can approach these conditions at low frequencies by artificially increasing the inductance per unit length L' of the line, which has the following results, from (3.25) and (3.26):

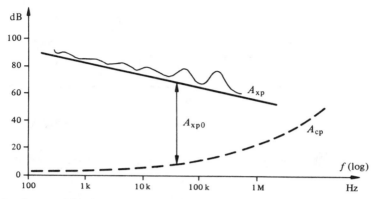

Fig. 3.20 Attenuation A_{xp} and near-end signal-to-crosstalk ratio A_{xp0} between two balanced pairs of a cable (0.6 mm φ, 1.5 km).

Table 3.21 Measured values of signal-to-crosstalk ratios between balanced pairs ($f \cong 1$ MHz).

Near-end signal-to-crosstalk ratios A_{xp0} between 2 pairs	Mean value	Minimum value
-of the exterior layer	62 dB	35 dB
-of the interior layer	65 dB	29 dB
-of adjacent layers	75 dB	50 dB
Far-end signal-to-crosstalk ratio A_{xt0} between 2 pairs (l = 1830 m)		
-of the exterior layer	50 dB	21 dB
-of the interior layer	57 dB	30 dB
-of adjacent layers	75 dB	51 dB

- the attenuation coefficient α diminishes and becomes independent of frequency;
- the phase constant coefficient increases and becomes proportional to the frequency.

While Krarup (1901) proposed sheathing each conductor with a magnetic material to increase L', Pupin (1900) had the simple and brilliant idea of inserting discrete inductance coils into the line at regular intervals. Such a line is called a *loaded line*. The hypothesis of uniformity (sect. 3.2.1) is nevertheless not satisfied in the case of loading. The line then behaves as a succession of low-pass LC elements, the schematic equivalent of which is presented in figure 3.22 (see also vol. III, sect. .8.8).

Rather than a homogeneous increase of L', the addition of discrete coils, of inductance L_p, only renders α constant in the frequency domain limited by the cut-off frequency f_c

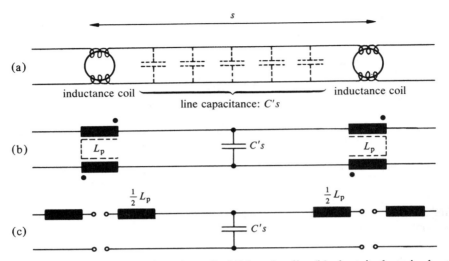

Fig. 3.22 Loaded line: (a) location of additional coils; (b) electrical equivalent; (c) decomposition into LC elements T type.

$$f_c = \frac{1}{2\pi} \frac{1}{\sqrt{\tfrac{1}{2}L_p \tfrac{1}{2}C's}} = \frac{1}{\pi\sqrt{L_p C's}} \tag{3.66}$$

Above this frequency, the attenuation increases rapidly (a low-pass filter behavior, fig. 3.23).

For telephony, we generally choose:

- $s = 1830$ m (proposed by Pupin, equal to 2000 yards)
- $L_p = 88.5$ mH

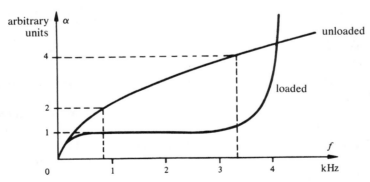

Fig. 3.23 Comparison of attenuation coefficient on lines with and without loading.

which, with a capacitance C' of 35 nF/km, gives the following characteristics (neglecting the inductance of the line itself and the parasitic capacitance of the coils).

- cut-off frequency: $f_c = 4228$ Hz
- characteristic impedance: $\underline{Z}_c = \sqrt{L_p/sC'} = 1175\ \Omega$ (real and constant)
- propagation time: $\tau_\varphi = \tau_g = \sqrt{L_p C'/s} = 41\ \mu s/km$.

3.5.7 Phantom circuits

The balance of the pairs allows construction of a third circuit per quad, the *phantom circuit*, in addition to the two *base circuits* on the two pairs. By way of balancing transformers, the phantom circuit borrows the two conductors of each pair in parallel (fig. 3.24). The result is that the phantom circuit has

- a resistance equal to half that of the base circuits
- a capacitance increased by a factor of 2.7 in star quads and 1.6 in twin quads

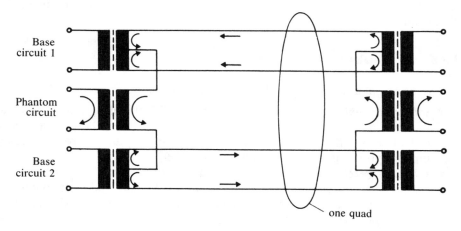

Fig. 3.24 Phantom circuit.

Crosstalk between the base and phantom circuits depends directly on the balance of the transformers and that of the pairs.

In the case of a loaded line, the phantom circuit must be supported by separated coils, each with four windings, in such a way that the magnetic flux due to the current in the four wires of the phantom circuit adds (fig. 3.25).

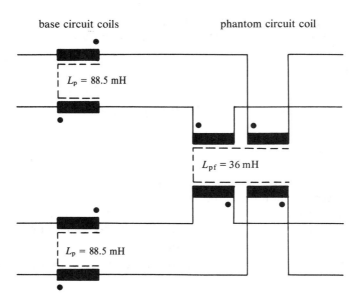

Fig. 3.25 Loading of the phantom circuit ($s = 1830$ m) (the black dot indicates the beginning of the winding).

3.5.8 Use of balanced pair cables

The principal applications of balanced pair cables are:
- *subscriber cables* with numbers of pairs up to 2400, even 3200, and conductor diameters from 0.3 to 1 mm, used principally at low frequencies (telephony), but also for broadcasting musical programs (wire broadcasting at high frequencies from 160 to 335 kHz);
- *urban interexchange cables* with several hundred pairs;
- *rural cables* of medium distance, generally loaded and with phantom circuits when carrying signals at voice frequencies. They can also be used for multiplexed carrier systems with a small number of channels, or, more recently, for digital systems at 2.048 Mbit/s;
- *long-distance transmission* on high-quality cables, at low frequencies (loaded), or by carrier systems of medium capacity (max. 120 channels). Loading permitted the realization of international telephone lines at the beginning of the century (e.g., Berlin-Milan on open-wire lines 1350 km long in 1914), before the invention of amplifiers.

The potential development of balanced pair cables for analog transmission is *limited in its useful frequency range by the rapid degradation of the signal-to-*

crosstalk ratio. Nevertheless, digital systems, although requiring a wider frequency range, but less sensitive to crosstalk interference, make advantageous use of older cables (sect. 9.8).

3.6 COAXIAL PAIR LINES AND CABLES

3.6.1 Definition

The *coaxial pair* is composed of two concentric conductors. The exterior conductor is grounded, while the interior conductor is insulated and centered by dielectric material. The critical dimensions for calculating electrical properties are (fig. 3.26):
- d_i, the external diameter of the interior conductor
- d_e, the internal diameter of the exterior conductor.

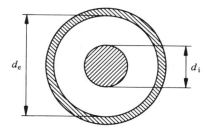

Fig. 3.26 Coaxial pair.

3.6.2 Construction of coaxial pair cables

The interior conductor is almost always a copper wire, and that of the exterior is a tube made of copper or aluminum ribbon wound lengthwise or spirally.

The problem of centering the interior conductor can be solved in many ways (fig. 3.27) by seeking to combine mechanical rigidity with minimization of the resulting relative permittivity ϵ_{res} of the dielectric (polyethylene + air).

Coaxial pairs are grouped in small bundles within a cable protected by a sheath and supported by an armoring of steel bands to resist mechanical constraints.

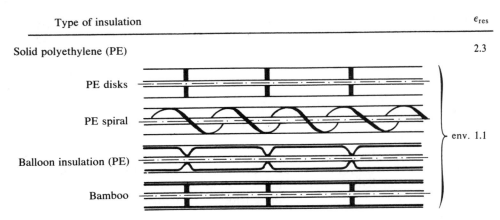

Fig. 3.27 Types of insulation.

3.6.3 Electrical properties of coaxial pairs

The concentric cylindrical structure easily lends itself to calculation of the capacitance and inductance (vol. III, sect. 8.4.1). With the hypothesis $\omega L' \gg R'$ (largely verified in the useful range of these cables), we can derive the expression for the characteristic impedance from (3.23):

$$\underline{Z}_c = \sqrt{\frac{L'}{C'}} = \frac{1}{2\pi} \sqrt{\frac{\mu_0 \mu_r}{\epsilon_0 \epsilon_{res}}} \ln \frac{d_e}{d_i} \qquad (3.67)$$

with $\mu_r = 1$ and $\sqrt{\mu_0/\epsilon_0} = 120\pi \ \Omega$ (characteristic impedance in vacuum, from vol. III, sect. 1.5.3):

$$\underline{Z}_c = \frac{60}{\sqrt{\epsilon_{res}}} \ln \frac{d_e}{d_i} \qquad (3.68)$$

The characteristic impedance of a coaxial pair is *real* and *constant* at high frequencies.

The attenuation coefficient α_R taken from (3.28), taking into account the skin effect to calculate R'_P according to (3.5), neglecting the leakage and assuming that the two conductors are made of the same metal ($\rho_i = \rho_e = \rho$) becomes,

$$\alpha_R = \frac{R'_p}{2\underline{Z}_c} = \sqrt{\pi\rho\epsilon_0\epsilon_{res}} \ \sqrt{f} \ \frac{1}{d_e} y \qquad \text{Np/m} \qquad (3.69)$$

with

$$y = \left(1 + \frac{d_e}{d_i}\right) \frac{1}{\ln \frac{d_e}{d_i}} \tag{3.70}$$

This function y (fig. 3.28) presents a minimum equal to approximately 3.6 for

$$(d_e/d_i)_{opt} \cong 3.6 \tag{3.71}$$

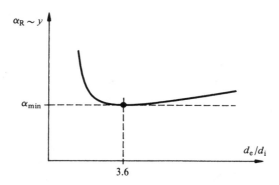

Fig. 3.28 Attenuation of a coaxial pair as a function of the ratio of its diameters.

For this optimal value of the diameter ratio, we have

$$Z_{c\,opt} = 77/\sqrt{\epsilon_{res}} \qquad \Omega \tag{3.72}$$

and

$$\alpha_{Rmin} = 3.6\sqrt{\pi\rho\epsilon_0\epsilon_{res}} \sqrt{f} \frac{1}{d_e} \qquad \text{Np/m} \tag{3.73}$$

It is important to note that:
- the attenuation coefficient α_R is inversely proportional to the diameter
- α_R depends directly on $\sqrt{\epsilon_{res}}$, hence the interest in reducing the proportion of polyethylene in favor of air as a dielectric
- the characteristic impedance does not depend on the absolute dimensions of the pair, but only on their ratio
- the condition (3.71) for minimizing α is in fact not critical, the optimum being very flat (fig. 3.28). It is not valid unless $\rho_i = \rho_e$

The phase change coefficient β is given by (3.26). With $1/\sqrt{\mu_0\epsilon_0} = c_0$ (velocity of light in vacuum, from vol. III, sect. 1.5.3) and $\mu_r = 1$, it remains

$$\beta = \frac{\sqrt{\epsilon_{res}}}{c_0} \omega \qquad (3.74)$$

and, from (3.12), the velocity of propagation in the cable is

$$v_\varphi = c_0/\sqrt{\epsilon_{res}} \qquad \text{m/s} \qquad (3.75)$$

Consequently,
- the coaxial pair exhibits no phase distortion at high frequencies ($f > 100$ kHz)
- the phase change and velocity of propagation do not depend on the geometry of the pair, but solely on the dielectric.

The effect of the attenuation coefficient loss is, from (3.28) and (3.1)

$$\alpha_G = \frac{1}{2} G' Z_c = \frac{\pi \sqrt{\epsilon_{res}} \tan\delta}{c_0} f \qquad \text{Np/m} \qquad (3.76)$$

This contribution to $\alpha = \alpha_R + \alpha_G$ is independent of dimension. As it increases in proportion to f, it only becomes significant with respect to α_R at very high frequencies.

3.6.4 CCITT standards for coaxial pairs

The advent of international coaxial cable links and the standardization of carrier systems have required that a limited number of types of coaxial pairs be standardized (table 3.29).

Table 3.29 Standardized coaxial pairs.

Type	2.6/9.5	1.2/4.4	0.7/2.9
CCITT recommendation	G.623	G.622	G.621
d_i	2.6 mm	1.2 mm	0.7 mm
d_e	9.5 mm	4.4 mm	2.9 mm
d_e/d_i	3.65	3.67	4.14
Usual type of insulation	PE disks	PE balloon bamboo	PE foam
ϵ_{res}	~ 1.1	~ 1.1	~ 1.1
Characteristic impedance 1 MHz	75 ± 1Ω	75 ± 1Ω	75 ± 1Ω

These pairs are generally grouped as 4, 6, 8, or 10 in the same cable.

The frequency behavior of the attenuation coefficient is given in figure 3.30. When presented as a double-logarithmic plot (log α as a function of log f),

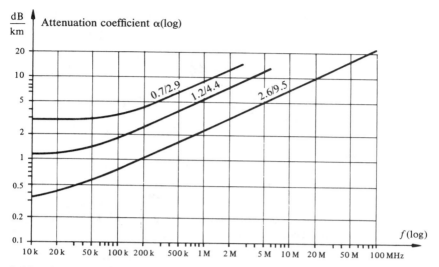

Fig. 3.30 Attenuation coefficient of coaxial pairs.

the law in $\sqrt{f}$ is reduced to a straight line of 1/2 slope. Nonetheless, at low frequencies, the influence of R' is no longer negligible before $\omega L'$. Therefore, $\alpha(f)$ approaches a horizontal asymptote determined by the resistance for direct current.

3.6.5 Crosstalk between coaxial pairs

The structure of the coaxial pair itself practically eliminates the possibility of capacitive or inductive coupling. Nevertheless, complex coupling, of an indirect and galvanic nature, appears due to the intermediary of the two exterior conductors [12]. As a result of the skin effect, these couplings diminish rapidly as the frequency increases, which leads to excellent crosstalk properties at high frequencies, while crosstalk attenuation can become critical below approximately 60 kHz (fig. 3.31).

3.6.6 Use of coaxial pair cables

These cables constitute the long-distance transmission medium *par excellence*. They are used for carrier systems of large and very large capacity (up to 10,800 telephone channels per pair, given a 60 MHz maximum frequency), and high bit-rate digital systems (several tens of Mbit/s), or other wideband

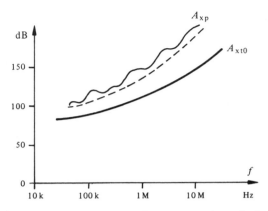

Fig. 3.31 Near-end crosstalk attenuation A_{xp} and typical far-end signal-to-crosstalk ratio A_{xt0} between two coaxial pairs.

signals (cable television). They are generally not used for frequencies lower than 60 kHz due to
- phase distortions which they exhibit at these frequencies
- degradation of their crosstalk properties.

3.7. OPTICAL FIBERS

3.7.1 Principle of guided light propagation

The direct transmission of information by means of light across the atmosphere is limited by meteorological conditions on which the transparency of the medium depends. It is preferable to be free of these restrictions by making the light propagate through multiple reflections in the interior of a strictly delineated channel of known and stable characteristics. Much better than the reflection on the internal face of a tube whereby only a part of the light would be reflected, the phenomenon of total internal reflection (vol. III, sect. 6.6.2) gives rise to optimal transmission of light across a dielectric guide or *fiber*, with a refractive index n greater than that of the surrounding medium. A detailed study of this mode of propagation is found in vol. XIII, sect. 2.10.

3.7.2 Structure of optical fibers

The fiber must be able to enter into contact with supports or a protective tube without disturbance of the internal reflections. To this end, the *core* of the fiber is surrounded by a *cladding* with a lower refractive index. The total reflections are thus carried back to the interface between the two media, i.e., at the interior of the fiber. The difference in the refractive index between the core (n_1) and the cladding (n_2) is chosen to be very low (some %0) so that a small difference of path length (and thus of propagation time) between the direct axial ray and the limit of obliquity leading to total reflections is obtained. One can further diminish this difference by replacing the discontinuity between core and cladding, which exists in the *step index fiber* (fig. 3.32(a)) through a continuous variation of the index of refraction, giving rise to a *graded index fiber* (fig. 3.32 (b)).

The finer the fiber, the less detectable is the difference between the paths. With a core diameter of a few microns (2–8 μm), there is practically only one possible mode of propagation, and the fiber is thus termed *single mode* or *monomode* (fig. 3.32(c)) as opposed to *multimode* fibers of larger diameter (25–100 μm), which are much more easily manipulated, but subject to significant group delay distortions that ultimately limit the bit rate transmissible by the fiber.

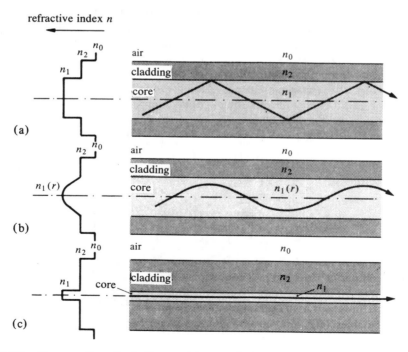

Fig. 3.32 Types of optical fibers: (a) step index; (b) graded index; and (c) single mode.

The material generally utilized for optical fibers is an extremely pure quartz glass (silica), with additions of boron, phosphorus, or germanium to control the index of refraction.

The fibers are placed in a protective tube, then bound in a cable with many fibers and armored with metal to withstand pulling.

The CCITT has specified a graded index fiber of parabolic profile, and with core and cladding diameters of 50 μm and 125 μm, respectively (Rec. G.651).

3.7.3 Intrinsic attenuation

The intrinsic attenuation of the fiber is essentially due to the absorption and diffusion of the light in the dielectric medium of the fiber.

The impurities and inhomogeneities in the material are sources of loss for the optical power transmitted. The reduction of the intrinsic attenuation is thus entirely a matter of technology and production methods. From values over 1000 dB/km in 1966, we have arrived at only a few tenths of dB/km (impurity proportion on the order of 10^{-6}). Due to the fact that the diffusion increases with frequency, the attenuation is more favorable in the near-infrared range (0.7–1.6 μm) than in that of visible light (0.4–0.7 μm).

A typical variation curve of the attenuation as a function of wavelength is presented in figure 3.33. The figure also shows the technological progress realized since 1973.

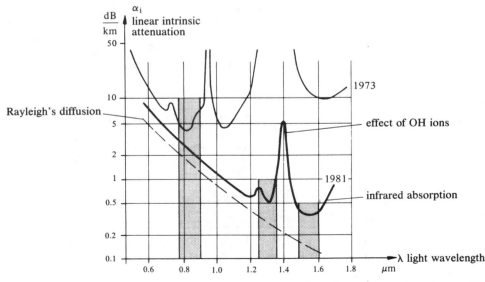

Fig. 3.33 Variation in intrinsic attenuation as a function of wavelength.

The lower limit of the attenuation is given by Rayleigh's diffusion law, which states that the optical power absorbed decreases with λ^{-4}. The peaks of the curve are due to resonances of OH radicals (ions), present as impurities in the fiber.

We will note that three "windows" exist, regions in which the attenuation is minimal. Until now, the majority of systems used the first window ($\lambda \cong 0.85\mu m$), historically justified at a time when fibers presented minimum attenuation for this wavelength. At present, the second ($\lambda \cong 1.3\mu m$) offers lower values for attenuation, but assumes a more sophisticated technology, notably in the area of opto-electronic transducers. Concerning the third ($\lambda \cong 1.55\mu m$), it is still the subject of study. Figure 3.34 pinpoints these ranges in their physical context.

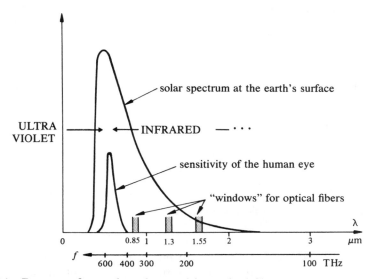

Fig. 3.34 Ranges of wavelengths used in optical fibers.

3.7.4 Numerical aperture. Definition

The *numerical aperture* of an optical fiber expresses its faculty for capturing the light from a source at its forward cross section. It is directly related to the aperture of the cone of acceptance of critical light rays leading to a total internal reflection (fig. 3.35).

The numerical aperture N of a step index fiber is defined by

$$N = n_0 \sin\vartheta_{0max} \tag{3.77}$$

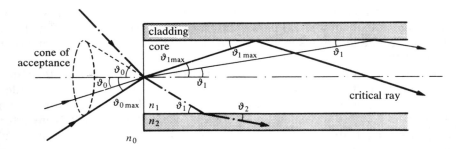

Fig. 3.35 Conditions at the input of a step index fiber.

According to the laws of geometrical optics:

$$n_0 \sin\vartheta_0 = n_1 \sin\vartheta_1 \tag{3.78}$$

$$h_1 \cos\vartheta_2 = n_1 \cos\vartheta_2 \tag{3.79}$$

At the limit of total reflection, ϑ_2 is equal to 0°. We, therefore, have

$$\cos\vartheta_{1max} = \frac{n_2}{n_1} \tag{3.80}$$

and

$$\sin\vartheta_{0max} = \frac{1}{n_0}\sqrt{n_1^2 - n_2^2} \tag{3.81}$$

The numerical aperture is thus given by

$$N = \sqrt{n_1^2 - n_2^2} < 1 \tag{3.82}$$

In graded index fibers, the refractive index of n_1 of the core varies along the diameter. The numerical aperture is thus maximum at the center of the fiber and zero at the periphery of the core.

3.7.5 Modal dispersion

The light pulses which traverse the fiber have a tendency to stretch, principally because of different propagation times. This phenomemon is generally called *dispersion*. One of its principal forms, *modal dispersion*, is caused by the different path lengths of the axial ray and the critical reflected ray (fig. 3.35). In a step index fiber, their lengths are related by $\cos\vartheta_{1\,max}$. The result is a difference in group delay $\Delta\tau_g$ between these two paths

TRANSMISSION MEDIA   119

$$\Delta\tau_g = \frac{n_1}{c_0}\left(\frac{1}{\cos\vartheta_{1\max}} - 1\right) \qquad \text{s/m} \qquad (3.83)$$

where c_0 is the velocity of light in vacuum. With (3.80), we obtain

$$\Delta\tau_g = \frac{1}{c_0}\frac{n_1}{n_2}(n_1 - n_2) \qquad \text{s/m} \qquad (3.84)$$

Modal dispersion and numerical aperture thus come together: small values of $\Delta\tau_g$ (e.g., 50 ns/km) requires a small relative difference in the refractive index between the core and the cladding (e.g., 1%). The result of this is a small numerical aperture (e.g., 0.14), which leads to geometric adaptation problems of opto-electronic transducers to the fiber.

Modal dispersion can be reduced by a suitable progressive reduction of the refractive index of the core along the radius (graded index fiber). In this way, the velocity increases when the light ray is farther from the axis, which compensates for the increase in path length. The theoretically optimal index profile is a parabola. This permits the reduction of the value of $\Delta\tau_g$ by a factor of approximately 3000 with respect to a step index fiber. The mastery of the index profile poses a delicate manufacturing problem. The slightest deviation with respect to the ideal profile brings about a noticeable increase of $\Delta\tau_g$.

Only the single mode fiber is exempt from modal dispersion.

3.7.6 Chromatic dispersion and guide dispersion

The refractive index n_1 of the core depends on the wavelength. It is, therefore, necessary to take into account the delay distortion in the spectral range covered by the available opto-electronic sources. This linear distortion results in a *chromatic dispersion*, i.e., a broadening of the transmitted pulses.

In the materials used for optical fibers, the derivative of the index of refraction with respect to the wavelength cancels for $\lambda \cong 1.3$ μm. The result is a minimum of chromatic dispersion in the second "window", which further increases its technical value.

Finally, a guide effect (vol. XIII, sect. 2.10.13) appears in the fiber. The result is a *guide dispersion*, which is only significant in single mode fibers.

3.7.7 Mode coupling

Unfortunately for theory, but fortunately for practical results, secondary effects particularly complicate the phenomenon of propagation in a fiber. Minimal irregularities (variation in diameter, microscopic curves, *et cetera*) lead to the appearance of new modes, coupled with the modes injected intentionally.

120   TELECOMMUNICATION SYSTEMS

The result is hybrid mode propagation, with supplementary losses due to light leakage into the cladding. This phenomenon is very difficult to calculate rigorously. Empirically, it has been established that
- a supplementary attenuation adds to the intrinsic attenuation of the fiber material
- the resulting attenuation only increases linearly with length for short fibers. Over a certain length, its increase is smaller

3.7.8 Impulse response

If we speak of impulse response $h(t)$ of an optical fiber, we are referring to the *envelope* of the optical signal at the output of the fiber, when an extremely brief (supposedly rectangular) light pulse is emitted at the input, and not the optical signal itself (contrary to the pulse response of the line).

Figure 3.36 gives an idea of the form of $h(t)$, normalized with respect to its maximum, as a function of the length l of a step-index fiber. It has been calculated from simplifying theoretical hypotheses [16].

In order to generally characterize the impulse response $h(t)$, we can define its *root mean squared* rms *pulse width* $d_{h\,eff}$ to be the standard deviation of a distribution, which has a function $h'(t)$ proportional to $h'(t)$, as a probability density:

$$d_{h\,eff} = \sqrt{\int_0^\infty t^2\, h'(t)\mathrm{d}t - \left[\int_0^\infty t\, h'(t)\mathrm{d}t\right]^2} \tag{3.85}$$

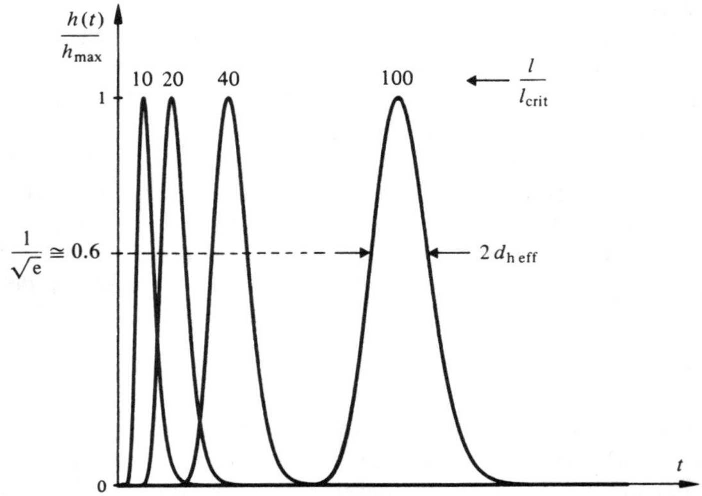

Fig. 3.36 Normalized impulse response of a step index fiber.

with $h(t)$ such that

$$\int_0^\infty h'(t)dt = 1 \qquad (3.86)$$

Experimental observations of the pulse response of optical fibers allows us to draw the following conclusions (fig. 3.36).

- conforming to the calculation for the delay distortion $\Delta\tau_g$ according to (3.84), the modal dispersion leads to an impulse response whose rms duration $d_{h\ eff}$ **increases linearly** with the length l of the fiber for short fibers;
- over a certain critical length $l > l_{crit}$, whose value varies enormously from one fiber to the next, and even according to the condition of the fiber (from a few meters to a few kilometers), the mode coupling results in the rms pulse width $d_{h\ eff}$ **increasing less rapidly** and approaching a dependence on $\sqrt{l}$ (fig. 3.37);
- when $l \gg l_{crit}$, the form of the pulse reponse $h(t)$ becomes approximately **Gaussian**. It can, therefore, be approached by

$$h(t) = \frac{1}{\sqrt{2\pi}d_{h\ eff}} \exp\left(-\frac{t^2}{2d_{h\ eff}^2}\right) \qquad (3.87)$$

Its rms pulse width $d_{h\ eff}$, equal to the standard deviation of the Gaussian distribution, corresponds to half of the measured duration at $h_{max}/\sqrt{e} \cong 0.6\ h_{max}$ (fig. 3.36).

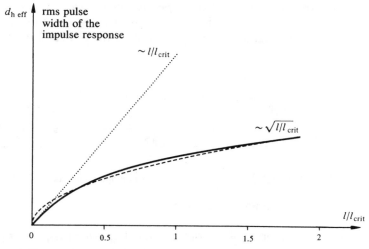

Fig. 3.37 Effect of length on the rms pulse width of the impulse response.

It has been established that the coupling has a favorable effect on the time behavior of long fibers, at the cost, however, of a supplementary attenuation.

3.7.9 Frequency response

The *optical* frequency response is given by the variation of the intrinsic attenuation as a function of the wavelength (fig. 3.33). Once the wavelength has been chosen in one of the optimal "windows," the interest lies rather in the frequency response relative to the *optical signal envelope*, because, contrary to the case of metallic lines, but in a manner analogous to radio transmissions, it is this envelope, and not the optical signal, which in fact contains the information to be transmitted.

The frequency response, understood in this way, is determined by the dispersion (notably modal) phenomena and mode coupling. The *transfer function* $H(f)$, which characterizes it, is the Fourier transform of the impulse response $h(t)$. In particular, for a long fiber, where $h(t)$ becomes Gaussian according to (3.87), $H(f)$ is also a Gaussian function

$$H(f) = H(0) \exp(-2\pi^2 d_{h\,\text{eff}}^2 f^2) \tag{3.88}$$

The logarithmic expression of $1/H(f)$ gives the attenuation distortion $\Delta A(f)$ of the optical signal envelope as a function of the frequency of this envelope, after a great length l of the fiber.

$$\Delta A(f) = -20 \lg[H(f)] \sim d_{h\,\text{eff}}^2 f^2 \quad \text{dB} \tag{3.89}$$

This attenuation distortion *increases proportionally with the square of the frequency* of the envelope. It is added to the intrinsic attenuation $A_i = \alpha_i l$ of the optical signal itself (fig. 3.38), characterized by the optical flux Φ_o (optical power).

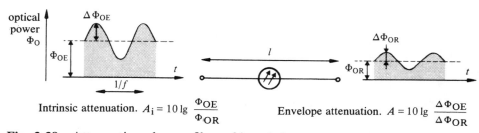

Intrinsic attenuation. $A_i = 10 \lg \dfrac{\Phi_{OE}}{\Phi_{OR}}$ Envelope attenuation. $A = 10 \lg \dfrac{\Delta\Phi_{OE}}{\Delta\Phi_{OR}}$

Fig. 3.38 Attenuation along a fiber of length l.

Finally, the attenuation A of the optical signal envelope is given by

$$A(f) = A_i(\lambda) + \Delta A(f) = \alpha_i(\lambda)\, l + k\, d_{h\,\text{eff}}^2 f^2 \quad \text{dB} \tag{3.90}$$

where λ is the optical wavelength and f is the frequency of the envelope.

In summary, let us note that:
- for $f \ll 1/d_{h\ eff}$, the attenuation is practically given by the intrinsic attenuation; therefore, it depends on λ but not on f;
- the rms pulse width $d_{h\ eff}$ of the pulse response depends **nonlinearly on the length l**, from figure 3.37. It, therefore, similarly varies to the attenuation distortion ΔA, by virtue of (3.89);
- the fiber has a very pronounced low-pass filter behavior, with a very high cut-off frequency $\approx 1/d_{eff}$. For a given length, the attenuation, compared with that of a coaxial cable (fig. 3.30), is illustrated by figure 3.39.

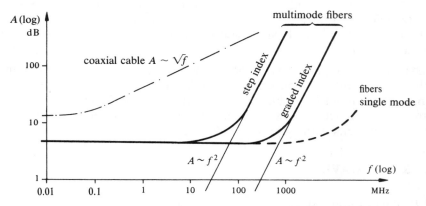

Fig. 3.39 Qualitative comparison between the attenuation of a fiber and coaxial cable (arbitrary length).

3.7.10 Applications of optical fibers in telecommunications

The advantages of optical fibers as a transmission medium are the following:
- very large useful bandwidth, determined by the dispersion (several tens of MHz in multimode fibers, several GHz in single mode fibers), allowing very high bit rates;
- minimal attenuation coefficent permitting the spacing of repeaters at several km (sect. 14.4);
- immunity to electromagnetic influences, i.e., notably elimination of crosstalk and allowing for the possibility of use in a very cluttered electromagnetic environment;
- very low sensitivity to temperature variations;
- electrical insulation between the emitter, the fiber (insulating in itself), and the receiver. This point, however, poses new problems for supplying direct current power to intermediate repeaters;

- very small dimensions, minimal weight, and mechanical flexibility;
- favorable economic prospects with respect to the development of integrated optics and opto-electronics.

The principal problems posed by this new medium are:
- the opto-electronic transducers, their optical adaptation to the fiber, and their own reliability;
- the splices and connectors, which have very high micromechanical requirements (especially for single mode fibers), and from which losses due to mismatching, poor alignment, or damage to the fiber can play a significant part in transmission attenuation. Furthermore, operating constraints (splicing over land) and reliability problems arise;
- the derivations which would allow coupling of light signals from one fiber to another.

Experimental applications of optical fibers or those envisaged for the future, with a few exceptions, concern digital transmission (telephony, music, television, or data) at very high bit rates or under very specific ambient conditions. They are the subject of study for long-distance (intercity network) links, and for local networks with large bandwidths (chap. 14).

3.8 RADIO WAVES

3.8.1 Earth's atmosphere

The atmosphere constitutes a complex medium for the propagation of electromagnetic waves. Unlike lines, the atmosphere cannot be optimized, but must be accepted as it is. Schematically, the atmosphere is composed of three overlapping principal regions (fig. 3.40):
- the *troposphere* (altitude lower than 15 km), which is characterized by turbulence (winds), the presence of water vapor (clouds), and diminishing temperature with altitude. There is a gradient of the refractive index which causes the trajectory of electromagnetic waves to curve in the direction of the ground;
- the *stratosphere* (altitude from 15 to 40 km), which is practically devoid of water vapor and has a temperature that increases with altitude before stabilizing;
- the *ionosphere* (altitude from 40 to 500 km) which is composed of ionized layers whose ionization density depends strongly on the time of day, the season, and the activity of sunspots (cycle of approximately 11 years). This region plays an important role in the propagation of radio waves. It is the locus of refraction, absorption, and reflection phenomena.

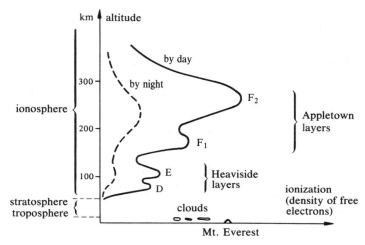

Fig. 3.40 Ionized layers of earth's atmosphere.

3.8.2 Classification of wave ranges

The collection of frequencies of electromagnetic waves is conventionally divided into ranges, which roughly correspond to modes of propagation and to very different types of use (telecommunications, radar, medicine, industrial applications, *et cetera*).

Since the beginning of radio communications, the assignment of wave ranges to particular services and users has been the object of international conventions. At present this role is filled by the CCIR (International Radiocommunications Consultative Committee, or *Comité consultatif international des radiocommunications*), and the IFRB (International Frequency Registration Board), both in Geneva. The overloading of frequency bands and the very specific propagation conditions for certain of them require absolute worldwide coordination on a global scale [14].

Figure 3.41 summarizes the conventional designations of frequency bands (classified by decades of wavelength) and their principal applications in telecommunications.

3.8.3 Wave bands assigned to radio broadcasting

- Longwave (150-285 kHz).
- Medium wave (525-1605 kHz): propagation is essentially by groundwaves (the ground is, in effect, a better conductor for longwave; hence they

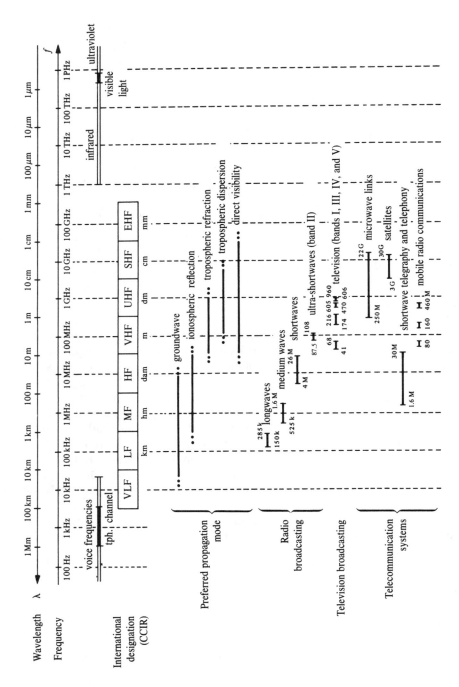

Fig. 3.41 Frequency ranges. The ranges attributed to the different services are in fact not continuous within the indicated domains.

reach longer distances). The reflection on the ionospheric layers O is accompanied by stronger absorption during the day; because of this fact, transmission over long distances is better at night than during the day.
- Shortwave (4-26.1 MHz): the groundwave predominates at short distances, while multiple reflections between the ground and the ionosphere permit transmission at very long distances (especially during the day). Nevertheless, reception can be seriously disrupted, on the one hand, by interference between direct and reflected waves, and, on the other, by variations in propagation conditions in the ionosphere. The result is significant fluctuations (fading) in the received signal level. Long, medium, and short waves are used for national and international radio broadcasting with amplitude modulation (AM, sect. 8.2).

The curvature of propagation paths, due to variation in the refractive index in the ionosphere, causes reflection only if the angle of incidence is sufficiently oblique. As a result, a skip zone whose radius increases with frequency around the emitter, is inaccessible by reflection (fig. 3.42).

The increase of the skip zone (table. 3.43) imposes an upper limit on the range of usable frequencies for this mode of transmission.

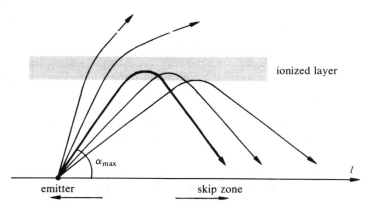

Fig. 3.42 Limit of reflection on the ionosphere.

- Ultra-shortwave: five discrete bands have been defined in the range from 41 MHz to 0.96 GHz, of which one is reserved for radio broadcasting (with frequency modulation FM, sect. 8.6) and four for television (with vestigial sideband modulation VSB, sect. 8.5). Transmission requires practically direct visibility between emitter and receiver, which limits the use of ultra-shortwave to regional broadcasting.

Table 3.43 Skip zone with shortwave.

Frequency	Radius of the skip zone
15 MHz	2000 km
25 MHz	5000 km
> 30 MHz	no more reflection

3.8.4 Wave bands assigned to radio communications

Besides the broadcasting of audio or video programs, radio waves are frequently used in telecommunications for single (radiotelephone, radiotelegraphy) or multiplexed (microwave links) communications, or for long-distance signaling (radio paging, air security, *et cetera*). It deals with the following types of services:

- point-to-point links at long distances
- fixed links with inaccessible areas
- mobile ground, airborne, or maritime links.

The frequency bands assigned to these different services are numerous, varied, and often very narrow. It would be impossible to try listing them all here, so only a few large categories will be given.

Shortwave was the first means used for a transatlantic telephone conversation (1927), but lost its importance for this use because of its mediocre quality and instability. It remains widely used for intercontinental radiotelegraphy.

The bands at 80 MHz, 160 MHz, 460 MHz and 900 MHz are reserved for short distances (line-of-sight) ***mobile radiocommunications***. The higher the frequency, the smaller the antenna, but the requirements for precision and stability of the oscillators are greater. On the other hand, the Doppler effect becomes noticeable.

Microwave links (chapter 12), which, along with coaxial cables, constitute an important part of the long distance telecommunications network, occupy frequencies between 250 MHz and 22 GHz. The antenna dimensions become prohibitive below this band, while absorption due to water vapor in the troposphere provokes very disturbing losses starting at around 10 GHz. Microwave links require direct visibility between emitter and receiver, taking into account tropospheric refraction. Leaps over the ground of from 50 to 200 km are possible.

When geographic or political reasons allow neither visibility nor installation of relay stations, we sometimes have recourse to a particular mode of propagation, based on a phenomenon of wave ***diffusion*** in the higher regions of the troposphere, which is still poorly explained. These inhomogeneous zones become the site of diffuse secondary emissions, of which a very small part return to the earth's surface and, quite randomly, to the receiver (fig. 3.44).

Such a link, whose efficiency is low and of erratic quality, is called *troposcatter link*. Its maximum reach (approximately 1000 km) is limited by the altitude at which diffusion is observed.

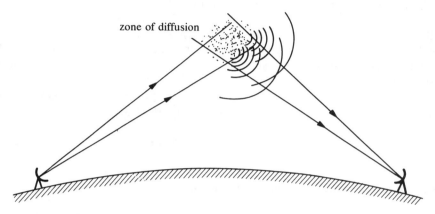

Fig. 3.44 Troposcatter link.

Satellite links (chapter 13) constitute a special type of microwave link, beyond the atmosphere that they must cross while undergoing a slight refraction. They are assigned pairs of frequencies at 4 and 6 GHz, and, more recently, at 11 and 14 GHz, and even 20 to 30 GHz. They also share a portion of the same bands as terrestrial microwave links, which can be a source of interference.

3.8.5 Reception in diversity

When propagation conditions are critical and unstable (fading, interference, troposcatter link), the probability of satisfactory quality can be increased by doubling the link and selecting, at each instant, the receiver whose output gives the best signal-to-noise ratio. This process is termed reception in *diversity* and takes two principal forms, which can be combined:
- *space diversity*, which consists of spacing two receivers at a distance of greater than 150 times the wavelength apart, with the idea that the wave propagation conditions will be different on the two neighboring paths.
- *frequency diversity* on the same antenna, but with two emission frequencies and a receiver tuned to each. The relative difference between the two frequencies must be on the order of 1%. Here too, we are counting

on the fact that fluctuations will probably not affect both frequencies at the same time.

3.9 ANTENNAS

3.9.1 Isotropic source: definition and properties

If a radio source radiates a total power P_E uniformly in all directions, it is said to be isotropic. There is no practical antenna that exhibits this property perfectly, but the ideal of an isotropic source nevertheless serves as a point of reference.

At a distance l from this isotropic source, the power flux density (per unit surface) is uniform, and, in a lossless medium, equal to

$$p_{iso} = \frac{P_E}{4\pi l^2} \qquad \text{W/m}^2 \qquad (3.91)$$

3.9.2 Characteristics of an emitting antenna

Generally, the power flux density p_α at a distance l of an antenna is dependent on the direction α, defined in azimuth and elevation. The **gain** g_α, **in the direction** α is defined in comparison with the isotropic source, by

$$g_\alpha = \frac{p_\alpha}{p_{iso}} \qquad (3.92)$$

or

$$G_\alpha = 10 \lg \frac{p_\alpha}{p_{iso}} \qquad \text{dB} \qquad (3.93)$$

The (abstract) surface in space defined by g_α for all directions of radiation around the antenna constitutes the *radiation diagram* of the antenna. The majority of antennas possess an axis, or a plane of symmetry, which is also found in this diagram and facilitates its plane representation (fig. 3.45).

The maximum value of g_α (direction of maximum radiation) is termed antenna *gain*.

In a lossless medium, and provided that the antenna does not dissipate power, all the power P_E provided by the emitter must be distributed (unequally, in general) over the sphere of radius l. The result is a condition for the radiation diagram.

$$P_E = \iint\limits_{\text{sphere}} p_\alpha dS = \iint p_{iso} g_\alpha dS = \frac{P_E}{4\pi l^2} \iint g_\alpha dS \qquad (3.94)$$

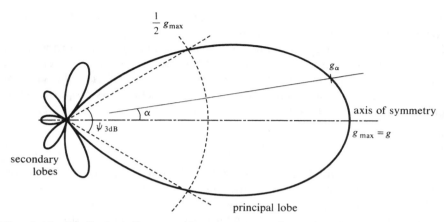

Fig. 3.45 Radiation diagram of an antenna with axial symmetry (of revolution). Example: parabolic reflector.

By introducing the solid angle Ω, we have

$$d\Omega = dS/l^2 \qquad (3.95)$$

and

$$\iint_{\text{sphere}} g_\alpha \, d\Omega = 4\pi \qquad (3.96)$$

In conclusion, if the gain g_α is high in one direction it must, in accordance with (3.96), be low in others. ***An antenna with high gain is thus a directional antenna and vice versa.***

In reality, the power effectively radiated by the antenna is lower than the power P_E that is provided by the emitter because of losses in the antenna itself (section III.7.2). These losses will be taken into account in the yield factor r introduced in relation (3.101).

3.9.3 Characteristics of a receiving antenna

The power P_R available at the output of an antenna terminated with its characteristic impedance (adapted load) is proportional to the power flux density p_β radiated to this location by an emitting antenna located in the direction β with respect to the receiving antenna being considered. The proportionality coefficent, which exhibits the dimension of a surface, is termed *effective surface* $A_{e\beta}$, or *effective area in the direction* β.

$$A_{e\beta} = \frac{P_R}{p_\beta} \qquad \text{m}^2 \qquad (3.97)$$

Its maximum value (preferred direction of reception) is the *effective area* A_e, or *effective aperture* of the antenna.

The reciprocity theorem, valid if the transmission medium is isotropic (i.e., if its local properties do not depend on the direction under consideration), confirms the fact that *the directive properties of an antenna are the same at reception as emission*. The antenna gain g_β in the direction β is proportional to its effective area $A_{e\beta}$ in this direction

$$\frac{g_\beta}{A_{e\beta}} = \text{const.} \quad \forall \beta \tag{3.98}$$

In particular, the directions of maximum radiation and preferred reception coincide. In the same way, the (maximum) gain g and the (maximum) effective area A_e are linked by a proportionality coefficient, which is identical for all antennas (vol. III, sect. 7.2.7) and depends only on the wavelength λ

$$\frac{g_\beta}{A_{e\beta}} = \frac{g}{A_e} = \frac{4\pi}{\lambda^2} \tag{3.99}$$

The result is that the ideal isotropic antenna, whose gain is by definition equal to 1, has an effective area equal to

$$A_{eiso} = \frac{\lambda^2}{4\pi} \qquad \text{m}^2 \tag{3.100}$$

3.9.4 Properties of large surface antennas

When the geometrical surface of an antenna is very large with respect to λ^2, its effective area A_e can be derived from its plane geometric area perpendicular to the direction of radiation A by

$$A_e = rA \qquad \text{m}^2 \tag{3.101}$$

where r is a dimensionless *yield factor*, which takes into account the nonuniform radiation on the surface, the effects of edges and of losses in the antenna. In practice, the value of r is around 0.5.

As a consequence, the effective area of such an antenna is independent of frequency. Further, from (3.99), its gain g increases in proportion to f^2. In logarithmic form, G increases by 6 dB when the frequency doubles.

Notably, the *parabolic reflector antenna*, frequently used by microwave links and satellite links, falls in this category. In the case, often encountered in practice, of a reflector with a circular aperture of diameter d, the gain is expressed by

$$g = r \frac{\pi d^2}{4} \frac{4\pi}{\lambda^2} = r \left(\frac{\pi d}{\lambda}\right)^2 \tag{3.102}$$

The radiation diagram presents a highly accentuated and narrow main lobe as well as undesirable, but weak, sidelobes (fig. 3.45). The aperture of the main lobe, defined by the angle ψ between the directions presenting a gain reduced by 3 dB, is given approximately by the empirical relation

$$\psi_{3dB} \cong 70 \frac{\lambda}{d} \qquad \text{in degrees} \qquad (3.103)$$

3.9.5 Wire antennas

There are a great variety of wire antennas, i.e., antennas composed of wires, or carefully arranged conducting bars, each with its own advantages from the economic, directive, bulk, bandwidth, and other points of view. The properties of some of these are presented in vol. III, sect. 7.3 and in [18]. Table 3.46 summarizes their principal characteristics as well as their applications. Often, particularly with emitters, these antennas are grouped in periodic arrays in order to increase directivity.

3.10 WAVE TRANSMISSION

3.10.1 Hypotheses

We will hereafter make the following assumptions:
- the medium is *isotropic*, which validates the reciprocity theorem;
- the medium is *lossless*; notably, the losses due to absorption in the troposphere (water vapor) are, nevertheless, only negligible at frequencies lower than approximately 10 GHz;
- transmission takes place without obstacles, and without reflection (direct visibility). It deals with *free-space propagation*. Precautions must be taken to guarantee that this hypothesis is valid (sect. 12.2).

3.10.2 Attenuation of the link

The attenuation A expresses in logarithmic form the relationsihp between the emitted power P_E and the received power P_R at the ouput of the adapted receiving antenna. Thus,

$$A = 10 \lg \frac{P_E}{P_R} \qquad \text{dB} \qquad (3.104)$$

Table 3.46 Some wire antennas

Type	Form	Z_c	Radiation pattern	Gain $g(G)$	Utilization
Dipole (or Hertz doublet)	$\ll \lambda$			1.5 (1.8 dB)	Longwave, medium wave, and shortwave
Dipole quarter wave	$\lambda/2$	~ 73 Ω		1.64 (2.1 dB)	Ultra-shortwave
Dipole $\lambda/4$ replicated	$\lambda/2$	~ 300 Ω			
Yagi	$>\lambda/2$ $\lambda/2$ $<\lambda/2$	< 300 Ω		8 ... 9 dB	Ultra-shortwave reception (narrow band)
Tourniquet		~ 70 Ω			Omnidirectional emission
Dihedral		~ 130 Ω		~ 9 dB	Ultra-shortwave emission
Butterfly				5 dB	Ultra-shortwave, TV emission
Cigar				16 dB	Special microwave links
Helix		90 ... 220 Ω			Pursuit and telecommand of satellites
Rhombic	~50°, 3 ... 10λ			15 ... 22 dB	Intercontinental radiotelegraphy (shortwaves)

If the two antennas were replaced by isotropic sources, we would have, from (3.91), (3.97), and (3.100):

$$P_{Riso} = A_{eiso} p_{iso} = \frac{\lambda^2}{4\pi} \frac{P_E}{4\pi l^2} \tag{3.105}$$

and the attenuation, termed *free space attenuation* would be

$$A_{iso} = 20 \lg \left(\frac{4\pi l}{\lambda}\right) \qquad \text{dB} \tag{3.106}$$

As the result of a regrettable conflict of international symbols (IEC), using the same letter A, it is necessary to carefully distinguish between

- A attenuation (in dB) and
- A_e effective area (in m^2).

The value of the attenuation calculated by (3.106) is an *optimistic minimum*, given the assumption of a lossless medium. In reality, meteorological or geographic conditions along the path, as well as interference between signals received by different paths, can noticeably increase the attenuation and lead to a momentary *fading* of the signal (12.2.4).

In the case of antennas with known radiation patterns, and with the relative orientation indicated by figure 3.47, the power received is, from (3.91), (3.92), (3.97), and (3.99):

$$P_R = A_{eR\beta} g_{E\alpha} \frac{P_E}{4\pi l^2} = A_{eR\beta} A_{eE\alpha} \frac{P_E}{\lambda^2 l^2} = g_{R\alpha} g_{E\alpha} \frac{\lambda^2 P_E}{(4\pi)^2 l^2} \tag{3.107}$$

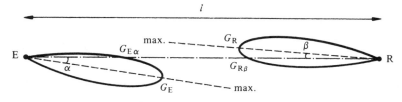

Fig. 3.47 Relative orientation of antennas.

and the attenuation becomes

$$A = 20 \lg \left(\frac{4\pi l}{\lambda}\right) - 10 \lg g_{R\beta} - 10 \lg g_{E\alpha} \qquad \text{dB} \tag{3.108}$$

$$A = A_{iso} - G_{R\beta} - G_{E\alpha} \qquad \text{dB} \tag{3.109}$$

If, as is the case of microwave links, the two antennas are *exactly pointed toward each other*, i.e., if their axes of maximum radiation coincide, (3.109) becomes

$$A = A_{iso} - G_R - G_E \quad \text{dB} \quad (3.110)$$

where G_R and G_E are the respective gains of the receiving and emitting antennas (in dB).

Antenna gain is thus quite analogous to amplifier gain and corresponds (in dB) to a reduction of attenuation with respect to a link between two omnidirectionally radiating (isotropic) antennas.

3.11 CRITICAL COMPARISON OF TRANSMISSION MEDIA

3.11.1 Comparison of transmission characteristics

Table 3.48 summarizes the respective properties of lines (balanced pairs or coaxial cables) and radio wave links (radio broadcasting or radiocommunications). It should be stressed that the laws which dictate how attenuation A

Table 3.48 Attenuation behavior

	Lines	Radio waves
Ratio of power emitted P_E and received P_E	$\dfrac{P_E}{P_R} = 10^{\alpha l / 10}$	$\dfrac{P_E}{P_R} = \dfrac{\lambda^2 l^2}{A_{eE} A_{eR}} = \dfrac{(4\pi)^2 l^2}{\lambda^2 g_E g_R} = 10^{A/10}$
Dependence on length l	$A = \alpha l \sim l$ $\dfrac{A_2}{A_1} = \dfrac{l_2}{l_1}$	$\dfrac{P_E}{P_R} \sim l^2$ $A_2 - A_1 = 20 \lg \dfrac{l_2}{l_1}$
If l doubles	A doubles	A increases by 6 dB
Dependence on frequency	$A \sim \sqrt{f}$ $\dfrac{A_2}{A_1} = \sqrt{\dfrac{f_2}{f_1}}$	$\dfrac{P_E}{P_R} \sim \dfrac{1}{f^2}$ * $A_2 - A_1 = -20 \lg \dfrac{f_2}{f_1}$ *
If f doubles	A is multiplied by $\sqrt{2}$	A decreases by 6 dB *

*Valid only if the effective area is independent of frequency (large surface antenna) and to the extent that the medium can be considered lossless.

varies as a function of the distance or frequency are *very different*, according to whether transmission is by line or radio waves.

3.11.2 Case of point-to-point links

Cables (balanced pair or coaxial) and directed radio links (terrestrial microwave links, troposcatter links, or satellite links, shortwave radiocommunications) are two competing transmission media in the intercity and international telecommunications networks and sometimes even in the local network.

The advantages of cables are:

- stable transmission medium, known and able to be optimized
- possibility of re-use of the same frequency band on distinct pairs of the same cable (paying attention to crosstalk if the pairs are balanced!)
- unlimited length and density of the network, without risk of interference between cables
- protection against interception or malicious interference
- long duration of operation (40 to 50 years) and high reliability, except for catastrophic breakdowns (rupture or damage by excavation).

The advantages of radio links are following:

- they permit links with isolated locations, or those with difficult access (deserts, high mountains, islands)
- the equipment can be transportable (temporary links, television reporting, electronic news gathering ENG)
- the construction is faster, provided the civil engineering infrastructure already exists (tower, pylons, buildings)
- the investments are independent of the distance, except for step increments (if relay stations are necessary)

3.11.3 Case of broadcasting networks

Radio waves by nature lend themselves to omnidirectional or relatively directional broadcasting (radio, television). The number of receivers is unlimited and the range of coverage often very extended, even global (in the case of shortwave).

Nevertheless, a cable network can also serve as a means of information broadcasting, with the following advantages:

- greater immunity to electromagnetic interference (atmospheric, industrial, *et cetera*)
- optimal receiving conditions

- increase in the number of programs offered
- elimination of individual receiving antennas

In regions where the population density is sufficient to justify the significant investment that this represents, coaxial cable can serve to support a large number of audio and television programs (cable or community antenna television, CATV). Optical fibers open new perspectives for the development of broadband local networks (sect. 15.8.5).

Chapter 4

Transmission Processes

4.1 LIMITATIONS DUE TO THE CHANNEL

4.1.1 Introduction

If it is, in principle, possible to transmit any type and quantity of information across any channel, **provided that we devote the time**, the transmissible information **rate** in "real time" (sect. 4.1.8) across a given channel is **strictly limited** by the characteristics of the channel, in particular by the channel's **capacity**, which represents the absolute theoretical limit of what it is possible to transmit.

Seeking to exceed the capacity of a channel is as vain as refusing to believe in the impossibility of perpetual motion.

These limitations, which are inherent in the channel, are equally valid for analog and digital transmission. The highly summarized and heuristic approach which will be taken in this section is based on digital information transmission, a case in which the idea of capacity is more concrete. Nevertheless, this idea retains all its significance in the case of analog information transmission, which can be considered as a limit case of digital information in which the subtlety of the nuances approaches infinity.

4.1.2. Characteristics of digital information flow

With respect to the definitions given in section 1.5, all digital transmission is characterized by:
- the bit rate $\dot{D}$ (in bit/s), which principally concerns the user of the channel
- the symbol rate $\dot{M}$ (in baud), which represents the physical speed of transmission of elementary signals or moments
- the number m of specific states per symbol characterized, for example, by m levels, m frequencies, or m phases of elementary signals, according to the agreed upon convention between the sender and the receiver.

These values are linked by the relation (1.10):

$$\dot{D} = \dot{M} \text{ lb } m \qquad (4.1)$$

Physically, the receiver must be able to distinguish between two successive symbols whose values are, in the worst case (fig. 4.1), immediately next to each other on the scale of m appropriate values.

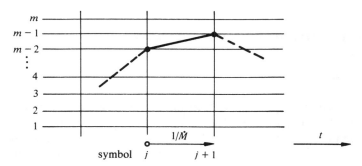

Fig. 4.1 Required resolution in time and amplitude.

Transmitting a bit rate $\dot{D}$ requires that the channel have

- a certain *time resolution* to permit a variation in the characteristic value of the signals from one signal to the next, given during the time $1/\dot{M}$
- a certain *amplitude resolution* to be able to distinguish m values effectively from one another, even in the presence of imprecision, interference, and noise.

4.1.3 Relationship between bandwidth and rise time of the step response

Any real transmission channel includes reactances which oppose sharp variations of the signal. The channel has a certain inertia. This effect can be shown

- in the *frequency domain*: the attenuation of the channel increases more or less suddenly as a function of the frequency. The channel behaves as a kind of low-pass filter;
- in the *time domain*: the step response of the channel (response to a unit step of infinite slope) presents a non-infinite slope.

Exact knowledge of the complex frequency transfer function permits deduction of the step response by inverse Fourier transform and integration, and *vice versa* (vol. IV, chap. 2 and vol. VI, chap. 4).

However, for transmission channels used in telecommunications, the transfer function is not known in a sufficiently precise and general way. In all cases, it is very different from classical mathematical models of communications theory (ideal low-pass filter, Gaussian filter, *et cetera*). Generally, the only known indication in the frequency domain is that of the **bandwidth B**, with all the uncertainty of the definition given in section 2.4.8.

In the same way, the description of the step response (highly complex) is generally reduced, in boldly abridged form, to knowledge of the *rise time t_m*, defined as the time taken for the step response to pass from 10% to 90% of its final value (asymptote) (figure 4.2).

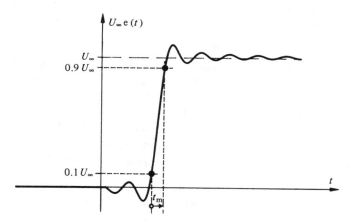

Fig. 4.2 Definition of the rise time t_m.

Clear transmission of a vertical slope (unit step, $t_m = 0$) would require a channel transmitting all the frequencies from zero to infinity without distortion. The more the bandwidth of the channel is narrowed, the more the rise time of the step response increases.

Despite the arbitrary character of their definitions, it has been shown that the bandwidth B and the rise time t_m of known theoretical or real channels are related by the following empirical relation:

$$Bt_m = 0.35 \ldots 0.45 \tag{4.2}$$

4.1.4 Resolution in time: Nyquist's theorem

In a channel given by its bandwidth B, by virtue of relation (4.2), it is not possible to vary the parameters of the signal to any speed because the rise time t_m represents a limit to this speed.

Consequently, the minimum time $1/\dot{M}$ of the symbols is tied to t_m, and symbol rate $\dot{M}$ (sect. 1.6.6) is inversely proportional to the rise time, thus it is proportional to the bandwidth

$$\dot{M}_{\max} \sim 1/t_m; \text{ that means, } \dot{M} \sim B \tag{4.3}$$

In 1928, Nyquist [19] mathematically studied the problem of digital transmission in the model case of a channel with ideal low-pass filter characteristics and thereby determined the proportionality coefficent between bandwidth and maximum symbol rate in order to guarantee the independence between consecutive symbols (no intersymbol interference, sect. 5.3) after transmission:

$$\dot{M}_{\max} = 2B \quad \text{(ideal low-pass)} \tag{4.4}$$

This result, known as the *Nyquist's theorem*, is essentially of theoretical interest. Nevertheless, it gives an absolute limit which channels may approach but will never reach.

This theorem is in fact a corollary of the sampling theorem (sect. 4.3), stated later by Shannon. Figure 4.3 summarizes this duality by showing
- the sampled transmission of a time continuous signal;
- continuous transmission of a time discrete signal (symbols).

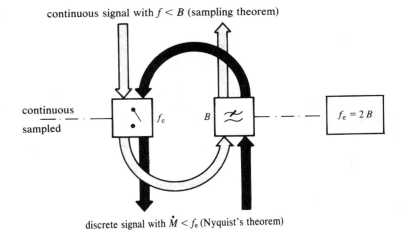

Fig. 4.3 Relation between the Nyquist's and sampling theorems.

The sampling frequency f_e and the cut-off frequency $f_{\max} = B$ of the ideal low-pass, filter which performs the inverse operation, are related by

$$f_e = 2B \tag{4.5}$$

The result is that

- restoring the continuous signal after it has been sampled is possible if it is contained in the pass band of the filter, thus, if

$$f \leq B = \tfrac{1}{2} f_e \quad \text{(sampling theorem)} \tag{4.6}$$

restoring the time discrete signal after its transmission in a channel with frequencies strictly limited at B is possible if each symbol is sampled at least once, thus, if

$$\dot{M} \leq f_e = 2B \quad \text{(Nyquist's theorem)} \tag{4.7}$$

In the case of real channels, if it is assumed that the rise time t_m corresponds to more than half the duration of a symbol, we have

$$\dot{M}_{max} = \frac{1}{2t_m} \tag{4.8}$$

Taking into account (4.2), we obtain

$$\dot{M}_{max} = \cong 1.25B \quad \text{(real channel)} \tag{4.9}$$

This relation, albeit empirical and approximate, gives a more realistic idea about the behavior of normal transmission channels than relation (4.4). It means that *we can transmit 1.25 Bd per Hz of bandwidth, or, inversely, that it requires devoting 0.8 Hz for each Bd to be transmitted.*

4.1.5 Amplitude resolution

Independently of the inertia of the channel, considered in section 4.1.3, another phenomenon, also unavoidable, disturbs transmission: noise, which is inevitably superimposed on the signal in the course of transmission, rendering reception more difficult and uncertain the larger the number m of values to be distinguished per symbol. Because the power of the signal P_S is limited, it is not possible to recognize an infinite number of different values in the presence of noise whose power at the same point of the channel is equal to P_N.

Reasoning geometrically in a multidimensional space in which signals and noise can be represented [20, 21], Shannon demonstrated in 1948 that the theoretical limit for m, with the hypothesis of white Gaussian noise, is given by

$$m_{max} = \sqrt{1 + \frac{P_S}{P_N}} = \sqrt{1 + \xi} \tag{4.10}$$

where ξ is the signal-to-noise ratio defined in section 2.4.15.

The maximum transmissible decision rate per symbol results from (4.10):

$$D_{Mmax} = \text{lb } m_{max} = \tfrac{1}{2} \text{lb}(1 + \xi) \tag{4.11}$$

4.1.6 Channel capacity: definition and significance

Starting from the two theoretical relations (4.4) and (4.11), it is possible to express the maximum bit rate $\dot{D}_{max}$ that an ideal channel can transmit, which we characterize by
- its bandwidth B (ideal low-pass filter!)
- its signal-to-noise ratio ξ (white Gaussian noise)

This theoretical maximum rate is called the channel *capacity* C. According to (4.1), it is equal to:

$$C = \dot{D}_{max} = \dot{M}_{max} \text{ lb } m_{max} = B \text{ lb}(1 + \xi) \tag{4.12}$$

In fact, Shannon, the author of this relation (1948), showed that the transmission of a bit rate $\dot{D}$ equal to the capacity C is not only theoretically possible, but can be done with an error probability approaching zero, provided that we find the optimal mode for representing these signals.

Relation (4.12) gives an absolute theoretical limit which can only be slightly approached by real technical transmissions systems. It has been established in practice that the bit rate realizable $\dot{D}$ across a channel characterized by its capacity C according to (4.12) is lower than C the lower the tolerable error probability (always remaining non-zero).

Nevertheless, relation (4.12) allows us to compare and evaluate systems.

We should note that the same capacity C can be obtained with different values of B and ξ. This interchangeability of bandwidth and signal-to-noise ratio brings up the following remarks:
- it is accomplished through the intermediary of a logarithm. It is, therefore, necessary to considerably increase ξ in order to be able to reduce B (system 1 of figure 4.4), or, on the other hand, to accept a very large bandwidth if we wish to use a very noisy channel (low ξ, system 2 of figure 4.4);
- it is not automatic, but requires processing of the signal in order to adapt it to conditions (B and ξ) of the channel by means of **modulation** and **coding**. Certain procedures prove to be more effective than others and allow us to use a channel with a capacity which better approaches the bit transmission rate;
- the capacity C is not entirely proportional to the bandwidth B. In fact, in the presence of "white" noise (e.g., thermal noise, sect. 2.4.12), doubling

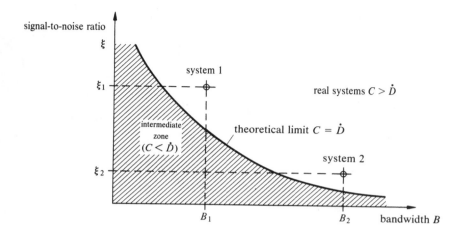

Fig. 4.4 Comparison of channel capacities C required by systems transmitting the same bit rate $\dot{D}$.

B means also doubling the received noise power P_N. For a given signal power P_S, the capacity increases less when the bandwidth is increased, the smaller the signal-to-noise ratio ξ.

Figure 4.4 shows the channel capacity theoretically required (minimum $C = \dot{D}$, independent of the error probability) and practical (strongly dependent on the tolerable error probability) for transmission systems with the same bit rate $\dot{D}$.

4.1.7 Summary of relations between message and channel

The fundamental relations (4.4), (4.11), and (4.12) can be represented geometrically (fig. 4.5) by conditions that allow us to pass a parallelepiped (of volume equal to the total decision content D carried by the message) during a time T, through a rectangular opening (of cross section equivalent to the capacity C of the channel). Not only must the parallelepiped section D be lower than the surface C, but its two dimensions must be compatible with the corresponding dimensions of the channel.

4.1.8 Transmission in real time and differed time

The idea of *real time* is relative and somewhat ambiguous. Two conditions must be satisfied for a transmission to be declared to be in real time:

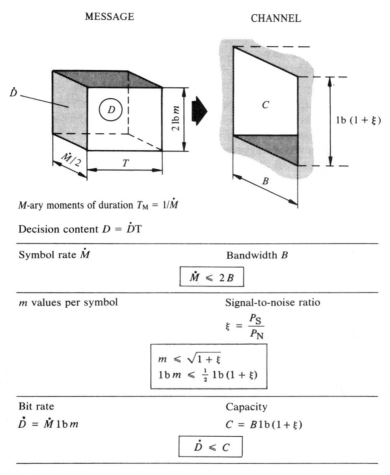

Fig. 4.5 Reciprocal fitting of channel and messages.

- the *duration* of the received message must be identical to that of the emitted message
- the *delay* between emission and reception must not noticeably exceed the propagation time on the transmission media used.

In all other cases, we speak of transmission in *differed time*. The same idea can be applied to coding and is thus related to the information produced by the source and received by the sink.

When the available channel capacity is low with respect to the real bit rate of the source, with appropriate coding we can, as long as the transmission

TRANSMISSION PROCESSES

need not be made in real time, stretch the parallelepiped of figure 4.5 by increasing its duration T so as to adapt its cross section $\dot{D}$ to the narrowness of the channel. This is notably the case for the transmission of still pictures from space probes.

4.2 OBJECTIVES AND PRINCIPLES OF MODULATION

4.2.1 Baseband transmission

Transmission is said to be *baseband* if the signals are transmitted in the same way as they exit the source transducer, i.e., in their original frequency band.

Baseband transmission is used whenever the available transmission medium lends itself, and when economic conditions allow us to provide physical support to each communication.

In particular, this is the case for:

- telephony: systematically in the local network and often in the network at medium distances, for voice signals issued from a microphone (approximately 100 Hz to 5 kHz) and transmitted on balanced pairs (eventually loaded or supported by amplifiers);
- television: for video signals (50 Hz to 5 MHz) produced by the camera and transmitted at short distance on a coaxial cable;
- data transmission: for signals coded and put into transmissible form, but not transposed in frequency, emanating directly from terminal equipment.

4.2.2 Symbolic representation of the baseband

By convention, the frequency range occupied by the baseband is represented by a shaded right triangle (fig. 4.6) with its hypotenuse indicating the direction of increasing frequencies.

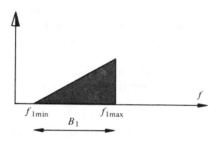

Fig. 4.6 Symbolic representation of the baseband.

It must be strongly emphasized that the symbolic triangular form has nothing to do with the envelope of the amplitude spectrum of the baseband signal!

This representation, trivial in the case of baseband transmission, becomes all important when the band is transposed into another frequency range, and eventually inversed, by way of modulation. The transposed band will be represented by the same triangular symbol, unshaded.

4.2.3 Definition and goals of modulation

Modulation is an operation which consists of transforming a signal representing information into another signal without noticeably modifying the information which it carries.

This is intended for the following goals:

- *adaptation* to the particular conditions of a transmission medium. By way of appropriate modulation, we can use a very noisy channel and guarantee high-quality transmission in spite of it. For radio wave transmission, it is necessary to transpose the initial signal into a frequency range in which the propagation conditions (range, usable bandwidth) are adapted to the problem that is to be solved;
- *multiplexing*, i.e., simultaneous use of the same transmission medium by several communications (sect. 4.4).

These two goals cannot always be achieved in a single stage, nor with the same modulation technique. We must often have recourse to multiple modulations and mixed procedures (sect. 4.5)

Modulation is necessary in all cases in which, for technical or economic reasons, baseband transmission is not possible.

4.2.4 Principle of modulation

By modulation, a signal $u_1(t)$, called the *primary signal* (modulating signal), is transformed into a *secondary signal* $u_2(t)$ (modulated signal).

The operation of restoring the original primary signal is accomplished by demodulation.

Any modulator works according to a **convention**, which relates the characteristic values of the secondary signal $u_2(t)$ to the instantaneous values of the primary signal $u_1(t)$.

The demodulator examines the characteristic values of the secondary signal $u_2(t)$ that it receives and by a *reciprocal convention* deduces the instantaneous values of the reconstructed primary signal $u'_1(t)$ (figure 4.7).

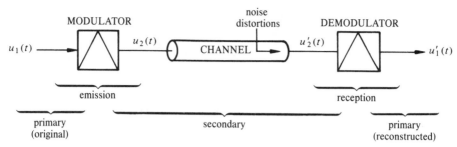

Fig. 4.7 Block diagram of a modulated transmission.

The reconstructed primary signal $u'_1(t)$ is not strictly identical to the original primary signal $u_1(t)$ due to the fact that, on the one hand, the distortions and interferences in the channel which render the received signal $u'_2(t)$ slightly different from the original signal $u_2(t)$ and, on the other hand, the imperfect reciprocity of the modulator and demodulator. The effect of these imperfections depends largely on the type of modulation and constitutes one of the comparison criteria that allows us to choose the most appropriate modulation in each case.

4.2.5 Types of modulation

We will distinguish two broad categories of modulation here, according to an unofficial terminology:
- *analog modulation* in which the modulation convention consists of varying one parameter (amplitude, frequency, phase, duration, *et cetera*) of the secondary signal in proportion to the instantaneous value of the primary signal. Such modulations ***do not modify the nature of the information*** (analog or digital) carried by $u_1(t)$ and $u_2(t)$. Chapter 8 is devoted to these modulations;
- *digital modulation* which performs analog/digital conversion between $u_1(t)$ and $u_2(t)$. The secondary signal is thus characterized by a bit rate $\dot{D}$ and the modulation convention becomes a *code* for representation of analog information in digital form. They are discussed in chapter 7.

Analog modulation interposes an auxiliary signal $u_p(t)$, called the *carrier*, one parameter of which is the object of modulation by the primary signal $u_1(t)$ (fig. 4.8).

Analog modulations are classified (table 4.9) by
- the form of the carrier: sinusoidal or pulse;

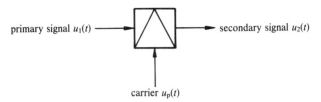

Fig. 4.8 Conventional symbol of an analog modulator.

- the parameter of the carrier which is the object of modulation: amplitude, frequency, phase, duration, *et cetera*;
- the nature of the information transmitted: analog or digital.

It must be noted that, according to this terminology, modulations involving digital information, particularly those used in data transmission on analog channels (OOK, FSK, PSK, *et cetera*) are treated as analog modulations, although the primary signal has a discrete nature in this case. There is, in fact, a definite analogy at the procedural level and concerning its properties between continuous analog modulations dealing with analog information (AM, FM) and discrete analog modulations (OOK, FSK, PSK, *et cetera*) involving digital information. These modulations are treated in sect. 8.11.

4.3 SAMPLING

4.3.1 Principles

Sampling of a signal $u_1(t)$ consists of replacing this signal with another signal $u_2(t)$, which is

- equal in instantaneous value to $u_1(t)$ during brief instants of duration τ, repeated periodically with a frequency f_e, which is called the sampling frequency;
- equal to zero between these instants.

Thus, this is the result of slicing the signal $u_1(t)$ with a switch activated periodically by a series of pulses of duration τ and period $T_e = 1/f_e$ (fig. 4.10).

In theory, sampling can be *ideal*, i.e., leading to pinpoint samples of zero duration ($\tau \to 0$) and mathematically representing these instantaneous values of the signal $u_1(t)$ at instant $t_0 + nT_e$.

Nevertheless, in practice, the duration of samples τ, although generally very brief with respect to their period T_e, is not zero. Therefore, we must deal with a *real* sampling.

Table 4.9 Modulation classification.

Information transmitted	Form of carrier	Modulated parameter	Type of modulation	
Analog modulation { analog -voice -audio -video	sinusoidal	amplitude frequency phase	AM SSB FM ΦM	amplitude modulation single sideband frequency modulation phase modulation
	pulse	amplitude frequency phase duration	PAM PFM PPM PDM	pulse amplitude mod. pulse frequency mod. pulse position mod. pulse duration mod.
digital -data -text	sinusoidal	amplitude frequency phase	ASK OOK FSK PSK	amplitude shift keying on-off keying frequency shift keying phase shift keying
Digital modulation { analog → digital	clock signal (bit rate $\dot{D}$)	code	PCM DPCM ΔM AΔM	pulse code mod. differential PCM delta modulation adaptive ΔM

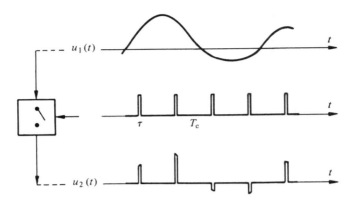

Fig. 4.10 Sampling principle.

The signal $u_2(t)$ can be considered the product of signal $u_1(t)$ and a *sampling function* $e(t)$, which is a periodic series of rectangular pulses of duration τ (fig. 4.11):

$$e(t) = \text{rep}_{T_e}\left[\text{rect}\left(\frac{t}{\tau}\right)\right] \tag{4.13}$$

Therefore, we have

$$u_2(t) = u_1(t)\, e(t) \tag{4.14}$$

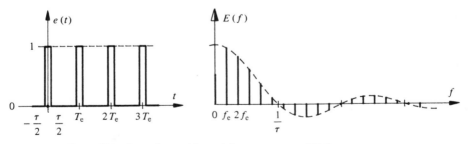

Fig. 4.11 Sampling function $e(t)$ and its spectrum $E(f)$.

4.3.2 Spectrum of a sampled signal

Applying the Fourier transform to a simple product of two terms in the time domain according to (4.14), we obtain a convolutional product in the frequency domain:

$$U_2(f) = U_1(f) * E(f) \tag{4.15}$$

$E(f)$ is the spectrum of the sampling function. $U_1(f)$ represents the spectrum of the original baseband signal, which we assume is limited between 0 and $f_{1\,\text{max}}$ and uniform between these limits.

As a result of (4.15) and with the symbolic baseband representation introduced in section 4.2.2, the (unilateral) spectrum of the sampled signal $u_2(t)$ is of the form given in figure 4.12 (see the convention for representation of spectra in the appendix, section 16.1).

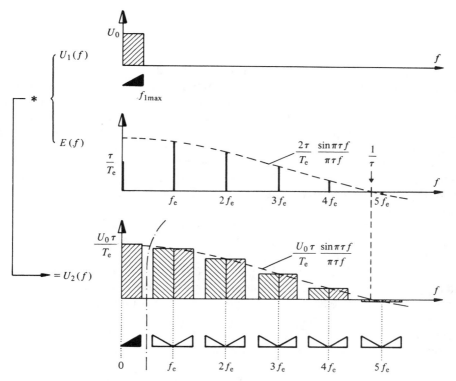

Fig. 4.12 Real sampling of a band-limited signal. Spectrum of the sampled signals; example = $\tau/T_e = 0.21$.

It is noted that

- the *baseband is present* in the spectrum of the sampled signal, with an amplitude reduced by a factor of τ/T_e. It is due to the $f = 0$ component of the sampling function spectrum;
- the convolutional product (4.15) causes *pairs of bands* to appear, equal by twos, similar to the baseband, but shifted with one of the two inverted,

located on either side of the multiples of the sampling frequency (sidebands);
- the amplitude of these pairs of bands is given by the values of a sin $\pi\tau f/\pi\tau f$ function at $f = nfe$, a function which cancels a first time for $f = 1/\tau$;
- if $\tau \gg T_e$, all the first pairs of sidebands and the baseband have approximately the same amplitude, which is that of the original signal multiplied by τ/T_e ($\gg 1!$).

4.3.3 Sampling as a form of modulation

The sampled signal $u_2(t)$ can also be considered a pulse train of frequency f_e and duration τ, whose amplitude is equal to the instantaneous values of $u_1(t)$ over the duration of the pulses. The amplitude of these pulses is thus **modulated** (in the sense of analog modulation as defined in sect. 4.2.5) by the original signal $u_1(t)$, which plays the role of primary signal. The **carrier** here is the series of pulses of frequency f_e and duration τ, i.e., the sampling function $e(t)$.

We should note that, in the spectrum of the sampled signal (fig. 4.12), the lines of the spectrum $E(f)$ of the sampling function are not present. This is why the sampling is also termed *pulse amplitude modulation-suppressed carrier* abbreviated PAM-SC.

This modulation is not used as a transmission technique *per se*. However, sampling is a fundamental preliminary stage for pulse modulation (sect. 8.10) and digital modulation (chap. 7).

Demodulation of the sampled signal $u_2(t)$ consists of reconstructing the original signal $u_1(t)$ from equidistant samples. This problem of interpolation in time can be solved more easily in the frequency domain. In effect, the presence of the baseband in the spectrum of the sampled signal (fig. 4.12) allows the restoration of $u_1(t)$ intact, but attenuated by a factor τ/T_e, by isolating the baseband by means of a low-pass filter which eliminates all the other bands.

4.3.4 Review and illustration of the sampling theorem

The sampling theorem, attributed to Shannon in its present form, constitutes the foundation of signal, information, and communication theory. It is the object of a rigorous and detailed presentation in vol. VI, chap. 9.

Only its statement is presented here:
- a primary signal $u_1(t)$, which only contains frequency components lower than $f_{1\,\text{max}}$ (bounded spectrum), can be entirely determined by equidistant samples, taken at frequency f_e, such that

$$f_e \geq 2f_{1\text{max}} \qquad (4.16)$$

TRANSMISSION PROCESSES

However, we must not forget that this fundamental theorem assumes three unrealistic hypotheses, which can only be approximated in practice:
- the spectrum of $u_1(t)$ is assumed to be bounded, thus the signal $u_1(t)$ must be stationary (with neither beginning nor end) and defined for $t = -\infty ... +\infty$;
- the samples are assumed to be infinitely brief ($\tau \to 0$);
- the reconstruction of the signal $u_1(t)$ from these samples assumes the existence of an ideal low-pass filter, which perfectly blocks all the components above $f_e/2$. Such an ideal filter cannot be constructed.

For reconstruction of $u_1(t)$ by filtering of the baseband in the spectrum of the sampled signal $u_2(t)$ to be possible (fig. 4.13(a)), the sidebands located on either side of the multiples of the sampling frequency must not run into each other.

This condition is fulfilled if $f_e \geq 2f_{1\,max}$, which illustrates the sampling theorem (without demonstrating it!).

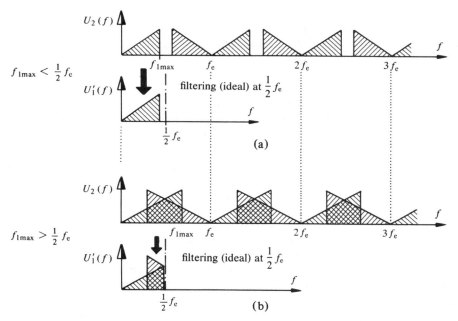

Fig. 4.13 Sampling with $\tau \ll T_e$.

4.3.5 Effect of a violation of the sampling theorem

If $f_{1max} > f_e/2$, the sidebands cross (fig. 4.13(b)). In particular, the lower sideband of f_e impinges on the baseband.

Filtering (even ideal) at $f_e/2$ does not restore the baseband, but an aliased band, composed of a part of the baseband ($f_1 < f_e/2$) over which other parts are superimposed, coming from the sidebands of f_e and its multiples.

In particular, if the primary signal $u_1(t)$ is a sinusoidal signal of variable frequency f_1, the reconstructed signal $u_1'(t)$ will also be sinusoidal, but with a frequency f_1', as given in figure 4.14.

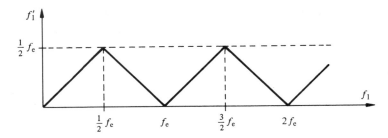

Fig. 4.14 Frequency f_1' of the signal reconstructed by filtering at $f_e/2$ for a sinusoidal primary signal of variable frequency f_1, sampled at f_e.

This phenomenon, called *aliasing*, is the cause of very disturbing nonharmonic distortions. It is illustrated in figure 4.15.

Consequently, it is absolutely necessary to assure by **preliminary** filtering that the primary signal does not contain any component higher than $f_e/2$ **before** sampling.

Because filtering cannot be ideal, the aliasing, even though weak, always exists.

4.3.6 Effect of imperfections in the demodulation filter

Because the attenuation of the band blocked by the demodulation filter is not infinite, the sidebands of f_e (and its multiples) are only imperfectly eliminated. This effect is particularly critical in the zone of transition between the pass band, which must allow $f_{1\ max}$ to pass, and the stop band, which must sufficiently attenuate $f_e - f_{1\ max}$ (fig. 4.16). The difference between these two frequencies, which defines the selectivity of the filter, is even smaller as $f_{1\ max}$ is closer to $f_e/2$ (theoretical limit).

The choice of the sampling frequency f_e with respect to the theoretical sampling minimum is finally dictated by the cost of this filter and that of the similar filter used before sampling.

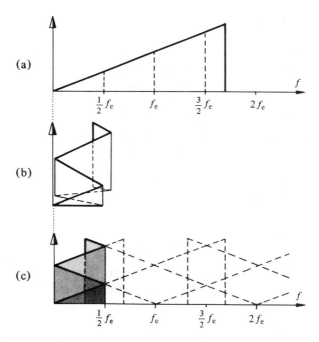

Fig. 4.15 Aliasing effect in the case of a violation of the sampling theorem: (a) original baseband (primary signal); (b) aliasing of the primary band; (c) result of the demodulation by filtering with an ideal low-pass filter at $f_e/2$.

For the conventional telephone band, a sampling frequency of

$$f_e = 8 \text{ kHz} \qquad (4.17)$$

has been almost universally chosen, which leaves a margin of 1.2 kHz for the range of filters between $f_{1\,\max} = 3.4$ kHz and $f_e - f_{1\,\max} = 4.6$ kHz.

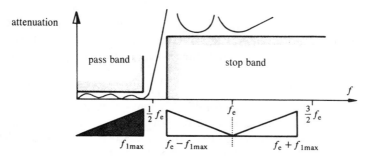

Fig. 4.16 Demodulation filtering process.

4.3.7 Sample-and-hold

Instead of the real sampling which follows the instantaneous variation of $u_1(t)$ over the duration τ of the sampling pulses, we can imagine an ideal sampling (pinpoint), followed by holding the pulse to a *constant* value during the entire time τ.

This procedure has two principal applications:

- it maintains a constant value during the *processing* of the sample, for example, in PCM modulation during the coding operation;
- post-transmission increase of the pulse ratio τ/T_e up to the maximum value equal to 1 ("staircase" signal, fig. 4.17) to improve the efficiency of demodulation filtering.

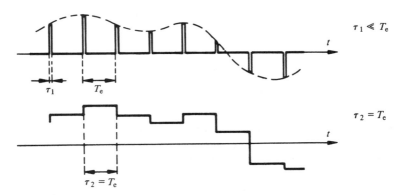

Fig. 4.17 Construction of a staircase signal by holding brief samples.

The held samples $u_2(t)$ can be considered as the convolutional product in the time domain of a rectangular pulse $\text{rect}(t/\tau)$ and pinpoint samples issued from an ideal sampling with frequency f_e.

According to the Fourier transform, this convolutional product becomes a simple product in the frequency domain: the spectrum of pinpoint samples ($\tau = 0$, hence sidebands all of the same amplitude) is multiplied by a function $\sin \pi\tau f / \pi\tau f$, which thus becomes the envelope of the spectrum of samples held (fig. 4.18).

It is important to note that

- in the case of sample-and-hold, low-pass filtering at $f_e/2$ restores the baseband with an amplitude in proportion to the duration τ of the hold, *but with an attenuation distortion* (linear distortion) corresponding to the part of the $\sin \pi\tau f / \pi\tau f$ curve between 0 and $f_e/2$:

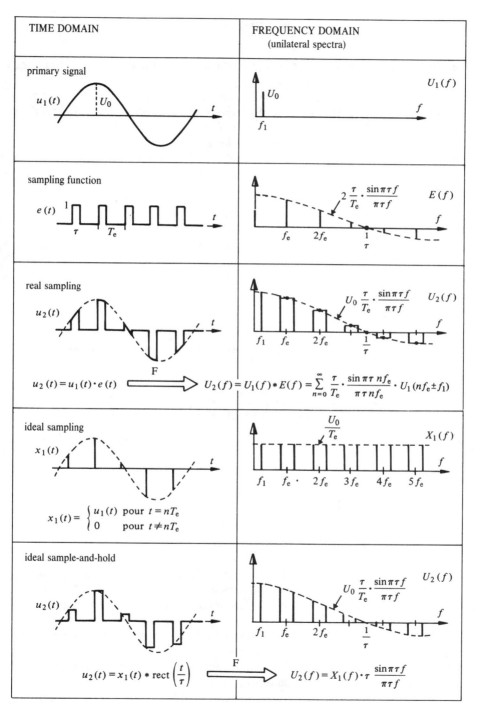

Fig. 4.18 Comparison of real sampling and sample-and-hold in the case of a sinusoidal primary signal; example: $T_e/\tau = 3.5$; $f_1 \cong 0.2 f_e$.

- filtering is facilitated by the fact that the sidebands at nf_e are attenuated with respect to the baseband;
- in the frequent extreme case of a *staircase signal* ($\tau = T_e$), the baseband distortion, i.e., the level difference with respect to $f = 0$, reaches $20 \lg \pi/2 = 3.9$ dB to $f_e/2$ (fig. 4.19). For a sampled telephone channel at 8 kHz, it is equal to 2.75 dB at 3.4 kHz;
- this distortion is generally unacceptable and must be compensated for after filtering or in the filter.

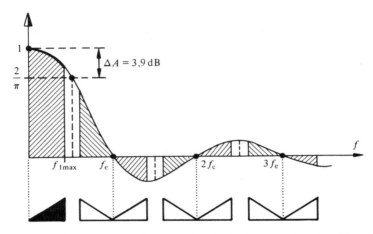

Fig. 4.19 Spectrum of a staircase signal (primary signal with spectrum bounded from 0 to $f_{1\max}$).

4.4 MULTIPLEXING

4.4.1 Definition and justification

Multiplexing is the operation which consists of grouping *several channels*, each assigned to a particular communication, in such a way as to transmit them *simultaneously* on the same physical medium (cable, carrier frequency of a radio link, satellite, *et cetera*) without mixing or mutual interference. At reception, a *demultiplexing* as perfect as possible allows these channels to be separated and restored to their original form. Economic imperatives require multiplexing. The cost of multiplexing equipment should be balanced by the economy obtained in the transmission medium by sharing its cost among z channels. In the case of line transmission, where this latter cost practically increases linearly with distance, there is a certain transmission distance above which multiplexing is justified (fig. 4.20).

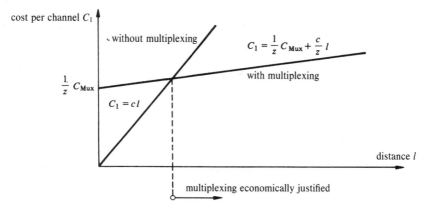

Fig. 4.20 Economics of multiplexing; C_{Mux} = total cost of multiplexing equipment; c = unit cost (per km) of the transmision medium (cable).

4.4.2 Frequency division multiplexing FDM

In a *frequency division multiplexing FDM*, the z channels are distributed along the axis of frequencies and assigned each to an individual frequency band, shifted with respect to those of its neighbors (fig. 4.21). Multiplexing thus consists of *frequency juxtaposition* of these channels, while their signals are *superimposed in the time domain*.

To accomplish frequency multiplexing, it is necessary to transpose the channel from its position in the baseband toward the place assigned to it in the frequency multiplex by means of a modulation.

The applications of this type of multiplexing are the following:

- *in telephony*, the procedure used is single sideband modulation (SSB) (sect. 8.4), the basic principle of *carrier systems* (sect. 10.1). In this case,

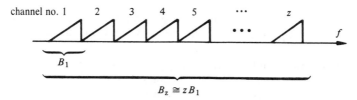

Fig. 4.21 Frequency multiplexing of z channels, each of bandwidth B_1.

the multiplex bandwidth B_z is notably equal to (in fact, slightly higher than) the sum of the channel bandwidths;

$$B_z = zB_1 \qquad (4.18)$$

- *voice frequency telegraphy* (sect. 11.3.5) is a process of frequency multiplexing several telegraph, or telex channels, in a telephone channel (OOK or FSK modulation);
- frequency distribution of several audio or television programs in the frequency band available for radio broadcasting, or on a distribution cable.

4.4.3 Time division multiplexing

Forming a *time division multiplexing TDM* consists of distributing the z channels periodically in time through the intermediary of pulse modulation, and the pulses correspond to a channel interleaved between those of other channels. The composition of time division multiplex is thus always by means of synchronous sampling of the channels with pulses shifted with respect to each other (fig. 4.22).

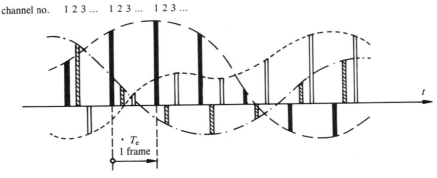

Fig. 4.22 Time division multiplexing (TDM).

The z interleaved time channels form a *frame* of a duration which corresponds to the sampling period T_e. The pulse modulation used can be analog (PAM, PPM) or digital (PCM).

In a time division multiplex, the channels are *juxtaposed in the time domain*, whereas their spectra are *superimposed in the frequency domain*.

There is a complete duality between frequency division multiplex and time division multiplex, except on one point: time is a relative value (in contrast to frequency), so time division multiplex requires a reference point in the frame cycle (fig. 4.23). The receiver must be synchronized (framing, sect. 9.2.2) in frequency *and in phase* with the aid of this reference repeated periodically, so as to be able to demultiplex the flux of pulses that it receives, i.e., to extract the signals concerning each channel individually.

As far as technical applications are concerned, time division multiplexing is becoming increasingly important, not only in transmission, but also (in contrast to frequency multiplexing) in switching, particularly in connection with the development of *digital PCM systems* (chap. 9).

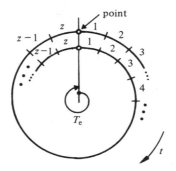

Fig. 4.23 Cyclical representation of a time division multiplex.

4.5 MIXED MODULATIONS

Very often several consecutive stages of modulation are necessary to construct a frequency multiplex (e.g., SSB modulation of a channel and of groups in carrier systems), or a time division multiplex (e.g., PAM, even PDM as preliminary stages of PCM).

On the other hand, the properties of different modulations with respect to adaptation to a transmission medium generally do not correlate with the interest value they represent for multiplexing.

This is why the majority of telecommunication systems make use of several consecutive modulations, often of different types. Table 4.24 gives some examples which will be discussed in the following chapters.

4.6 TWO-WIRE AND FOUR-WIRE LINKS

4.6.1 Bidirectional modes of communication: definitions

While two terminal stations A and B exchange information (conversation, dialogue), in principle two distinct unidirectional means of transmission, called

Table 4.24 Examples of modulation combinations.

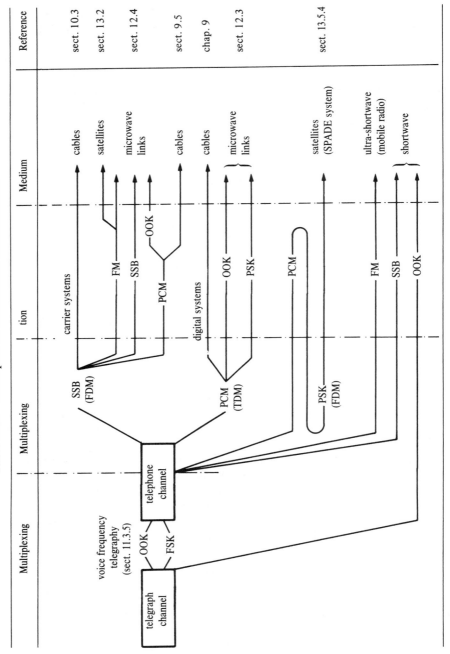

channels, are necessary, each capable of transmitting the information from the sender of one unit to the receiver of the other (fig. 4.25).

These two channels, generally identical, constitute a *transmission circuit*. The emitter and receiver are transducers; most often they are separate units:

microphone—receiver
keyboard—printer
camera—screen
sensor—actuator

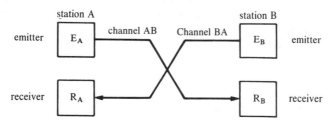

Fig. 4.25 Bidirectional communication (full duplex).

In the *full duplex* mode, the transmission of information takes place **simultaneously** in both directions. The two units can thus emit and receive simultaneously and, in particular, interrupt the emission of the other by sending it a message to this effect. The two channels of the circuit must always be available. The situation is that of figure 4.25. This is typically the case of telephone circuits.

In the *half duplex* mode, the transmission of information takes place **alternately** in each direction. Each unit is normally in a condition of reception, but leaves this mode in order to send. It, therefore, cannot be interrupted by the other unit (fig. 4.26).

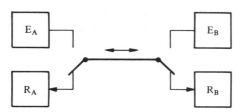

Fig. 4.26 Half duplex mode (switchover contacts are shown idle).

The means of transmission, therefore, must be bidirectional. This mode is used particularly for certain radiocommunications. It has the advantage of requiring only a single radioelectric channel, but requires a strict conversation protocol.

In the case of exclusively unidirectional communication, the term *simplex* is used.

4.6.2 "Two-wire" transmission

Passive transmission means (balanced or coaxial cables, radio channels) are bidirectional by nature. In the language used by telecommunications, they are called "2-wire" circuits, even if the "wires" are not actually wires (radio waves!) or have a tube form (coaxial cables).

In a "2-wire" circuit, the two channels (forward and return) are combined into a single unit.

4.6.3 "Four-wire" transmission

Bidirectional "2-wire" transmission dispenses with any active irreversible unidirectional elements such as amplifiers, envelope detectors, encoders, most modulators, *et cetera*

As soon as the distance or the transmission medium requires an amplification or modulation, it is necessary to separate the two directions and assign **one unidirectional channel to each direction**. The transmission is then said to be "4-wire".

4.6.4 "Frequency division pseudo-four-wire" transmission

The two unidirectional channels needed for "4-wire" transmission are not always physically distinct (two balanced pairs, two coaxial cables); they can also be composed of different frequency bands in the *same* physical medium. The system thus is only "2-wire" physically, but behaves exactly like a "4-wire" circuit. We, therefore, speak of "pseudo-4-wire" transmission. Strictly speaking, this is frequency division multiplex of both transmission directions, which has the effect of doubling the occupied bandwidth in the medium.

Two examples will illustrate this principle:
- microwave link with two different carriers but the same antennas for both directions (fig. 4.27);
- transatlantic link with multiple channels (FDM), on a single coaxial tube by frequency shifting the forward channels with respect to the return channels (fig. 4.28).

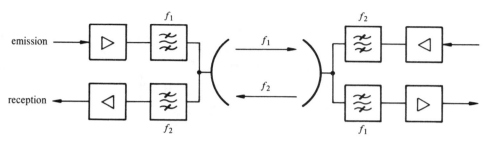

Fig. 4.27 "Frequency division pseudo-4-wire" radio circuit.

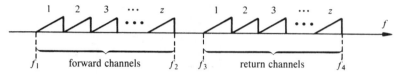

Fig. 4.28 Transatlantic system with a single coaxial tube.

The intermediate amplification (repeater) requires discrimination between the two forward and return frequency bands by means of filters and two separate amplifiers (fig. 4.29(a)), or a single amplifier of double bandwidth (fig. 4.29(b)).

4.6.5 "Time division pseudo-four-wire" transmission

Instead of frequency shifting the forward and return channels of the same circuit, it is possible to imagine transmission of periodically alternated samples of one or the other of these channels, on the same physical support (fig. 4.30). This procedure, which is in principle reminiscent of half duplex transmission, is, in fact, time division multiplexing in both transmission directions and on a macroscopic level gives the perfect illusion of a full duplex mode.

The duration of the samples d must decrease by a factor greater than 2 with respect to "4-wire" transmission, to take the propagation time into account (forward and return) on the line:

$$d < \tfrac{1}{2}(T_e - 2\tau) \tag{4.19}$$

This procedure is not widely used, but can be envisioned for certain cases involving digital PCM transmission (digital subscriber's line).

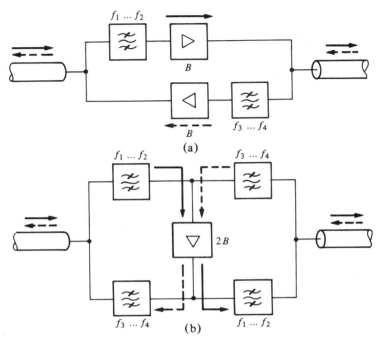

Fig. 4.29 Amplification in a "frequency division pseudo-4-wire" system.

4.6.6 Conversion from "two-wire" to "four-wire" and vice versa

Due to the irreversibility of most transducers, the terminal units are of the "4-wire" type (separate emitter and receiver). The same is true of the active elements (amplifiers, modulators, *et cetera*). However, economics work against the development of a "4-wire" transmission system from end to end. At least in the local network and everywhere that an unamplified baseband transmission is possible, "2-wire" circuits are used.

As a result, it is necessary to convert from "2-wire" to "4-wire" and back again several times, as the example of long-distance telephone links illustrated in figure 4.31 shows.

Toll exchanges switch the link with four contacts per circuit ("4-wire" switching), while the local exchanges require only two per circuit ("2-wire" switching).

The problem of conversion from a "4-wire" circuit to a "2-wire circuit" consists of preventing the receiver (e.g., R_A) from receiving signals sent by the emitter of the *same* unit (e.g., E_A).

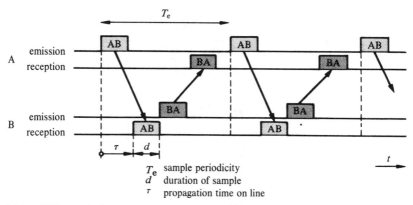

Fig. 4.30 "Time division pseudo-4-wire" transmission.

The solution is either half-duplex mode (not always practical), or the insertion of *hybrids*, which allow simultaneous transmission $E_A \to R_B$ and $E_B \to R_A$, while preventing $E_A \to R_A$ and $E_B \to R_B$.

4.6.7 Hybrid

The *hybrid* is a four-port device in which
- access 2 is connected to the "2-wire" circuit;
- the two accesses 4 and 4' are connected to the "4-wire" channels of the circuit;
- the fourth access is terminated with an impedance $\underline{Z}_E$, called the **balancing impedance**, which is as much as possible equal to the input impedance of the "2-wire" circuit.

In its simplest form, it consists of a differential transformer (fig. 4.32).
If $\underline{Z}_E = \underline{Z}_2$, the hybrid is balanced. In this case, it has the following properties:

- the composite attenuation A_{cp} (2 → 4') across the hybrid in the direction "2-wire" → "4-wire" is minimum for $\underline{Z}'_4 = 2\underline{Z}_2$ and is equal to 3 dB;
- the composite attenuation A_{cp} (4 → 2) across the hybrid in the direction "4-wire" → "2-wire" is minimum for $\underline{Z}_4 = \underline{Z}_2/2$ and also equal to 3 dB;
- the reflection attenuation $A_{cp}(4 \to 4')$ between the input of the "4-wire" side and the output of the "4-wire" side is infinite.

Consequently, in both directions, only half the power is transmitted, the other half being dissipated, either in the balancing impedance, (in the direction "4-wire" → "2-wire",) or in $\underline{Z}_4$ (in the direction "2-wire" → "4-wire").

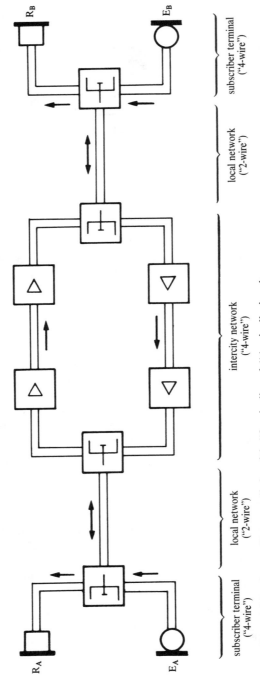

Fig. 4.31 Long-distance link with "2-wire" and "4-wire" circuits.

TRANSMISSION PROCESSES

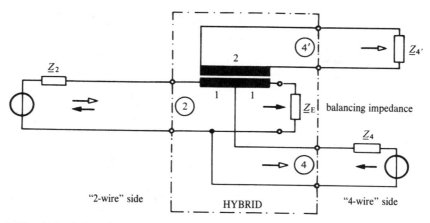

Fig. 4.32 Principle of a hybrid.

In real hybrids, the translators also dissipate some power ("copper" losses due to the resistance of windings, "iron" losses in the core). The crossing attenuation A_t in the two directions can thus be evaluated by

$$A_t = A_{cp}(2 \to 4') = A_{cp}(4' \to 2) \cong 3.5 \text{ dB} \qquad (4.20)$$

Numerous other more elaborate forms of hybrids have been proposed: balanced hybrids, active hybrids, microwave circulators, *et cetera*.

In practice, the balancing condition $\underline{Z}_E = \underline{Z}_2$ is practically impossible to achieve perfectly, since $\underline{Z}_2$ is the input impedance (complex and varying widely with frequency) of the line. A part of the power received from the "4-wire" side is thus re-emitted from the same side (reflection).

The importance of this imbalance can be expressed by the reflection factor ρ (or reflection coefficent, vol. III, chap. 8.5) of the "2-wire" side:

$$\underline{\rho} = \frac{\underline{Z}_2 - \underline{Z}_E}{\underline{Z}_2 + \underline{Z}_E} \qquad (4.21)$$

or by the *balance return loss* $A\rho$

$$A\rho = 20 \lg \left| \frac{1}{\underline{\rho}} \right| \qquad (4.22)$$

Seen from the "4-wire" side, the reflected signal undergoes a supplementary attenuation of $2A_r$ by its double crossing of the hybrid (fig. 4.33). The *transhybrid attenuation* A_r between the incident signal and the reflected signal is, thus,

$$A_r = A_{cp}(4' \to 4) = A_\rho + 2A_t \qquad (4.23)$$

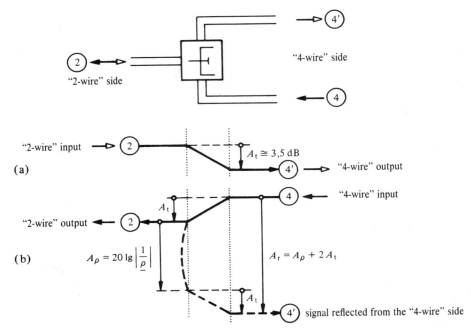

Fig. 4.33 Reflection in a hybrid: (a) Level diagram of the path "2-wire" → "4-wire"; (b) level diagram of the path "4-wire" → "2-wire".

In the "4-wire" telecommunications network, these reflections give rise to *echoes* and the risk of *instability* (sect. 15.3).

On the other hand, in the telephone subscriber's unit, it is psychologically desirable that reflections, weak but present, allow the subscriber to hear his own voice in the receiver. The reference equivalent (sect. 2.4.2) of this *sidetone* must be higher than 17 dB to make the subscriber speak loudly enough (CCITT, Rec. G.121).

Chapter 5

Digital Transmission

5.1 CHARACTERISTICS OF DIGITAL TRANSMISSION

5.1.1 Principles

Digital transmission is intended for communication of *discrete information* from one point (the transmitter) to another (the receiver), coming from a source which has a *finite* number of characters ("alphabet") available.

According to a *prearranged convention*, the receiver knows the alphabet used by the source. It can then *interpret* the received information as a function of this alphabet. It compares the signals received (deformed and disturbed by transmission in the channel) to the list of possible characters and deduces by a *decision* which of them is *most probably* the origin of the received signal.

Thus, in principle, the digital information transmitted can be *regenerated* integrally upon reception. However, if the distortion of the received signals is such that it simulates the presence of another character, irreversible *regeneration errors* appear. The probability ϵ of regeneration errors is the principal criterion for judging the quality of digital transmission.

5.1.2 Representation of digital information

The characters produced by the source are represented for transmission by physical signals composed of a series of *symbols*, or elementary signals, of duration T_M. Generally, each symbol can take m values (m-ary transmission). The bit rate $\dot{D}$ (in bit/s) is thus tied to the symbol rate $\dot{M} = 1/T_M$ (in baud) by relation (1.10):

$$\dot{D} = \dot{M} \, \text{lb} \, m \tag{5.1}$$

A particularly interesting case is that of binary transmission. We therefore have

$$m = 2 \rightarrow \dot{D} = \dot{M} \tag{5.2}$$

The number μ of symbols necessary to represent one of the n characters arranged by the system varies with m according to the relation:

$$m^\mu \geq n \tag{5.3}$$

i.e.,

$$\mu \geq \frac{\text{lb} n}{\text{lb} m} \tag{5.4}$$

For example, to represent an alphabet of $n = 60$ characters, requires at least

$\mu = 6$ binary symbols ($m = 2$)

$\mu = 4$ ternary symbols ($m = 3$)

$\mu = 3$ quaternary symbols ($m = 4$), etc.

In the rest of this chapter, we will consider only symbols, without taking into account their grouping into characters.

5.1.3 Clocked transmission: definition

Digital transmission is said to be *clocked* (or isochronous) if the symbols have a fixed and identical duration T_M. The succession of symbols is thus dictated by a *clock* of frequency $\dot{M} = 1/T_M$.

At reception, a local clock is necessary to identify the different symbols in time. This clock must be **synchronized in frequency and phase** (isochronism) with the clock that determines the series of symbols. This is why it is generally extracted from the received signal by appropriate filtering and processing.

5.1.4 Transmission mode: definition

The law of correspondence between the m possible values of each symbol and the characteristics of the physical signal which represents them constitutes the *mode* of transmission.

The signal emitted is composed of a random train of *elementary signals*, each corresponding to a symbol.

It will be assumed that, for **baseband** digital transmission, the elementary signals are multiples of an *elementary base signal* $u_{BE}(t)$, in the form of an impulse at emission, limited to a duration close to T_M.

Thus, the elementary signal corresponding to a symbol of value k is

$$u_{Ek}(t) = a_k\, u_{BE}(t) \tag{5.5}$$

or

$k = 0 \ldots m - 1$

where a_k is a coefficient which can take m values.

The mode is thus characterized by

- the form $u_{BE}(t)$ of the elementary base signal at emission;
- the coefficients a_k;
- an eventual sequential law, imposing certain constraints on the determination of coefficents a_k as a function of the signals previously emitted (memory modes).

Knowledge of these characteristics and the statistical distribution of the m values allows us to calculate the autocorrelation function $\varphi_E(t)$ by the usual methods in order to obtain by Fourier transform the *power spectral density* $\Phi_E(f)$ of the random signal emitted in this mode.

5.1.5 Unipolar and antipolar modes: definitions

According to the numerical values attributed to the coefficent a_k, and especially their signs, we can distinguish two particular categories of modes:

- *unipolar* modes, in which the m possible values of a_k are *positive* integers: $a_k = 0, 1, 2, \ldots, m-1$;
- *antipolar* modes characterized by m values of a_k equal pairs, but *of opposite signs*. If m is odd, we have: $a_k = 0, \pm 1, 2, \ldots, (m-1)/2$; and if m is even: $a_k = \pm 1/2, 3/2, \ldots, (m-1)/2$.

5.1.6 Examples of clocked modes

- a binary unipolar mode with rectangular signals of duration equal to $T_M/2$ (return to zero mode, RZ);

  elementary base signal: $u_{BE}(t) = U_{BE} \operatorname{rect}(2t/T_M)$

  state 0: $a_0 = 0$

  state 1: $a_1 = 1$

- ternary restrictive mode (pseudo-ternary), antipolar with alternating polarity (memory mode) and rectangular elementary signals of duration equal to T_M (AMI-NRZ mode, alternate mark inversion non-return to zero). The restriction consists of making ternary symbols (with three coefficients: a_0, a_1, a_2), carry binary information only (0/1), according to the following convention:

  binary 0 represented by $a_1 = 0$

  binary 1 represented by $a_0 = -1$ or $a_2 = +1$, alternatively

  elementary base signal: $u_{BE}(t) = U_{BE} \operatorname{rect}(t/T_M)$

- antipolar quaternary mode (2 bit per symbol) with signals in $\cos^2$
  elementary base signal: $u_{BE}(t) = U_{BE} \, \text{rect}(t/2T_M)\cos^2(\pi t/2T_M)$
  state 0: $a_0 = -3/2$
  state 1: $a_1 = -1/2$
  state 2: $a_2 = +1/2$
  state 3: $a_3 = +3/2$

Figure 5.1 gives the form of the emitted signal and its spectral power density for each of the three examples.

5.1.7 Notation

The following notation will be used in this chapter:

$u_{BE}(t)$ emitted elementary base signal
$U_{BE}(f)$ its Fourier transform
$u_E(t)$ random emitted signal, consisting of a random series of elementary signals according to (5.5), each corresponding to an m-ary symbol
$\Phi_E(f)$ spectral power density corresponding to $u_E(t)$
$u_{BR}(t)$ received elementary base signal

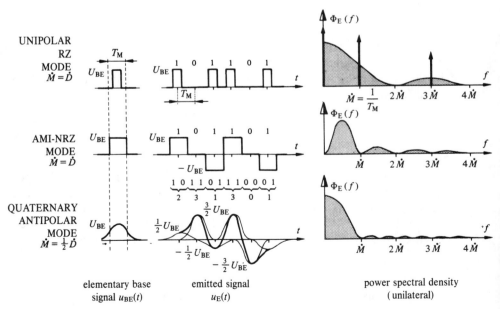

Fig. 5.1 Examples of clocked modes.

U_{BR} its maximum value
$u'_{BR}(t)$ received elementary base signal after equalization
$\underline{U}'_{BR}(f)$ its Fourier transform
$u'_R(t)$ received and equalized random signal, formed from the same random series as $u_E(t)$, but originating from the elementary base signal $u'_{BR}(t)$
$\Phi'_R(f)$ power spectral density corresponding to $u'_R(t)$

5.1.8 Influence of the channel

The channel can be characterized by
- its transfer function $\underline{H}(f)$ in the frequency domain or its impulse response $h(t)$ in the time domain;
- the spectral power density and the statistical distribution of the amplitudes of disturbing sources (noise, crosstalk) which affect it.

If the form $u_{BE}(t)$ of the emitted elementary base signals, or their Fourier transform $\underline{U}_{BE}(f)$ is known, the corresponding characteristics of the elementary base signals at reception can be derived, in the absence of interference:

$$u_{BR}(t) = u_{BE}(t) * h(t) \tag{5.6}$$

and

$$\underline{U}_{BR}(f) = \underline{U}_{BE}(f)\, \underline{H}(f) \tag{5.7}$$

Relation (5.6) allows us to evaluate the deformation of the base signal due to the linear distortions of the channel, in particular their duration and whatever influence they may have on the neighboring symbols. This *intersymbol interference, ISI*, is the subject of section 5.3.

The noise which affects the signal in the course of transmission makes discrimination among the m values by the receiver more dificult and can become the cause of *regeneration errors*, which have a probability that depends on:

- the statistical distribution of the disturbing sources (and thus their power);
- the amplitude of the useful signals received and their form;
- the number of m states.

The result is that the information received by the receiver is not exactly identical to that emitted by the source. The *transmission errors*, which express this difference between the characters emitted and received, have a probability of appearance which is not necessarily equal to that of the regeneration errors of the symbols. In fact, by appropriate coding and the addition of corresponding redundancy, it is possible to **detect** the presence of an erroneous symbol in a character and refuse its interpretation, or even to *correct* certain errors and to restore the correct character despite the fact that it contains erroneous symbols.

The analysis made in this chapter (section 5.4) concerns only regeneration errors of each symbol taken separately. In the same way, only *additive noise*, i.e., that which is superimposed on the signal, will be considered here.

In certain cases (especially in wave transmission), the transfer function of the channel is subject to random fluctuations. This is called ***multiplicative noise***. Its effect is discussed in section 12.2.4.

5.2 REGENERATION

5.2.1 Principle

Regeneration consists of reconstructing digital information transmitted by a signal as faithfully as possible after it has been attenuated, deformed, and disturbed by its passage along the transmission channel.

The information thus restored is given to the receiver (terminal regenerator) or retransmitted further by a signal emitted once again in the same form (intermediary regenerators).

The regenerator processes the received signals to remove as much as possible the disturbances which affect them, then it *interprets* the signals as functions of the m possible values, which are known according to a prearranged convention. Logical *decision thresholds* allow it to discriminate these m values, and thus to extract the digital information carried by the signal.

If the transmission is clocked, these decisions are taken only at certain moments, called *characteristic moments*, spaced by T_M, determined by the clock of frequency $\dot{M}$ with which the received signals are sampled.

5.2.2 Example of an unclocked regenerator: the relay

The relay, used since the beginning of telegraphy to regenerate the binary information being transmitted, reacts in a binary manner (operated or released) according to the value of the current (received signal) which traverses its coil. Due to the variable reluctance of its magnetic circuit according to its binary state, the thresholds of operation and release are different (hysteresis). The transitions of information emitted are only approximately restored in time. Nevertheless, the binary information is correctly transmitted, despite the distortion and noise that the signal undergoes (fig. 5.2).

DIGITAL TRANSMISSION

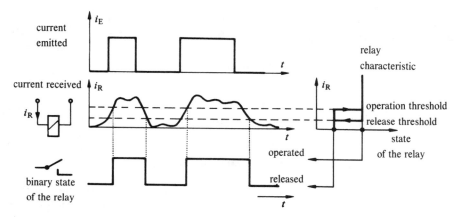

Fig. 5.2 Regeneration with a relay.

5.2.3 Functions of a clocked regenerator

The principal functions of a regenerator are (fig. 5.3):

- *equalization*, i.e., the preliminary restoration of the received signal in analog form in such a way as to reduce as much as possible the interference between symbols and the influence of noise;
- linear amplification to compensate for the atttenuation of the channel. This is necessary for technical reasons in order to facilitate sampling and discrimination, but does not influence the theoretical performance of the regenerator;
- sampling at moments iT_M with a clock frequency $\dot{M} = 1/T_M$ precisely extracted from the received signal;

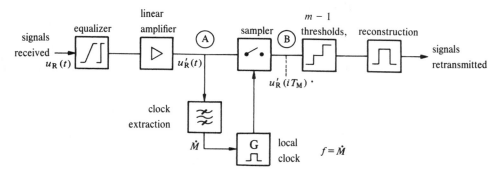

Fig. 5.3 Block diagram of a clocked regenerator.

- discrimination of the *m* discrete values by *m* − 1 thresholds;
- if necessary (intermediary regenerators), shaping of the signals into a new form (i.e., emission of new signals), with the original amplitude and form, and carrying reconstructed digital information.

5.3 EFFECTS OF DISTORTION: INTERSYMBOL INTERFERENCE

5.3.1 Time condition for zero interference between symbols

In the block diagram of the isochronous regenerator (fig. 5.3), the discrimination decision from among the *m* possible values per symbol is based on the signal sampled at frequency $\dot{M}$ (point B of the block diagram). Only the value of the signal at the characteristic moments spaced by T_M is considered and not its form between those moments.

The interference between symbols is zero if the value of the signal at the moment iT_M depends only on the *m*-ary information carried by the symbol *i* and not on that carried by its neighbors $i\pm1$, $i\pm2$, *et cetera*.

This condition can be expressed in the following way:

- the elementary base signal $u'_{BR}(t)$ received and equalized (point A of the block diagram, figure 5.3) must cancel for all the characteristic moments *except one* (fig. 5.4).

$$u'_{BR}(iT_M) = \begin{cases} U_{BR} \text{ for } i = 0 \\ 0 \text{ for integer } i \neq 0 \end{cases} \quad (5.8)$$

A signal which satisfies this condition is, for example, the following (fig. 5.6(a)).

$$u'_{BR}(t) = U_{BR} \frac{\sin(\pi t/T_M)}{\pi t/T_M} \quad (5.9)$$

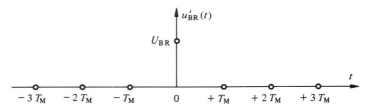

Fig. 5.4 Time condition for zero interference at the rate $\dot{M} = 1/T_M$.

5.3.2 Frequency condition for zero interference between symbols: first Nyquist criterion

The time condition (5.8) defines the response $u'_{BR}(t)$ only by means of its samples $u'_{BR}(iT_M)$. By virtue of the sampling theorem (sect. 4.3.4), $u'_{BR}(t)$ is only known unambiguously if the Fourier transform $\underline{U}'_{BR}(f)$ is strictly limited to a frequency range lower than half the sampling frequency, which here is equal to $\dot{M}$ (fig. 5.5). Thus, the elementary base signal takes the form given by (5.9).

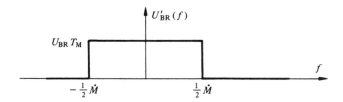

Fig. 5.5 First Nyquist criterion.

This allows us to state a frequency condition, called the *first Nyquist criterion*, after its author (1928):

- if $U'_{BR}(f) = \begin{cases} U_{BR}T_M & \text{for } |f| < \tfrac{1}{2}\dot{M} \\ 0 & \text{for } |f| > \tfrac{1}{2}\dot{M} \end{cases}$ (5.10)

- then, the interference between the symbols transmitted at a rate $\dot{M}$ is zero.

This condition is sufficient, but not necessary, as the examples of figure 5.6 show. Nevertheless, the first Nyquist criterion provides the solution with the minimum bandwidth:

$$B_{min} = \tfrac{1}{2}\dot{M} \qquad (5.11)$$

Nyquist's theorem (4.4), already roughly as presented in section 4.1.4, is found again here.

5.3.3 Limitation of the first Nyquist criterion: second Nyquist criterion

The elementary base response from (5.9) in the form $\sin x/x$ is impractical for several reasons:

- it extends from $t = -\infty$ to $t = +\infty$ and thus violates the causality principle;

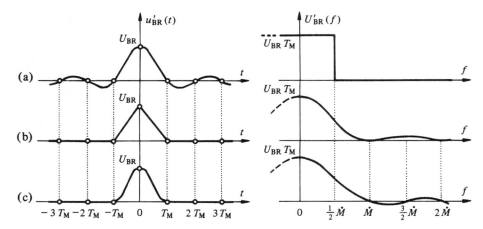

Fig. 5.6 Examples of elementary base signals without interference between symbols at the rate $\dot{M} = 1/T_M$: (a) signal in the form sin x/x; (b) triangular signal; (c) signal in the form $\cos^2$.

- considered as the result of a strict limitation of the pass band, it presupposes the existence of an ideal low-pass filter;
- a small error in the determination of the characteristic moment (sampling) would lead to significant intersymbol interference, because of the strong variations of this function around its zeros.

By restricting the time condition (5.8), the frequency condition (5.10) can be made more flexible, thus leading to the *second Nyquist criterion* which, itself, leads to practical solutions.

The new time condition is the following:

- the elementary base signal $u'_{BR}(t)$ received and equalized (point A of the block diagram, fig. 5.3) must not only present zeros to the non-zero multiples of T_M but also cancel once in the middle of the interval between each of the zeros (fig. 5.7).

$$u'_{BR}(iT_M) = \begin{cases} U_{BR} & \text{for } i = 0 \\ \frac{1}{2} U_{BR} & \text{for } i = \pm\frac{1}{2} \\ 0 & \text{for } i = \pm 1; \pm\frac{3}{2}; \pm 2; \pm\frac{5}{2} \ldots \end{cases} \quad (5.12)$$

A signal which satisfies this condition is, for example, the following (fig. 5.8):

$$u'_{BR}(t) = U_{BR} \frac{\sin(\pi t/T_M)}{\pi t/T_M} \frac{\cos(\pi t/T_M)}{1 - (2t/T_M)^2} \quad (5.13)$$

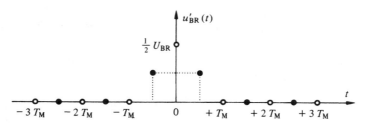

Fig. 5.7 Stricter time condition for zero interference at the rate $\dot{M} = 1/T_M$.

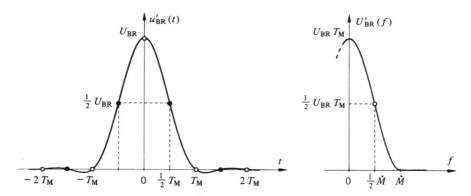

Fig. 5.8 Second Nyquist criterion.

This satisfies the *second Nyquist criterion*, which is expressed as follows:

- if $U'_{BR}(f) = \begin{cases} U_{BR}T_M \cos^2(\frac{1}{2}\pi T_M f) & \text{for } |f| \leq \dot{M} \\ 0 & \text{for } |f| > \dot{M} \end{cases}$ (5.14)

- then, the interference between symbols is zero for a rate $\dot{M}$ and the elementary signal satisfies the condition (5.12).

5.3.4 Compromise: expanded Nyquist criterion

The second Nyquist criterion leads to an elementary response $u'_{BR}(t)$ which, strictly speaking, is also non-causal but approaches zero much more rapidly than the sin x/x signal from (5.9), at the expense of **doubling the occupied frequency range** ($\dot{M}$ instead of $\dot{M}/2$).

184   TELECOMMUNICATION SYSTEMS

A compromise between the increase of the bandwidth and the time limitation of the elementary response can be found by the *expanded Nyquist criterion*[22], which is stated as follows:

- if $\underline{U}'_{BR}(f)$ presents a central symmetry around its value for $f = \dot{M}/2$ (fig. 5.9), in the interval $(0 \ldots \dot{M})$, i.e., if

$$U'_{BR}(\tfrac{1}{2}\dot{M} - f) + U'_{BR}(\tfrac{1}{2}\dot{M} + f) = U'_{BR}(0) = U_{BR}T_M \tag{5.15}$$

- then the interference between the symbols transmitted at a rate $\dot{M}$ is zero.

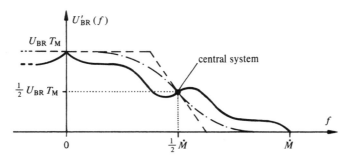

Fig. 5.9 Expanded Nyquist criterion.

Consequently, the signals occupy a frequency range (bandwidth) with an upper limit which is located between the minimum theoretical value of $\dot{M}/2$ from the first Nyquist criterion and twice this value (second criterion). A reasonable compromise for purposes of estimation is to assume that *the bandwidth necessary for transmission without interference at a rate $\dot{M}$* is equal to approximately $0.8\,\dot{M}$. This empirical value has already been given in relation (4.9).

5.3.5 Equalization

It would be purely by chance if the elementary base signals $u'_{BR}(t)$ at the output of the channel from (5.6) and (5.7) neatly satisfied the expanded Nyquist criterion and therefore presented zero interference.

To satisfy this condition with

- an elementary signal in the form $u_{BE}(t)$ given at emission
- a channel of transfer function $\underline{H}(f)$

it is necessary to place a quadripole in cascade with the channel, called the *equalizer* (fig. 5.10), one of the functions of which is to correct the form of the

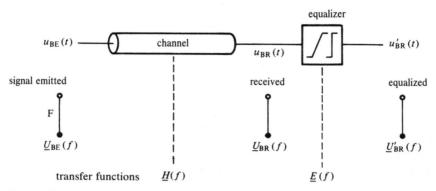

Fig. 5.10 Equalization and elementary base signals.

elementary signals $u_{BR}(t)$ received in order to make them zero-interference signals as much as possible.

Thus, we have

$$\underline{U}'_{BR}(f) = \underline{U}_{BE}(f)\,\underline{H}(f)\,\underline{E}(f) \tag{5.16}$$

The transfer function $\underline{E}(f)$ of the equalizer must be determined and adjusted so that $U'_{BR}(f)$ corresponds to the expanded Nyquist criterion (5.15).

It must be noted that:

- the transfer function $\underline{E}(f)$ of the equalizer is not in general equal to the inverse of that of the channel $\underline{H}(f)$;
- the relation (5.16) associated with the Nyquist criterion does not completely determine the function $\underline{E}(f)$.

5.3.6 Eye pattern

The elementary response $u'_{BR}(t)$ after equalization extends over a longer time than the elementary base signal $u_{BE}(t)$ at emission, due to the low-pass character of the usual channels (fig. 5.11). This response can be calculated by decomposing $u_{BE}(t)$ into elementary steps and summing the corresponding step responses, which are obtained, for example, by the method indicated in section 3.3 in the case of lines.

The emission of a ***random*** series of m-ary symbols from (5.5) leads, at reception and after equalization, to a signal $u'_R(t)$ composed of the elementary base signal $u'_{BR}(t)$, shifted by iT_M, each multiplied by a random m-ary coefficient a_k and in part overlapping one another due to the extension of these elementary signals.

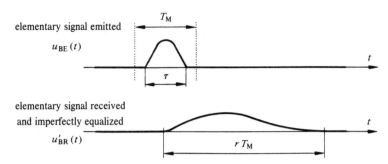

Fig. 5.11 Extension of the elementary signals after transmission.

The graphic superposition of all the intervals of duration T_M, taken over $u'_R(t)$, gives rise to an interesting diagram called the *eye pattern*. This pattern contains *all* the possible signals $u'_R(t)$ while respecting the rules of the transmission mode, and *all* the transitions possible between one symbol and the next, superimposed over a period T_M.

In assuming that $u'_{BR}(t)$ extends over a time rT_M, the signal corresponding to a symbol in the random series $u'_R(t)$ will be influenced by those of the $r - 1$ preceding symbols. Taking this influence into account, in an m-ary mode, there are m^r signal forms possible during the time T_M of a symbol. The eye pattern corresponds exactly to the graphic representation of these m^r combinations, on the same time interval T_M.

Figure 5.12 gives an example of the graphic construction of the eye pattern from an elementary response extending over $r = 3$ clock periods in the case of a binary unipolar mode.

The eye pattern presents "openings" in the form of rounded diamonds, which have given it its name. The form and dimension of these "eyes" are very meaningful indicators of the quality of an equalization.

Actually, the discrimination of m values after sampling at the characteristic moments iT_M is easier when the "eyes" of the diagram are more open. A perfect equalization, i.e., with zero interference between symbols, appears as a maximum vertical opening of the "eyes." The different lines of the eye pattern thus pass through m exact points at the characteritic moments (fig. 5.13).

Experimentally, the eye pattern can be obtained very easily on an oscilloscope by generating a (pseudo-) random series of m-ary symbols from the emitting side, emitted according to the mode used, and observing the received and equalized signal $u'_R(t)$ with a time base synchronized by the symbol clock (frequency $\dot{M} = 1/T_M$).

DIGITAL TRANSMISSION

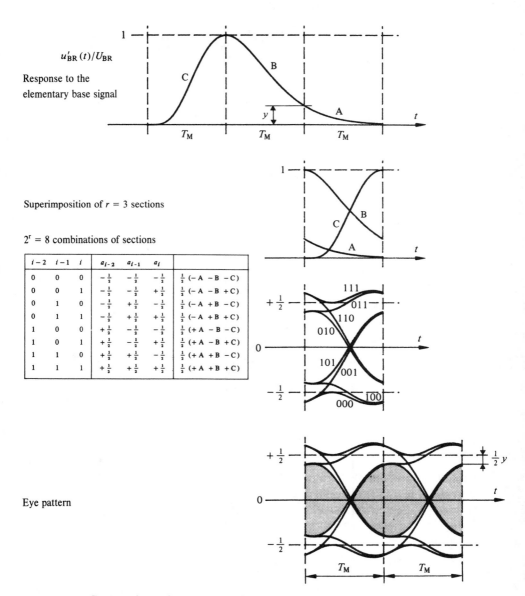

Fig. 5.12 Generation of the eye pattern (example).

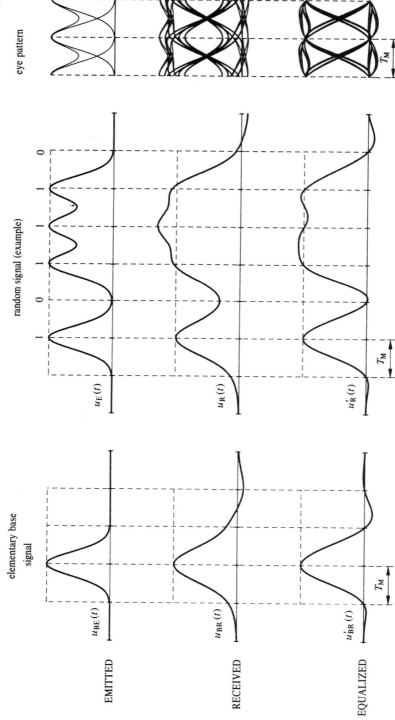

Fig. 5.13 Effect of equalization on the received signal and on the eye pattern (binary unipolar mode).

DIGITAL TRANSMISSION

The practical value of the eye pattern is very great. Through rapid overall examination with the oscilloscope, in the case of an installed digital transmission, it allows:

- fine tuning of the equalizer to cancel the interference between symbols;
- adjustment of the phase of the local sampling clock in the regenerator (fig. 5.3) so as to sample the equalized signals at the precise exact moment when the interference is zero;
- control of these adjustments while in service.

5.4 EFFECTS OF NOISE: REGENERATION ERRORS

5.4.1 Sources of errors

In the case of zero interference between symbols, the signals received and perfectly equalized $u'_R(t)$, then sampled at the frequency $\dot{M}$ (point B of fig. 5.3), can nominally take m values, which depend on the mode of transmission chosen:

$$u'_{Rk}(iT_M) = a_k U_{BR} \qquad (5.17)$$

where a_k is the same coefficient as in relation (5.5).

Whatever the mode, the deviation between two neighboring values among the m is, by hypothesis, equal to U_{BR}. The $m - 1$ logical decision thresholds for the discrimination of the m states are located between the nominal values of these states (fig. 5.14).

The noise $u_N(t)$ introduced in the course of transmission gives rise, after filtering by the transfer function $E(f)$ of the equalizer, to a contribution $u'_N(t)$ of which the samples $u'_N(iT_M)$ are superimposed on the nominal values of the information carrying signal (fig. 5.15).

The result is an uncertainty regarding the true value of the signal at the characteristic moments iT_M which can lead to interpretation errors.

5.4.2 Condition for regeneration without errors

There is no regeneration error when the disturbed values $u'_R(iT_M) + u'_N(iT_M)$ do not simulate the presence of *another* nominal value (fig. 5.16).

If the decision thresholds are located at $\pm U_{BR}/2$ with respect to the nominal values, in the case in which there is no interference between symbols, the condition can be expressed by:

$$|u'_N(iT_M)| < \tfrac{1}{2} U_{BR} \qquad (5.18)$$

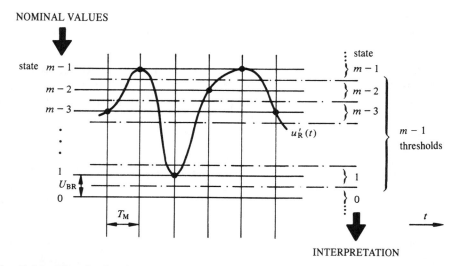

Fig. 5.14 Nominal values and decision thresholds.

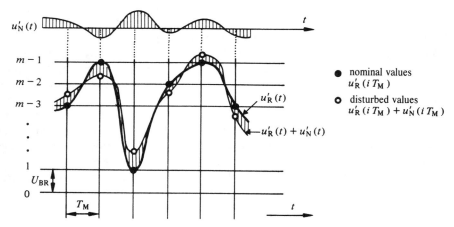

Fig. 5.15 Effect of noise on the values of the signal.

5.4.3 Evaluation of the error probability in the general case of an *m*-ary transmission.

The error probability per symbol ϵ can be estimated by the probability that the noise $u'_N(iT_M)$ crosses the decision threshold and causes confusion with a neighboring value.

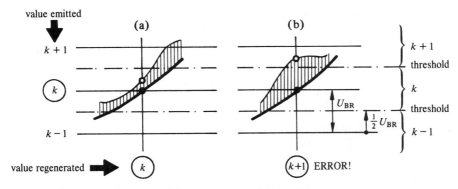

Fig. 5.16 Regeneration (a) without error and (b) with error.

This appears

- for the value $k = 0$: when $u'_N(iT_M) > U_{BR}/2$;
- for the value $k = m - 1$: when $u'_N(iT_M) < -U_{BR}/2$;
- for the intermediate values $k = 1...m - 2$; when $|u'_N(iT_M)| > U_{BR}/2$.

Since the m possible values per symbol are mutually exclusive and statistically independent of the instantaneous value of the noise, we can write:

$$\epsilon = \text{Prob}(k = 0) \cdot \text{Prob}(u'_N > \tfrac{1}{2}U_{BR}) + \text{Prob}(k = m - 1) \cdot \text{Prob}(u'_N < -\tfrac{1}{2}U_{BR})$$

$$+ \sum_{k=1}^{m-2} \text{Prob}(k) \cdot \text{Prob}(|u'_N| > \tfrac{1}{2}U_{BR}) \tag{5.19}$$

The probability $\text{Prob}(u'_N > U_{BR}/2)$ is equal to the integral between $U_{BR}/2$ and infinity of the ***probability distribution*** of the random continuous variable $u'_N(iT_M)$, designated here by $p(u'_N)$ (fig. 5.17).

With the following hypotheses:

- equal probability of the m values:

$$\text{Prob}(k) = 1/m \tag{5.20}$$

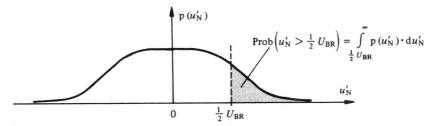

Fig. 5.17 Probability of exceeding and probability distribution.

- symmetry of the probability distribution of u'_N

$$p(-u'_N) = p(u'_N) \qquad (5.21)$$

i.e.,

$$\text{Prob}(u'_N > \tfrac{1}{2}U_{BR}) = \text{Prob}(u'_N < -\tfrac{1}{2}U_{BR})$$
$$= \tfrac{1}{2}\text{Prob}(|u'_N| > \tfrac{1}{2}U_{BR}) \qquad (5.22)$$

- perfect equalization (no interference)

relation (5.19) becomes:

$$\epsilon = \frac{2(m-1)}{m} \text{Prob}(u'_N(t) > \tfrac{1}{2}U_{BR}) \qquad (5.23)$$

5.4.4 Case of Gaussian noise

The calculation of the probability of error from (5.23) implies knowledge of the statistical distribution of the noise. Because equalization is a linear operation, it does not affect the form of this distribution.

An often used model, which is sufficiently realistic in the majority of cases, is the Gaussian distribution (given in the appendix, section 16.2) according to which the probability distribution is

$$p(u'_N) = g(u'_N/U'_{N\text{ eff}}) \qquad (5.24)$$

and the probability of exceeding it is

$$\text{Prob}(u'_N > \tfrac{1}{2}U_{BR}) = G_c(\tfrac{1}{2}U_{BR}/U'_{N\text{ eff}}) \qquad (5.25)$$

Since the integral complementary Gaussian function $G_c(x)$ is well known through tables and graphs (fig. 16.2), it permits a quantitative evaluation of the error probability ϵ.

Relation (5.23), illustrated in figure 5.18 (m equally probable values), in the Gaussian case, becomes

$$\epsilon = \frac{2(m-1)}{m} G_c(\tfrac{1}{2}U_{BR}/U'_{N\text{ eff}}) \qquad (5.26)$$

Examples:

- binary transmission ($m = 2$) with equally probable states with $U'_{N\text{ eff}} = U_{BR}/2$;

according to (5.26) and using figure 16.2, we evaluate the error probability at $\epsilon \cong 0.16$;

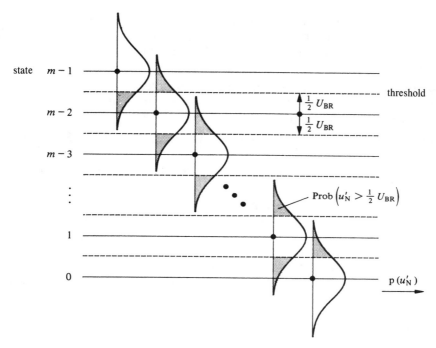

Fig. 5.18 *m*-ary transmission with Gaussian noise.

- quaternary transmission ($m = 4$) with equally probable states; to guarantee a error probability $\epsilon < 10^{-5}$, it is necessary that, from (5.26):

$$G_c(U_{BR}/2U'_{N\,eff}) < 10^{-5}$$

from which we obtain (from fig. 16.2):
$U_{BR} \geq 8.5 U'_{N\,eff}$

5.4.5 Role and optimization of the receiving filter (equalizer)

At the input of the regenerator, the signals $u_R(t)$ occupy a relatively limited frequency range, while the noise $u_N(t)$, particularly if it is caused by thermal noise, has a much wider spectrum.

We therefore have every interest in separating the signals from the noise as much as possible so as to accept $u_R(t)$ while rejecting a part of $u_N(t)$. This function is performed by the *receiving filter*, which can be considered as part of the equalizer and contributes to the transfer function $\underline{E}(f)$ of the latter.

Because the unilateral power spectral densities of the signals and noise at the input of the regenerator, $\Phi_R(f)$ and $\Phi_N(f)$, respectively, are given, the problem consists of optimizing the transfer function $\underline{E}(f)$ in such a way that simultaneously

- the noise power P'_N after the equalizer is at a minimum
- the amplitude U_{BR} of the elementary base signals, after equalization and sampling, are maximum.

Further, $\underline{E}(f)$ must satisfy the condition stated in section 5.3.5 to guarantee zero interference between symbols.

Generally, the solution to this problem is not simple. It can be simplified with the following hypotheses:

- Gaussian noise with uniform power spectral density at the input of the regenerator; for example, thermal noise (white) with $\Phi_N(f) = kT$ (unilateral);
- Fourier transform $U'_{BR}(f)$ of the elementary signals equalized in $\cos^2$ according to the second Nyquist criterion, from (5.14).

It can be shown [22] that, under these conditions, the optimal equalizer is of the form:

$$E(f) \sim \sqrt{U'_{BR}(f)/U_{BR}T_M} = \cos(\tfrac{1}{2}\pi T_M f) \tag{5.27}$$

This form is nevertheless not critical: a rectangular transfer function (ideal low-pass at $f = 1/2T_M$) or in $\cos^2$, such as $U'_{BR}(f)$, would lead to a value for the ratio between the power of the signals P'_R and that of the noise P'_N after equalizaton of only 2 dB lower than the optimum.

5.4.6 Error distribution in time

The error probability ϵ gives an idea of the *mean* symbol error rate, but says nothing about error distribution in time. The interval τ between these errors, measured in multiples of T_M in the case of a clocked transmission, is a random discrete variable.

While the errors are *statistically independent*, τ follows a geometrical distribution with the mathematical expectation (statistical mean) T_M/ϵ

$$\text{Prob}(\tau = iT_M) = \epsilon(1 - \epsilon)^{i-1} \tag{5.28}$$

The mean number N of errors expected during an observation period $T = kT_M$ is thus regulated by the *Poisson distribution*:

$$\text{Prob}(N, kT_M) = \frac{(k\epsilon)^N}{N!} \exp(-k\epsilon) \tag{5.29}$$

DIGITAL TRANSMISSION

The phenomenon is said to be *Poissonian*. Such errors appear when the source of disturbance is **white Gaussian noise** (e.g., thermal noise).

Nevertheless, transmission often is disturbed by **impulse noise**. This noise, with a much more complex statistical nature, is less well known than white noise, and is due to external influences (atmospheric disturbances, effects of heavy-current circuits), crosstalk couplings of straight-sided signaling pulses (e.g., selection pulses) on neighboring circuits, and switching effects (shaking of electromechanical contacts, poor contacts, *et cetera*). As a consequence, the errors are not independent, but appear grouped into bursts (several consecutive erroneous symbols), separated by intervals without errors.

5.5 PLANNING DIGITAL TRANSMISSION

5.5.1 Given situation

The specifications for the planned transmission are:
- the bit rate $\dot{D}$
- the transfer function of the channel $\underline{H}(f)$
- the unilateral power spectral density of the source of noise (in the case of thermal noise $\Phi_N(f)$ is constant and equal to kT)
- the probability distribution $p(u_N)$ of the noise
- the available and tolerable emission power P_E

5.5.2 Mode selection

The determination of the number m of values per symbol is the result of a compromise between
- the "bandwidth," i.e., the fitting of the power spectral density $\Phi_E(f)$ at emission with the channel characteristics. Increasing m means decreasing the symbol rate $\dot{M} = \dot{D}/\text{lb } m$ and, for a given channel, decreasing the intersymbol interference;
- the error probability ϵ per symbol, which increases very rapidly with m because, for a constant emission power, the difference U_{BR} at reception between the m states diminishes.

Often, however, the selection is imposed by the very pronounced low-pass or band-pass character of certain channels, which *a priori* limits the possible symbol rate (sect. 4.1.4). According to (4.9) and (4.1), we therefore have

$$\text{lb} m > \dot{D}/1.25B \tag{5.30}$$

Selection of the form of the elementary base signals $u_{BE}(t)$ and the coefficients a_k from (5.5) depend on the particular conditions imposed on the power spectral density $\Phi_E(f)$ and on the spectrum $U_{BE}(f)$ of the elementary signals.

Depending on the particular case, these conditions can be:

- mean value strictly zero: $U_{BE}(0) = 0$
- low content at low frequencies: $\Phi_E(0) = 0$ with possibly an additional zero derivative for $f = 0$
- high clock rate content ($f = \dot{M}/2$ or $\dot{M}$)

The antipolar modes cancel the long term mean value if the states are equally probable, but do not necessarily cancel $\Phi_E(0)$.

5.5.3 Outline of error probability calculation

The error probability ϵ per symbol is a determining criterion for planning digital transmission. Its evaluation proceeds from the general relation (5.23) in which the following factors are involved:

- the difference U_{BR} between the nominal m-ary values after equalization at the characteristic moments iT_M;
- the probability distribution $p(u'_N)$ of the noise *after* the equalizer. Because this is supposed to be a linear circuit, $p(u'_N)$ is of the same form as $p(u_N)$ although $u'_N(t)$ and $u_N(t)$ have different rms values. If, for example, $u_N(t)$ is Gaussian, $u'_N(t)$ is also.

Figure 5.19 illustrates symbolically an approach to the complete calculation with the following stages:

- calculation (followed by an adjustment according to the eye pattern) of the transfer function $\underline{E}(f)$ of the equalizer so as to cancel the intersymbol interference (Nyquist criterion);
- the only non-zero value of the elementary base signal $u'_{BR}(t)$ at the characteristic moments iT_M gives the voltage U_{BR};
- the effect of the equalizer on the power spectral density of the noise $\Phi_N(f)$ allows us to evaluate the power P'_N after equalization by integration of $\Phi'_N(f)$. From this we obtain the corresponding rms voltage $U'_{N\,eff}$ which determines the scale of the probability distribution $p(u'_N)$.

ϵ is thus determined, at least in principle.

It may be useful to express ϵ as a function of the signal-to-noise ratio ξ'_R defined at the output of the equalizer:

$$\xi'_R = \frac{P'_R}{P'_N} = \frac{\int_0^\infty \Phi_E(f)|H(f)|^2|E(f)|^2\,df}{\int_0^\infty \Phi_N(f)|E(f)|^2\,df} \tag{5.31}$$

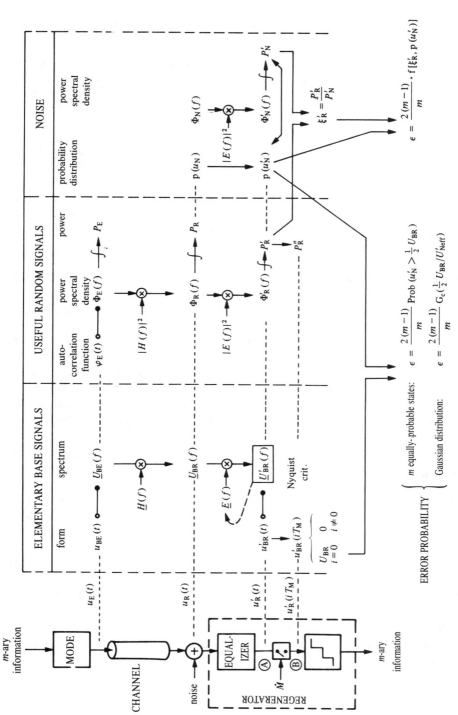

Fig. 5.19 Complete outline of the error probability calculation.

The function which relates ϵ to ξ'_R, not specifically expressed in the general case of figure 5.19, depends on the probability distribution of the noise, on the transmission mode, and on the transfer function of the equalizer. The exact form of this function is very complex.

5.5.4 Estimation of the error probability as a function of the signal-to-noise ratio in a given digital transmission

It is not always necessary, nor even possible with reasonable methods, to calculate exactly the probability of error per symbol ϵ because neither the characteristics of noise, nor those of the channel, are generally known with precision.

It is nevertheless useful to estimate at least roughly and quickly, ϵ as a function of the signal-to-noise ratio ξ'_R defined in section 5.5.3. The estimation will be made with the following hypotheses:

- *perfect equalization* with respect to canceling the intersymbol interference and nearly optimum with respect to noise filtering;
- *Gaussian* noise, but not necessarily with a uniform power spectral density $\Phi_N(f)$ at the input of the regenerator;
- *equal probability* of the m states, emitted in unipolar or antipolar mode;
- equalizer *without amplification*, i.e., such that it modifies the form of the elementary base signals $u_{BR}(t)$ received into $u'_{BR}(t)$ without changing their amplitude U_{BR}. This hypothesis is not restrictive, because amplification would not noticeably modify the ratio ξ'_R.

Under these conditions, the power P'_R of the equalized signals gives an approximate but realistic idea of the effective power P_R of the signals received at the ouput of the channel.

Furthermore, the power P''_R of the samples $u''_{BR}(iT_M)$ taken at the frequency $\dot{M}$ and supposed to be held until the following (staircase signal) is very close to P'_R. Thus, the signal-to-noise ratio

$$\xi'_R = \frac{P'_R}{P'_N} \cong \frac{P_R}{P'_N} \cong \frac{P''}{P'_N} \tag{5.32}$$

is a realistic image of the transmission quality insofar as the relative importance of the noise is concerned.

The power P''_R across a resistance R can be calculated from U_{BR}, from the probability of m states and from the coeffecints a_k which intervene in relations (5.5) and (5.17):

$$P''_R = \frac{1}{R} \sum_{k=0}^{m-1} \text{Prob}(k) \cdot a_k^2 \cdot U_{BR}^2 \tag{5.33}$$

In the two categories of modes envisaged and with Prob(k) = $1/m$:

- unipolar modes (any m) $a_k = 0,1,2,...m-1$

$$P''_R = \frac{1}{R} \frac{(m-1)(2m-1)}{6} U^2_{BR} \tag{5.34}$$

- antipolar modes, even m: $a_k = \pm 1/2; \pm 3/2, \ldots \pm(m-1)/2$; odd m: $a_k = 0; \pm 1; \pm 2, \ldots \pm(m-1)/2$

$$P''_R = \frac{1}{R} \frac{m^2 - 1}{12} U^2_{BR} \tag{5.35}$$

In both cases, ξ'_R becomes, respectively:

- unipolar modes

$$\xi'_R \cong \frac{(m-1)(2m-1)}{6} \frac{U^2_{BR}}{U'^2_{N\text{eff}}} \tag{5.36}$$

- antipolar modes

$$\xi'_R \cong \frac{m^2 - 1}{12} \frac{U^2_{BR}}{U'^2_{N\text{eff}}} \tag{5.37}$$

By substituting the ratio $U_{BR}/U'_{N\text{ eff}}$ in (5.26) with its expression from (5.36) or (5.37), we obtain the desired relation between the error probability ϵ per symbol and the signal-to-noise ratio ξ'_R

- unipolar modes

$$\epsilon \cong \frac{2(m-1)}{m} G_c \left(\sqrt{\frac{3}{2(m-1)(2m-1)}} \cdot \sqrt{\xi'_R} \right) \tag{5.38}$$

- antipolar modes

$$\epsilon \cong \frac{2(m-1)}{m} G_c \left(\sqrt{\frac{3}{m^2 - 1}} \cdot \sqrt{\xi'_R} \right) \tag{5.39}$$

These relations bring up the following remarks:

- taking into account the properties of the integral complementary Gaussian function G_c (fig. 16.2), an increase of ξ'_R causes ϵ to decrease more rapidly the lower ϵ was to begin with. For example, for $m = 2$ (binary transmission), an increase in $10 \lg \xi'_R$ of 6 dB causes ϵ to go from 10^{-2} to 2.10^{-6}, or from 10^{-3} to 3.10^{-10};
- the influence of the number of m states on ϵ is much more subtle.

In fact, at constant emission power P_E and at constant bit rate $\dot{D}$, ξ'_R depends on m by the intermediary of $\dot{M}$ and of the power spectral densities $\Phi'_R(f)$ and $\Phi'_N(f)$. If m increases, M and P'_N diminish, which leads to an increase of ξ'_R to be balanced with the decrease of its coefficient in the argument G_c in relations (5.38) and (5.39). In this way it can be shown that for a given channel, there is an optimal value of m with respect to the error probability. This optimum is nevertheless not always realistic, given the technical difficulty of constructing a regenerator capable of discriminating among a great number of states in an economical and reliable manner.

5.5.5 Specific case: binary transmission

In the particular interesting case in which $m = 2$ (binary transmission), relations (5.38) and (5.39) become:

$$\epsilon \cong G_c(\sqrt{\tfrac{1}{2}\xi'_R}) \qquad \text{(unipolar mode)} \qquad (5.40)$$

$$\epsilon \cong G_c(\sqrt{\xi'_R}) \qquad \text{(antipolar mode)} \qquad (5.41)$$

Figure 5.20 shows these relations. It is noted that, all other things being equal, the unipolar mode demands a signal-to-noise ratio some 3 dB higher than the antipolar mode, because it represents twice the power for an equal difference U_{BR} between the two states (fig. 5.21).

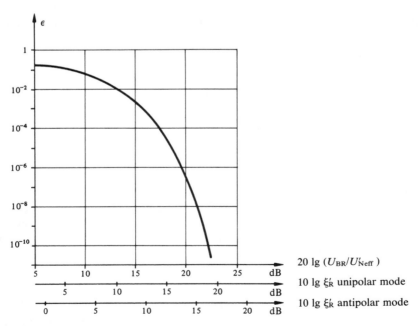

Fig. 5.20 Error probability ϵ in binary transmission.

DIGITAL TRANSMISSION

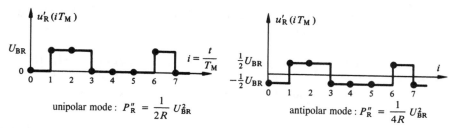

Fig. 5.21 Binary transmission modes.

5.6 COMBINED EFFECTS OF INTERSYMBOL INTERFERENCE AND NOISE

5.6.1 Effect on the eye pattern

If the equalization is not perfect, the elementary base signal $u'_{BR}(t)$ does not cancel for the characteristic moments iT_M. An example is given in figure 5.22, where a symbol produces relative residual signals (with respect to U_{BR}) $y_{-1}, y_1, y_2, y_3, \ldots$ on the neighboring symbols.

From (5.17), the random useful signals $u'_R(t)$ are the sum of the elementary base signals shifted by iT_M and multiplied by the coefficients a_k of which the m possible values are dependent on the mode.

If, as in section 5.3.6, it is assumed that $u'_{BR}(t)$ is significantly different than zero during a limited period of time, corresponding to rT_M, the influence of the $r - 1$ symbols surrounding a given symbol appears as an interference Y relative to U_{BR}, which can take m^{r-1} different values according to the status

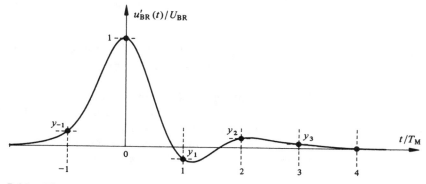

Fig. 5.22 Elementary base signal imperfectly equalized at reception.

of the neighboring symbols. These values are the algebraic sum of the residues y_j of the neighboring symbols, multiplied by the coefficents a_{kj} corresponding to their state.

The m exact characteristic points of the ideal eye pattern at iT_M (perfect equalization) are shifted by these interference values Y. The result is that the eye pattern undergoes a reduction of the vertical aperture Y_{max} (fig. 5.23). The aperture of the eye is, thus,

$$U_0 = U_{BR}(1 - 2Y_{max}) \tag{5.42}$$

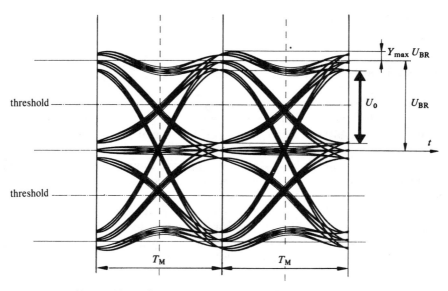

Fig. 5.23 Effect of interference on the eye pattern (example: $m = 3$, bipolar).

While $U_0 > 0$, an errorless discrimination between the states is possible (in the absence of noise). Otherwise, the intersymbol interference leads to regeneration errors.

Each of the lines of the eye pattern has a definite character. Nevertheless, the appearance of any one of them at a given moment is unpredictable because it is related to the flux of the digital information being transmitted.

The effect of random noise is superimposed on this line diagram and encloses each line with a sort of fuzziness, which makes the edges of the "eyes" less clear.

5.6.2 Evaluation of the error probability in the presence of noise and interference

When the imperfectly equalized signal undergoes the effect of intersymbol interference the difference between the representative values of the m states at the characteristic moments is reduced from U_{BR} to U_0 according to (5.42).

The decision thresholds are situated at the midpoint of the nominal values, and a noise value greater than $U_0/2$ is sufficient to give rise eventually to an error. Relation (5.23), which gives the probability of error ϵ, must be modified into

$$\epsilon = \frac{2(m-1)}{m} \operatorname{Prob}(u'_N(t) > \tfrac{1}{2}U_0) \tag{5.43}$$

With respect to a transmission with perfect equalization, it is necessary to increase U_{BR} (and thus U_{BE}) by a factor $1/(1 - 2\ Y_{max})$, i.e., to increase the signal-to-noise ratio in the same proportion to the square of this value, to obtain the same error probability ϵ. It must be noted, nevertheless, that this evaluation is pessimistic, in the sense that the minimal aperture U_0 of the "eyes" of the diagram represents an extreme which can be rare. Nevertheless, the experimental observation of the eye pattern with an oscilloscope provides a realistic value of U_0, as the rare combinations only appear on the screen with a weak luminosity.

5.7 CUMULATIVE REGENERATION

5.7.1 Digital transmission with distributed regenerators

While transmission must be made at a certain distance so that the attenuation, distortion, and noise of the channel give rise to a signal-to-noise ratio ξ'_R that is too low to guarantee a sufficiently low error rate, it is necessary to proceed to intermediary regeneration at regular intervals along the length of the channel. The spacing of the regenerators depends greatly on the characteristics of the medium and the disturbances. Some examples of such planning will be presented in sections 9.8 and 9.9.

The cascade arrangement of several regeneration sections gives rise to the following new problems:

- accumulation of regeneration errors
- increasing imprecision of the restored clock at each regenerator, based on the signals coming from the preceding regenerator (phase jitter)

5.7.2 Accumulation of errors after several sections in cascade

We can reasonably assume that the noise affecting the different sections is statistically independent from one section to the next. On the other hand, the probability that the same symbol is the object of two errors which compensate each other on two different sections is negligible with respect to the simple error probability.

As a consequence, the error probability ϵ per symbol after N sections is

$$\epsilon = 1 - \prod_{i=1}^{N} (1 - \epsilon_i) \tag{5.44}$$

ϵ_i being the error probability of section i considered separately.

Relation (5.44) can be approached by

$$\epsilon \cong \sum_{i=1}^{N} \epsilon_i \qquad \text{if } \epsilon_i \ll 1^{-i} \tag{5.45}$$

In particular, if all the sections present the same error probability ϵ_i, these relations become

$$\epsilon = 1 - (1 - \epsilon_i)^N \tag{5.46}$$

$$\epsilon \cong N\epsilon_i \qquad \text{if } \epsilon_i \ll 1 \tag{5.47}$$

5.7.3 Sensitivity to the noisiest section

In the case of Gaussian noise, we have already mentioned in section 5.5.4 that the error probability depends very strongly on the signal-to-noise ratio.

Thus, in a chain of regenerators, it is sufficient that the signal-to-noise ratio of a single section is degraded by only several decibels so that the error probability ϵ_i of this section is increased by several orders of magnitude, which has the effect of rendering all the other terms of the sum negligible according to (5.45) and gives rise to an overall error probability ϵ, which is in all cases as high as that of the noisiest section.

For example, a binary transmission (unipolar mode), composed of 20 similar sections, each having a signal-to-noise ratio of 17 dB, presents an overall error probability:

$$\epsilon \cong 20\epsilon_i = 20 \cdot 2.9 \cdot 10^{-7} = 5.8 \cdot 10^{-6}$$

If the signal-to-noise ratio of a single section is diminished by 3 dB, its error probability becomes $2 \cdot 10^{-4}$ and the overall error probabiltiy becomes

$$\epsilon' \cong 2 \cdot 10^{-4} + 19\epsilon_i \cong 2 \cdot 10^{-4}$$

DIGITAL TRANSMISSION

This sensitivity to the worst case is typical of digital transmission. In analog transmission, on the contrary, the noise increase of a particular section is drowned in the overall noise and has only a slight influence on the performance of the overall system (6.4.6).

5.7.4 Phase jitter: definition

Each clocked regenerator extracts the frequency clock $\dot{M}$ from the flux of symbols it receives, and uses this to sample and retransmit these signals (fig. 5.3). The phase shifts at the reconstruction of the clock thus have a tendency to propagate from one regenerator to the other.

The phase of the restored clock varies with the sequence of symbols received and according to an inertia which depends on the quality factor (and thus selectivity) of the extraction apparatus. The clock thus does not always exactly coincide with the maxima of the apertures of the eye pattern.

Phase jitter $\Delta\varphi$ is the term for the (variable) deviation between the pulse of the internal clock of the regenerator and the optimal characteristic moments of the received and equalized signal $u'_R(t)$.

In the case of sequences of random symbols, the phase jitter is a random continuous variable with a zero mean. Its standard deviation (or rms value) $\Delta\varphi_{\text{eff}}$ and frequency distribution (spectral density) can be determined.

5.7.5 Phase jitter in a chain of regenerators

After a long chain of N similar regenerators ($N \gg 1$) it is found that

- the statistical distribution of the phase jitter $\Delta\varphi$ approaches a Gaussian law.
- the standard deviation $\Delta\varphi_{\text{eff}}$ of this distribution increases proportionally to $\sqrt{N}$
- the frequency range occupied by the spectrum of $\Delta\varphi$ is even more limited to low frequencies when N is large.

5.7.6 Consequences of phase jitter for error probability

Here also, the eye pattern provides at least a qualitative indication of the effect of phase jitter on the discrimination of the m states (fig. 5.24).

If the local sampling clock is slightly shifted with respect to the moments at which the eye pattern presents its maximum apertures, the margin between the received signal $u'_R(t)$ and the decision thresholds narrows, thus increasing the error probabilty. An evaluation of this probability ϵ, pessimistic as with

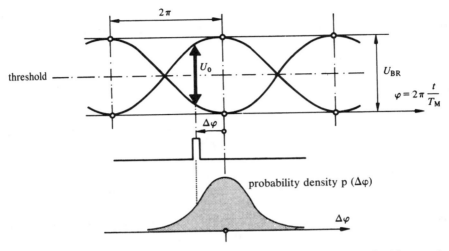

Fig. 5.24 Effect of phase jitter on the eye pattern (example of a binary signal $m = 2$).

the case of interference between symbols (fig. 5.23), can be made by a relation identical to (5.42) in which U_0 represents the opening of the eye at shifted moments, for example of $\Delta\varphi_{\text{eff}}$ with respect to the optimum.

5.8 APPLICATIONS OF DIGITAL TRANSMISSION

5.8.1 Scope of the preceding results

The methods and principles discussed in the preceding section are sufficiently general to cover any digital transmission with respect to the following points:
- the type of digital information transmitted: data, telegraphic characters, digital representation of analog information (speech, music, pictures)
- bit rate $\dot{D}$
- nature of the channel: balanced pair or coaxial cable, radiocommunications, guided optical transmission (optical fiber)
- nature of the additive noise in the channel, particularly its statistical distribution. The Gaussian model, often used as an illustration, is not a limitation. The general relations, such as (5.23), apply to other distributions as well.

In this chapter, the input and output of the channel are assumed to be in **baseband**. If, however, as is often the case, a discrete analog modulation (OOK,

FSK, PSK) is necessary to adapt the signals to a given medium, the modulators and demodulators are then assumed to be part of the channel. The effect of noise actually undergone by the modulated signal in the course of transmission is, in this case, again modified by the properties of the demodulator. These properties are the subject of section 8.11. Another restriction of this chapter is the *exclusion of multiplicative noise* which appears, notably in radiocommunications in the form of random fluctuations of the received signal's amplitude. These particular effects are treated in section 12.2.4.

5.8.2 Technical applications

Baseband digital transmission, as it has been presented in this chapter, is found in a large number of telecommunication systems in which the evaluation of the error probability is always an essential concern. For example:
- telegraphic baseband transmission, with relays as regenerators
- telex transmission on subscriber lines (rate: 50 baud)
- data baseband transmission (without modem) on local lines
- digital transmission of analog information (for example, speech) by means of a digital modulation (either PCM or another type). The error probability ϵ is the principal criterion for the planning of these systems installed on balanced pairs or coaxial cables (sections 9.8 and 9.9).

Chapter 6

Analog Transmission

6.1 CHARACTERISTICS OF ANALOG TRANSMISSION

6.1.1 Principles

In analog transmission, the information coming from a continuous source (analog information) is represented by the ***continuous*** variation of one parameter of the transmitted signal.

Each detail concerning this parameter of the received signal is considered by the receiver to be carrying information. This is why all modifications of this parameter in the course of transmission, no matter how minimal, cause an irreversible alteration of the analog information it carries.

Analog transmission can be considered as a limit case of digital transmission, in which the number of m states per symbol would approach infinity. Since the available power is limited, the difference between these states becomes infinitely small and it is no longer possible to discern them with the aid of thresholds.

The essential difference between analog transmission and digital transmission resides in the means with which the receiver interprets the signals it receives:

- as a function of a finite ***alphabet***, prearranged with the emitter, in the case of ***digital*** transmission
- according to ***all the details*** of a parameter, prearranged with the emitter, in the case of ***analog*** transmission.

6.1.2 Characteristics of the transmitted signals

At emission, the signals $u_E(t)$, as unpredictable as the messages they represent, are principally characterized by

- the bandwidth B (in the sense defined in section 2.4.8) occupied by their (unilateral) power spectral density $\Phi_{SE}(f)$
- their power P_{SE} or their level L_{SE}, given

$$P_{SE} = \int_0^\infty \Phi_{SE}(f) \cdot df \cong \int_B \Phi_{SE}(f) \cdot df \qquad (6.1)$$

6.1.3 Characteristics of the transmission channel

As with the case of digital transmission, the channel is defined by
- the *transfer function* $\underline{H}(f)$, which expresses the linear distortions (attenuation and phase distortion) of the channel and from which we obtain the "bandwidth,"
- the power spectral density $\Phi_N(f)$ of the *noise* which affects it,
- the statistics of this noise.

There is an additional characteristic of typical significance to analog transmission:
- the *linearity* of the input-output characteristics of the system. Nonlinear distortion is, in effect, the cause of intermodulation products (section 2.4.10) which impair the transmission with additional noise.

6.1.4 Criteria for analog transmission quality

Consistent with the quality objectives stated in section 2.4, the principal evaluation criteria in the case of an analog transmission are:
- the absolute *level* of signal reception L_{SR} at the *output* of the transmission system,
- the *signal-to-noise ratio* ξ_R at the same point

$$10 \lg \xi_R = L_{SR} - L_{NR} \tag{6.2}$$

- the *distortion factor d*, which expresses the effect of system nonlinearities on a sinusoidal signal of frequency f_1 by the ratio between the effective value of the collection of harmonics $U_n (n \geq 2)$ and that of the distorted signal (which can be practically approached by the effective value of the fundamental frequency U_1):

$$d = \frac{\sqrt{\sum_{n=2}^{\infty} U_{n\,\text{eff}}^2}}{\sqrt{\sum_{n=1}^{\infty} U_{n\,\text{eff}}^2}} \cong \frac{\sqrt{\sum_{n=2}^{\infty} U_{n\,\text{eff}}^2}}{U_{1\,\text{eff}}} \quad \text{if } d \ll 1 \tag{6.3}$$

The distortion factor d depends on the amplitude of the signal. It can also be expressed in decibels by $20 \lg d$.

These three quantities are defined for the *sinusoidal test signal* (nominal signal, section 2.4.14), i.e., for the *nominal level* of the system. They do not necessarily correspond to the values which would be measured in the presence of real signals (e.g., telephone conversations) when the system is in service.

6.2 AMPLIFICATION

6.2.1 Ideal conditions for amplification

The purpose of amplification is to compensate for losses and attenuation of the medium used as a transmission channel in order to obtain a level of reception L_{SR} sufficient for the output of the system.

So that the ideal conditions for conformal transmission stated in section 2.4.6 can be fulfilled, it is necessary that the amplification exactly compensate for the linear distortions (attenuation and phase) of the channel, at least in the whole frequency band occupied by the signal.

The gain $G(f)$ of the amplifier must thus follow the same law as a function of frequency, as the attenuation of the medium. Thus, we may consider the amplifier to be the cascade arrangement of a quadripole, called an *equalizer*, which compensates for the linear distortions of the channel and of an ideal amplifier with constant gain (fig. 6.1).

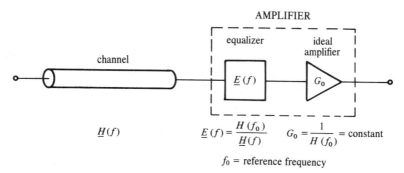

$$\underline{E}(f) = \frac{\underline{H}(f_0)}{\underline{H}(f)} \qquad G_0 = \frac{1}{H(f_0)} = \text{constant}$$

f_0 = reference frequency

Fig. 6.1 Block diagram of an amplifier.

It is important to note that here the role of the equalizer is different from that in digital transmission (sect. 5.3.5). In the case of analog transmission, its transfer function $\underline{E}(f)$ must be as similar as possible to the inverse of the channel transfer function, $\underline{H}(f)$, in such a way that the arrangement of the channel and the equalizer brings about a conformal transmission, a condition which is unnecessary in digital transmission.

6.2.2 Notation

The quantities relative to the input of the amplifier are designated by an index x, and those relative to the output by an index y. The quantities concerning the signal are indexed S, and those concerning the noise N. The indices E and R are related respectively to the input (emission) and output (reception) of the transmission.

6.2.3 Sources of background noise in an amplifier

Each amplifier introduces noise into the transmission system. This noise comes primarily from three sources:
- the most important source is the thermal agitation within the *input resistance* of the amplifier. It is responsible for a background noise source called *thermal noise* (section 2.4.12);
- the active components inside the amplifier are the source of random phenomena (e.g., granular effect), which make a noise contribution called *internal noise* and expressed by the noise factor;
- the *intermodulation products*, due to amplification nonlinearities and causing *intermodulation noise* in the presence of a real signal.

6.2.4 Thermal noise contribution

This noise source, which is the major contributor to background noise, is principally located at the *input* of each stage of amplification. It generates a noise level which is directly dependent on the bandwidth B of the amplifier and the absolute ambient temperature T_a according to (2.41):

$$L_{Nthx}(\text{re } 1\text{mW}) = 10 \lg \frac{kT_a B}{1\text{mW}} \quad \text{dBm} \tag{6.4}$$

In the absence of any other source of noise, it defines a signal-to-thermal noise ratio at the *input* of the amplifier:

$$10 \lg \xi_x = L_{Sx} - L_{Nthx} \quad \text{dB} \tag{6.5}$$

6.2.5 Internal noise contribution. Noise figure: definition

The second source of noise is located in the active components of the amplifier and constitutes its *internal noise*. Consequently, the signal-to-noise ratio ξ_y at the *output* of the amplifier is lower than the signal-to-thermal noise ratio ξ_x at the input, as calculated from (6.5).

The *noise figure* is the difference between the logarithms of these ratios, when the ambient temperature is at the conventional value of $T_a = T_0 = 290$ K (fig. 6.2). We therefore have

$$F = 10 \lg \xi_x - 10 \lg \xi_y \quad \text{dB} \tag{6.6}$$

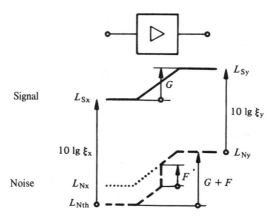

Fig. 6.2 Expression of the internal noise by the noise figure F.

It is convenient to refer the internal noise contribution to the *input* of the amplifier. In effect, at $T_a = T_0 = 290$ K, everything occurs as if the noise level at the input of the amplifier were not that of the thermal noise L_{Nthx} only, but rather this value *increased* by the noise figure F:

$$L_{Nx} = L_{Nthx} + F \quad \text{(at } T_0 = 290 \text{ K)} \quad \text{dBm} \tag{6.7}$$

6.2.6 Noise temperature: definition

Another way of expressing the contribution of the internal noise consists of *increasing the ambient temperature* T_a by a hypothetical value ΔT_N, so as to obtain a level of *hypothetical thermal noise* equivalent to the combined effect of the real thermal noise *and* the internal noise. The resulting hypothetical temperature $T_a + \Delta T_N$ is called the *[equivalent] noise temperature* T_N. It is given by

$$T_N = \frac{P_{Nx}}{kB} = T_a + \Delta T_N \quad \text{K} \tag{6.8}$$

where P_{Nx} represents the total power of the noise (thermal and internal) referred to the input of the amplifier.

Relation (6.7), valid by definition for $T_a = T_0 = 290$ K, permits us to establish the link between the noise factor F and the increase of noise temperature ΔT_N which expresses the *internal noise temperature* (table 6.3).

$$F = L_{Nx} - L_{Nthx} = 10 \lg \frac{k(T_0 + \Delta T_N)B}{1\text{mW}} - 10 \lg \frac{kT_0 B}{1\text{mW}}$$

$$= 10 \lg \left(1 + \frac{\Delta T_N}{T_0}\right) \tag{6.9}$$

Table 6.3 Equivalence between the noise factor and the internal noise temperature.

F	1	2	5	10	15	dB
ΔT_N	75	170	627	2610	8880	K

6.2.7 Case of a multi-stage amplifier

If the amplifier, as is often the case, is composed of a cascade arrangement of several stages with gains (input/output power ratios) which are respectively $g_1, g_2, ..., g_i, ..., g_n$ from figure 6.4, where each stage constitutes a noise source whose power referred to the input of the stage is equal to $P_{Ni} = kT_{Ni}B = k(T_a + \Delta T_{Ni})B$.

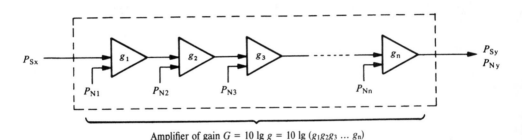

Amplifier of gain $G = 10 \lg g = 10 \lg (g_1 g_2 g_3 ... g_n)$

Fig. 6.4 Multistage amplifier.

We therefore have for the signal

$$P_{Sy} = P_{Sx} g_1 g_2 ... g_n = P_{Sx} g \quad (6.10)$$

The noise at the output is the sum of the noise contributions from each of the n stages, each amplified by the stages which separate it from the output. This noise is analyzed on the basis of two terms: the thermal noise determined by the ambient temperature T_a (assumed to be identical for all the stages), and the internal noise expressed by the internal noise temperature ΔT_i, which can be different at each stage.

$$P_{Ny} = kT_a B[(g_1 g_2 ... g_n) + (g_2 g_3 ... g_n) + ... + g_n]$$
$$+ kB[\Delta T_{N1}(g_1 g_2 ... g_n) + \Delta T_{N2}(g_2 g_3 ... g_n) \quad (6.11)$$
$$+ ... + \Delta T_{Nn} g_n]$$

Let

$$P_{Nn} = kT_aBg\left[1 + \frac{1}{g_1} \ldots + \frac{1}{g_1g_2\ldots g_{n-1}}\right]$$
$$+ kBg\left[\Delta T_{N1} + \Delta T_{N2}\frac{1}{g_1} + \ldots + \Delta T_{Nn}\frac{1}{g_1g_2\ldots g_{n-1}}\right] \tag{6.12}$$

by introducing the total gain $g = g_1g_2\ldots g_n$,

If, as is generally the case, $g_i \gg 1$, we can write approximately

$$P_{Ny} \cong k(T_a + \Delta T_{N1})Bg \tag{6.13}$$

In conclusion, this signifies that

- the equivalent noise temperature of an amplifier with several stages is determined by the ambient temperature T_a and the noise temperature of the *first stage* only;
- the contribution of the other stages, as much with the thermal noise as with the internal noise of the amplifier, can be ignored.

6.2.8 Effect of nonlinearities

In an ideal amplifier, the instantaneous voltage of the output $u_y(t)$ must be exactly proportional to that of the input $u_x(t)$. Real amplifiers can fulfill this ideal condition only in a limited voltage range, beyond which $u_y(t)$ is no longer proportional to $u_x(t)$, but tends to remain constant (saturation, fig. 6.5). The result is a distortion rate which increases with the signal level.

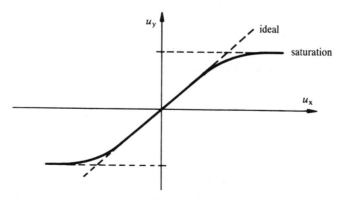

Fig. 6.5 Real amplification characteristics.

If the input signal is not purely sinusoidal, *intermodulation products* appear, part of which fall in the frequency band of the original signal (section 2.4.10). In the case of a random signal (speech, music, video) the consequence of this is *self-interference* of the signal, i.e., additional noise, called *intermodulation noise*, with a relative importance which increases when the level of the signal increases. The third-order intermodulation products, provoked by a nonlinear characteristic with central symmetry around the origin, are the most important in actual amplifiers.

6.2.9 Compromise between background noise and intermodulation

The total noise at the output of an amplifier is finally the sum of two components:
- the background noise (thermal and internal) of level L_N which is independent of that of the signal
- the self-interference due to the intermodulation products, which is very strongly dependent on the level of the signal

If we increase the nominal level L_S of the signal, the background noise level L_{N0} referred to zero (2.4.18) decreases:

$$L_{N0} = L_N - L_S \qquad \text{dB} \qquad (6.14)$$

and thus its power referred to zero P_{N0} decreases also.

However, as the level approaches the saturation point of the amplifier, the intermodulation products increase in importance and their power referred to zero P_{IM0} increases rapidly.

It follows that the total noise level referred to zero $L_{N0\,tot}$, which results from the logarithmic expression of the sum of these two powers, reaches a minimum for a certain nominal signal level (fig. 6.6).

This conflict between background noise and intermodulation products is typical of all analog systems. It is encountered as often in transmission problems as in measurement devices and recording procedures.

6.3 NOISE BUDGET

6.3.1 Effect of a single noise source

In an analog transmission system dealing with thermal noise, the principal sources of noise are the amplifiers and more precisely their input resistance. These sources are precisely located at defined points of the system.

ANALOG TRANSMISSION

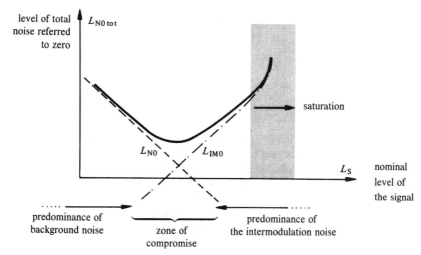

Fig. 6.6 Compromise between background noise and intermodulation noise (third-order distortions).

If such a source, *considered separately*, introduces a noise power P_{NX} at point X of a system, what is its effect on the output of the system?

The noise introduced at point X undergoes the same attenuation and amplification as the signal, and, as it is assumed by hypothesis that there are no other sources of noise in the system, the signal-to-noise ratio ξ_R at reception (output of the system) will be the same as at point X (fig. 6.7).

$$\xi_R = \frac{P_{SR}}{P_{NR}} = \xi_x = \frac{P_{SX}}{P_{NX}} \qquad (6.15)$$

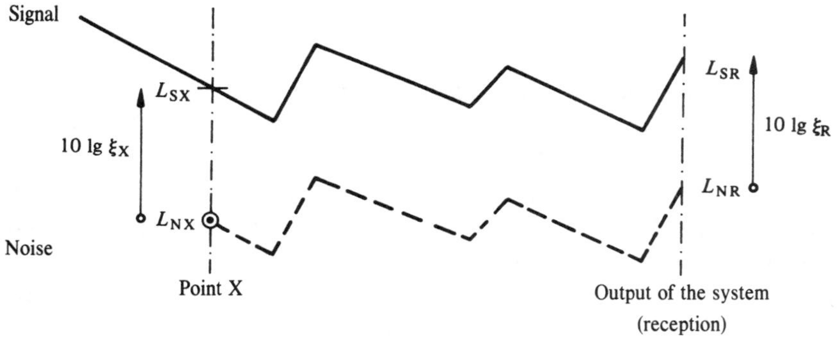

Fig. 6.7 Level diagrams of the signal and noise in the case of a single noise source.

The result is that the power P_{NR} of the reception noise, due to this single source, is

$$P_{NR} = P_{SR} \frac{P_{NX}}{P_{SX}} \tag{6.16}$$

6.3.2 Cumulative effect of several noise sources

Very frequently, several noise sources, with different powers P_{N1}, P_{N2}, ..., P_{Ni}, ..., affect transmission. Because they intervene at different places in the system, i.e., characterized by a signal power equal to P_{S1}, P_{S2}, ..., P_{Si}, ..., respectively, they cannot be simply added together. However, as these sources are assumed to be incoherent (statistically independent), it is possible to evaluate their cumulative effect at the *same* point of the system, in particular at the output, by the sum of the powers that they develop there.

$$P_{NR} = \sum_i P_{NRi} = P_{SR} \sum_i \frac{P_{Ni}}{P_{Si}} \tag{6.17}$$

The noise-to-signal ratio $1/\xi_R$ proceeds at the output of the system by this relation:

$$\frac{1}{\xi_R} = \sum_i \frac{1}{\xi_i} \tag{6.18}$$

with

$$\xi_R = \frac{P_{SR}}{P_{NR}} \tag{6.19}$$

at the output of the system (receiver), and

$$\xi_i = \frac{P_{Si}}{P_{Ni}} \tag{6.20}$$

at the point where the noise source i intervenes in the system.

As a consequence of (6.18), the sum of the effects of several noise sources is regulated by the following rule:

- in a transmission system in which several noise sources intervene, the *noise-to-signal ratio at the output of the system is the sum of the noise-to-signal ratios of several noise sources* considered separately.

In conformity with the definition of the noise level referred to zero given in section 2.4.18, the noise power P_{Ni0} of the source i referred to zero is defined by

$$P_{Ni0} = \frac{P_{Ni}}{P_{Si}} 1mW = \frac{1mW}{\xi_i} \tag{6.21}$$

ANALOG TRANSMISSION

P_{Ni0} is a hypothetical noise power. It is that noise power which would be found if the signal power at the point where the source i intervenes were not P_{Si} but 1 mW (given a level of 0 dBm), while keeping the same signal-to-noise ratio ξ_R. Relation (6.18) leads to the following equivalent statement:

- the resulting noise power referred to zero P_{NR0} at the output of the system is the sum of the noise powers referred to zero P_{NR0} of the different noise sources.

$$P_{NR0} = \sum_i P_{Ni0} \qquad\qquad W0 \qquad\qquad (6.22)$$

This again states that the noise powers effectively injected into the system cannot be added *unless they have been first referred to a hypothetical signal level equal to 0 dBm*.

6.4 CALCULATION OF THE SIGNAL-TO-NOISE RATIO

6.4.1 Given situation

The transmission channel used is characterized by its attenuation A, which depends both on the type of medium (e.g., attenuation coefficient α of the line, or range of frequencies and gain of the antennas in radio transmission) and on the distance l of the transmission.

The sources of background noise are those identified in section 6.2.3. They are characterized by the level of thermal noise L_{Nth} at the input of the amplifiers and by their noise factor F.

Knowing the level of emission available L_{SE} (limited by the nonlinearities of the repeaters), it is thus possible to evaluate the signal-to-noise ratio ξ_R at reception, i.e., at the output of the channel.

The calculation is made with the hypothesis of a *baseband* transmission (without modulation) or with an SSB (single sideband) modulation, which does not modify the signal-to-noise ratio (sect. 8.8.8 and 8.9.3).

6.4.2 Planning for a nominal signal

The characteristics of the usual transmission channels (lines, radio) are dependent on frequency. This is why planning is always done on the basis of one or several *nominal sinusoidal signals* (section 2.4.14) with specified level and frequency.

The objectives for these signals are thus:

- the level of reception L_{SR}
- the signal-to-noise ratio at reception ξ_R

6.4.3 Repeaters distributed along the line: general case

In the general case of a transmission line composed of N sections, each terminated by an amplifier, called a *repeater*, according to the level diagram of figure 6.8, the background noise at reception results from the combination of N thermal noise sources located at the input of each repeater and increased by their internal noise.

As long as all the repeaters are at the same temperature ($T_0 = 290$ K) and have the same noise figure F (which is a reasonable assumption), the noise level injected into the system at each repeater is the same and equal to

$$L_{Nx} = L_{Nth} + F = 10 \lg \frac{kT_0 B}{1\text{mW}} + F = 10 \lg \frac{P_{Nx}}{1\text{mW}} \quad \text{dB} \quad (6.23)$$

However, the level of the signal L_{Sxi} is not necessarily the same at all these points. The combination of these N noise sources therefore must be done by the addition of the noise-to-signal ratios according to relation (6.18). The noise-to-signal ratio $1/\xi_R$ at the ouput of the system thus becomes

$$\frac{1}{\xi_R} = \sum_{i=1}^{N} \frac{1}{\xi_i} = \sum_{i=1}^{N} \frac{P_{Nx}}{P_{Sxi}} = P_{Nx} \sum_{i=1}^{N} \frac{1}{P_{Sxi}} \quad (6.24)$$

thus, in logarithmic form:

$$10 \lg \frac{1}{\xi_R} = 10 \lg \frac{P_{Nx}}{1\text{mW}} + 10 \lg \sum_{i=1}^{N} \frac{1\text{mW}}{P_{Sxi}} \quad \text{dB} \quad (6.25)$$

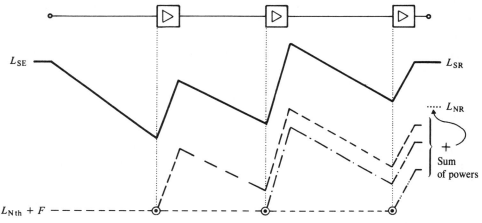

Fig. 6.8 Level diagram of a link with distributed repeaters: ⊙ noise source.

ANALOG TRANSMISSION

from which we obtain the logarithmic signal-to-noise ratio at the output with (6.23):

$$10 \lg \xi_R = -(L_{Nth} + F) - 10 \lg \sum_{i=1}^{N} \frac{1\text{mW}}{P_{Sxi}} \quad \text{dB} \quad (6.26)$$

It must be noted that it is the level of the signal *at the input* of the repeaters (L_{Sxi}) which is the determining factor for the calculation of ξ_R.

6.4.4 Identical and equidistant repeaters

If we have the following restricting but realistic hypotheses:
- *uniform* line of total length l and attenuation coefficient α (at the nominal signal frequency)
- *N equidistant* repeaters, including the repeater at the end of the last section
- same nominal level $L_{Sy} = L_{SE} = L_{SR}$ at the output of all the repeaters, including the last (reception level of the system) and identical to the emission level L_{SE} at the input of the line

then, the level diagram of the link takes the form presented in figure 6.9.

The result of the above hypotheses is that
- the gain G of each repeater must compensate exactly for the attenuation of the preceding section

$$G = \alpha \frac{l}{N} \quad \text{dB} \quad (6.27)$$

- the input level L_{Sx} is identical for each repeater

$$L_{Sx} = L_{SR} - \alpha \frac{l}{N} \quad \text{dB} \quad (6.28)$$

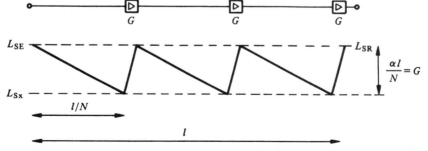

Fig. 6.9 Level diagram of a link with identical and equidistant repeaters.

Consequently, the signal-to-noise ratio ξ_R at the output of the system, calculated by (6.26), becomes

$$10 \lg \xi_R = -(L_{Nth} + F) + L_{Sx} - 10 \lg N \qquad \text{dB} \qquad (6.29)$$

or, with (6.28):

$$10 \lg \xi_R = L_{SR} - \alpha \frac{l}{N} - (L_{Nth} + F) - 10 \lg N \qquad \text{dB} \qquad \mathbf{(6.30)}$$

It is noted that the expression $L_{SR} - (\alpha l/N) - (L_{Nth} + F)$ represents the signal-to-noise ratio $10 \lg \xi_i$ at the output of *one* of the repeaters considered separately. Logically, there is indeed, from N identical and equidistant repeaters, an addition of noise powers according to (2.25), i.e.,

$$10 \lg \xi_R = 10 \lg \xi_i - 10 \lg N \qquad \text{dB} \qquad (6.31)$$

The noise power referred to zero P_{NR0} at reception, from (6.22), is

$$P_{NR0} = N P_{Ni0} \qquad \text{W0} \qquad (6.32)$$

6.4.5 Influence of the number of repeaters

Figure 6.10 compares the level diagram and the signal-to-noise ratios in the same link with, on the one hand, a single terminal repeater compensating for the total attenuation αl of the line and, on the other hand, $N = 5$ equidistant repeaters along the line.

The significant increase in signal-to-noise ratio brought about by the solution with five repeaters has been established. This improvement justifies the interest and the frequent necessity of distributing the amplification between several repeaters along the length of the line, despite the technical and economic problems that this poses. An extreme case is that of repeaters submerged along a transoceanic cable.

The number of repeaters N intervenes in relation (6.30) with two antagonistic effects:

- by dividing the total attenuation αl of the link, it contributes to raising the input level of the repeaters and consequently improving the signal-to-noise ratio ξ_R. This is the principal effect of an increase in the number of repeaters;
- secondarily, however, the number of noise sources is increased, which has the effect of decreasing the final signal-to-noise ratio ξ_R.

There is in fact an optimal number N_{opt} of repeaters for a given link, which leads to a maximum signal-to-noise ratio, and from which the second effect

ANALOG TRANSMISSION

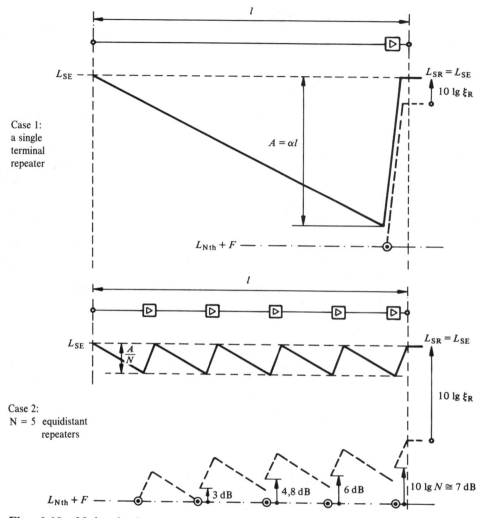

Fig. 6.10 Noise budget on a line with one or several ($N = 5$) repeaters: —— signal; - - - noise; ⊙ source of noise.

mentioned above predominates. This optimal number, obtained by calculating the highest ξ_R from (6.30), is

$$N_{opt} = \frac{\alpha l}{10 \lg e} \cong \frac{\alpha l}{4.34} \qquad (6.33)$$

where α is expressed in dB per unit of length.

N_{opt} cannot generally be usefully raised, and for economic reasons we attempt rather to install only the minimum number of repeaters compatible with quality requirements.

6.4.6 Sensitivity to the noisiest section

If, for some reason (lowering of the signal level, disturbing external influences, *et cetera*) the signal-to-noise ratio of one of the sections is degraded by several decibels with the others remaining unchanged, expression (6.30) is no longer applicable, since the sections are no longer identical. The addition of noise powers referred to zero according to (6.22), or of the noise-to-signal ratios from (6.24), shows that the effect of this deterioration concerns only one of the terms of this sum and has little influence on the final result, especially if the number of sections is high.

By analogy with the example mentioned in section 5.7.3 concerning the same, but then catastrophic, effect in digital transmission, if the 20 identical sections each have a signal-to-noise ratio of 77 dB (a more realistic value for analog transmission than the 17 dB assumed in section 5.7.3), the signal-to-noise ratio at the output of the line is equal to

$$10 \lg \xi_R = 77 \text{ dB} - 10 \lg 20 = 64 \text{ dB}$$

from (6.31).

If the signal-to-noise ratio of one of them is degraded by 3 dB, the noise power referred to zero rises to 40 pW0, while that of the 19 others remains equal to 20 pW0 (corresponding to $L_{N0} = -77$ dBm0). The total noise power referred to zero at the output thus increases from 400 pW0 to 420 pW0, which decreases the final signal-to-noise ratio by only 0.2 dB.

6.5 APPLICATIONS OF ANALOG TRANSMISSION

6.5.1 Scope of the preceding results

The calculation of the final signal-to-noise ratio ξ_R from the preceding section concerns analog transmission with the following characteristics:
- signal of given bandwidth B, carrying any analog information (speech, music, images, *et cetera*)
- transmission medium with any transfer function, the calculation always being made for a nominal signal of given frequency
- *background noise* of power spectral density assumed to be uniform

("white" noise), the level of noise being calculated in the bandwidth B of the signal
- ***baseband*** transmission or in a band transposed by single modulation sideband, SSB

Often, an ***analog modulation*** of type AM, FM, ΦM, PPM, *et cetera* is necessary to adapt the signal to the available channel. This is notably the case for radio transmission. The signal-to-noise ratio at the output of the demodulator is thus different from that at the output of the transmission channel, according to the type of modulation used. These properties are discussed in chapter 8.

If ***digital modulation*** (PCM, ΔM, *et cetera*) intervenes in the link, the transmission is no longer analog, but digital (chapter 5). In any case the (analog) output of the demodulator (decoder) is also affected by noise of a different nature which proceeds partly from quantizing distortion (section 7.2) and partly from the effect of regeneration errors (section 7.7). The calculation of ξ_R is thus completely different.

On the other hand, the disturbances due to ***crosstalk*** or other external causes, as well as the self-interference due to ***intermodulation*** are ***excluded*** from the preceding calculations. They must be the subject of separate evaluations in the planning of analog links (section 10.5).

Finally, the effect of ***attenuation fluctuations*** (multiplicative noise), frequent in radiocommunications and occasionally found in lines (as an effect of temperature variations), must also be considered separately as a variation of the nominal signal level. This has repercussions for the signal-to-noise ratio.

6.5.2 Technical applications

Analog transmission, while being increasingly replaced by digital sections, remains the principal form of telecommunication. It is found in particular in the following cases for which the evaluation of the final signal-to-noise ratio ξ_R from (6.26) or (6.30) is of primary importance:

- baseband telephone transmission with terminal amplifiers
- carrier systems for telephony (frequency multiplexing by SSB modulation) installed on coaxial cables or, more rarely, balanced pairs, with distributed repeaters
- television distribution network by coaxial cable (CATV) with distributed repeaters
- by further considering the effect of supplementary modulations, these calculations can also be adapted to the case of analog radiocommunications by shortwave or ultra-shortwave, by microwave links, or by satellite

6.6 COMPANDING

6.6.1 Dynamic range of the signal: definition

All the calculations made in the preceding sections of this chapter refer to a conventional signal level, called the nominal level. In real operation, the channel is occupied by unpredictable non-stationary signals, with an instantaneous level which varies enormously over the course of time and from one communication to the next.

The *dynamic range* of real signals is the difference between

- the maximum level determined by the limit of linearity (saturation) of the amplifiers
- the minimum level of the useful signals which must still be transmitted. This level is subjectively defined according to the quality of transmission required.

For telephone signals, a dynamic range of 50 dB is assumed, i.e., real levels of -45 dBm0 (45 dB lower than the nominal level) to $+5$ dBm0 (5 dB above the nominal level) with a mean real level of -15 dBm0 (section 2.4.19).

6.6.2 Principle of dynamic range compression

The inconvenience of a significant dynamic range is threefold:

- the high levels could lead to saturation of the amplifiers, and thus strong intermodulation
- the low levels are more exposed to noise (poor signal-to-noise ratio)
- the risks of crosstalk between neighboring circuits are increased (interference of low levels by high levels)

In order to reduce these effects, we proceed *before* transmission at a relative amplification of the low levels in such a way as to *compress the dynamic range*. To avoid too substantial an increase in the mean power of the signal thus modified, the high levels must be attenuated. Reciprocally, the original levels are re-established *after* transmission through amplification and attenuation, respectively. Together these effect an *expansion* of the dynamic range. The collection of circuits which brings about these two operations in an exactly reciprocal manner is called a *compandor*.

To each input level L_S, the compressor assigns a level L_{SC} according to a generally linear rule, chosen so that the nominal level (by definition 0 dBm0) remains unchanged (fig. 6.11).

ANALOG TRANSMISSION

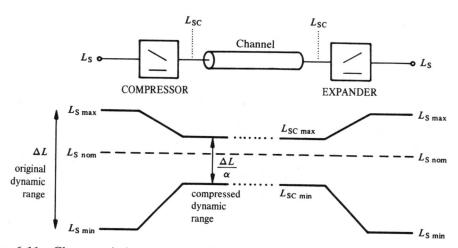

Fig. 6.11 Characteristics of companding.

The compression rule can be expressed by

$$L_{SC} - L_{Snom} = \frac{1}{\alpha}(L_S - L_{Snom}) \tag{6.34}$$

where α is the *compression factor* that the CCITT recommends be chosen equal to 2 (Recommendation G.162).

The ideal expansion law is exactly reciprocal:

$$L_S - L_{Snom} = \alpha(L_{SC} - L_{Snom}) \tag{6.35}$$

6.6.3 Effect on background noise

In the absence of any signal, a noise level L_N at the input of the expander is reduced, by the expansion rule, to a level L'_N at the output, such that

$$L'_N - L_{Snom} = \alpha(L_N - L_{Snom}) = -\alpha 10 \lg \xi_{nom} \tag{6.36}$$

It follows that the background signal-to-noise ratio (between the real power of the signal and that of the noise when the signal is absent) is increased from the value ξ *without* compression (fig. 6.12a):

$$\begin{aligned} 10 \lg \xi = L_S - L_N &= (L_S - L_{Snom}) + (L_{Snom} - L_N) \\ &= (L_S - L_{Snom}) + 10 \lg \xi_{nom} \end{aligned} \tag{6.37}$$

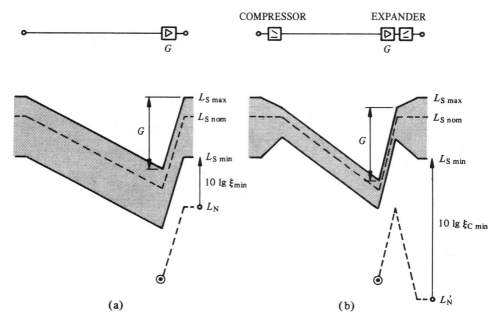

Fig. 6.12 Effect of companding: (a) transmission with the original dynamic range; (b) transmission with compressed dynamic range.

to a value ξ_C *with* compression (fig. 6.12b):

$$10 \lg \xi_C = L_S - L'_N = (L_S - L_{Snom}) + (L_{Snom} - L'_N) \tag{6.38}$$

with (6.36) and (6.37), let

$$10 \lg \xi_C = 10 \lg \xi + (\alpha - 1) 10 \lg \xi_{nom} \tag{6.39}$$

In conclusion, for $\alpha = 2$, companding according to (6.34) improves the background signal-to-noise ratio for all signal levels (including the nominal level) to a value equal (in dB) to the nominal signal-to-noise ratio (fig. 6.13).

It must be noted that companding is only effective with noise sources located *between* the compressor and the expander.

6.6.4 Effect of noise superimposed on the signal

While a noise level of L_N is superimposed on a signal level of L_{SC} at the input of the expander, the behavior of the latter is determined by the predominating level, i.e., practically by that of the signal.

The result is that the signal *and the superimposed noise* undergo the same attenuation or amplification across the expander. Thus, the superimposed signal-to-noise ratio is *the same* at the input and the output of the expander:

$$10 \lg \xi_C = L_S - L'_N = L_{SC} - L_N$$
$$= (L_{SC} - L_{Snom}) + (L_{Snom} - L_N) \qquad (6.40)$$

By using (6.37) to express the signal-to-noise ratio ξ without compression and the compression rule (6.34), we finally obtain

$$10 \lg \xi_C = 10 \lg \xi - (1 - \frac{1}{\alpha})(L_S - L_{Snom}) \qquad (6.41)$$

In conclusion, companding modifies the superimposed signal-to-noise ratio by a value which depends on the difference between the real signal level and the nominal level. If this difference is negative ($L_S < L_{Snom}$), ξ is increased. For $L_S = L_{Snom}$, ξ is unchanged. Less dramatic than the effect on background noise, the effect of companding on the superimposed noise is, however, more advantageous the lower the signal level (fig. 6.13).

6.6.5 Instantaneous companding

The compression law (6.34), valid for voltage levels (rms voltages), can also be applied to the instantaneous signal voltages $u_S(t)$

$$\frac{|u_{SC}(t)|}{U_{nom}} = \left(\frac{|u_S(t)|}{U_{nom}}\right)^{1/\alpha} \qquad (6.42)$$

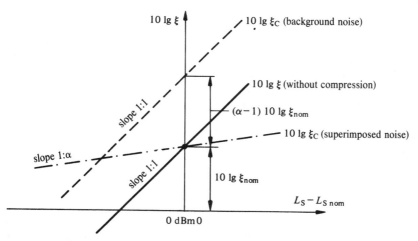

Fig. 6.13 Effect of companding on the signal-to-noise ratio.

The compressor (and reciprocally, the expander) thus present a *nonlinear* input/output characteristic (fig. 6.14).

This procedure has a major disadvantage which precludes its use in unsampled analog transmission: the distortion (essentially odd-ordered) introduced by the nonlinear compression characteristics gives rise to an *unacceptable broadening of the bandwidth occupied by the signal*.

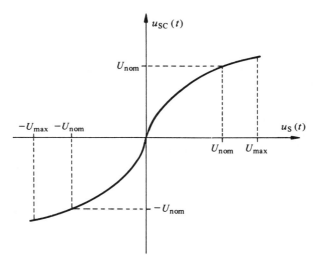

Fig. 6.14 Instantaneous compression characteristic.

Instantaneous companding is nevertheless currently used for *sampled* transmissions in pulse modulation. In this case, the broadening of the occupied signal bandwidth is not obstructive, despite an apparent violation of the sampling theorem, provided that the expansion takes place *before or during* demodulation (a signal with compressed dynamic range cannot be demodulated as such).

Nonuniform quantizing (section 7.4) is an example of instantaneous compression.

6.6.6 Syllabic companding

For a real signal of a random nature, the notion of level is ambiguous. It is tied to that of the rms value, which, in the usual case of a signal with a zero mean, is synonymous with the standard deviation of the random signal. For

ANALOG TRANSMISSION

telephone signals, this rms value varies considerably in time, with a rhythm corresponding to that of the syllables used in speech.

The principle of syllabic companding consists of varying the gain $G_C = L_{SC} - L_S$ of the compressor, and reciprocally of the expander, according to the compression law (6.34), by taking the *mean level* corresponding to the rms value of the signal as the signal level L_S, measured with a time constant equal to the mean duration of a syllable, approximately 20 ms.

The input/output characteristic of the compressor is linear, whereby its slope changes according to the mean signal level (fig 6.15).

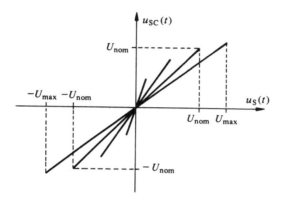

Fig. 6.15 Characteristic of syllabic compression.

The relatively slow variations of this characteristic cause only a very slight broadening of the signal bandwidth, being compatible anyhow with the conventional telephone band.

The inconvenience of this type of companding resides in the effect of trailing in the case of sharp variation in level and in its sensitivity to variations in channel attenuation. It is used for long-distance communications in very noisy media (e.g., shortwave radiotelephony).

Chapter 7

Digital Modulation

7.1 PRINCIPLES AND TYPES OF DIGITAL MODULATION

7.1.1 Principles

Digital modulations, generally defined in section 4.2.5, consist of converting analog information carried by a signal with continuous variations into a *sequence of discrete characters* issued from a finite alphabet of q characters, which are integers in this case (fig. 7.1).

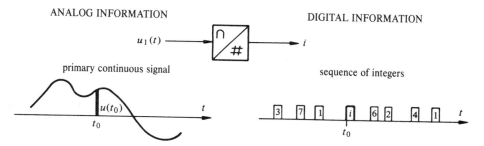

Fig. 7.1 Principle of digital modulation.

This conversion consists of two basic operations:
- the conversion of a continuous time signal into a discontinuous sequence implies *sampling*: only the values of the signal at certain discrete moments, not necessarily periodic, are the object of modulation;
- the reduction of the infinite number of possible analog values to a finite collection of numbers, which are supposed to represent them, is the result of *quantizing*. This operation constitutes the essential and fundamental element of digital modulation.

While sampling, under certain conditions described in section 4.3, does not alter the information carried by a signal, the restriction to a finite number of values does not allow expression of all the subtle nuances of the original analog information: certain details are irremediably lost, and *a systematic approximation is introduced*, which can be minimized, but never completely eliminated.

It is important to emphasize that digital modulations associate *integers*, without physical reality, to the exact values of a *physical* analog quantity (primary signal). In order to be transmitted, these integers must be represented by signals (secondary signals). For demodulation, these integers must be extracted from the secondary signals by regeneration (sect. 5.2). It is from these integers (and not directly from the secondary signals) that the *discrete* physical values of the demodulated signal are determined according to a pre-established convention. The signal thus reconstructed is not analog in the strict sense but can only be the physical translation of the *transmitted digital information* (fig. 7.2).

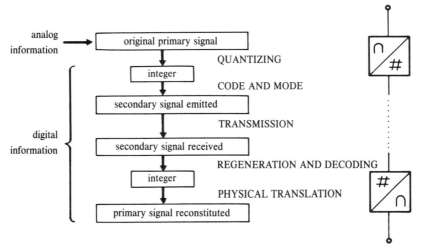

Fig. 7.2 Digital modulation and demodulation.

The decision rate $\dot{D}$, characteristic of digitally transmitted information, depends on the following:
- the specifications of the primary signal (bandwidth, frequency range, dynamic range)
- the convention (code) adopted for digital representation

Historically and technically, these modulations take on enormous importance. They form a bridge between the analog and digital domains. They also allow the principles of digital transmission to be applied to the transmission of continuous information (speech, music, continuous images, *et cetera*) and in large measure free this transmission from the problems of linearity, distortions, and noise interference.

DIGITAL MODULATION

7.1.2 Types of digital modulation

The principal digital modulation is *pulse code modulation* (PCM, sect. 7.5). To each analog sample, it assigns an integer (among q), which is then represented in an appropriate coded form. Announced in a patent in 1938 by A.H. Reeves (ITT), 10 years before the mathematical information theory by Shannon, and also 10 years before the invention of the transistor, the principle of PCM modulation had to wait for the development of semiconductors (transistors, integrated circuits) to be technically realized and commonly used with increasingly favorable economics. At present, PCM modulation forms the basis for new developments in telephone transmission and switching technology.

Differential digital modulation (sect. 7.8) does not encode the value of each sample, but the difference with respect to the preceding sample. In this category we find *differential PCM modulation* (DPCM), and its simplest form, *delta modulation* (ΔM).

Adaptive digital modulation (sect. 7.9) is that in which the correspondence between the analog values and their digital representation is dependent on the preceding sequence of events.

7.2 QUANTIZING

7.2.1 Definition

Quantizing is the approximation of the instantaneous value of a signal by the nearest value drawn from a a *finite* assortment of q *discrete* values, each designated by an integer.

In this way, each of these q integers represents a whole range of analog values called *quantizing intervals,* whose width Δ_i may be different from one interval to the next (fig. 7.3).

Proceeding to the quantizing of a signal $u_1(t)$, we must start by defining q intervals covering the entire presumed range of signal variation. Because q and Δ_i must be finite, only a limited range can be quantized. If the signal is outside of the range defined by the sum of all the intervals Δ_i, the quantizing results in a *clipping* of the signal.

7.2.2 Reconstruction of the quantized signal

The emitter transmits the identity of the interval i, in which the instantaneous value of the signal is found, in the form of an integer i of the q possible

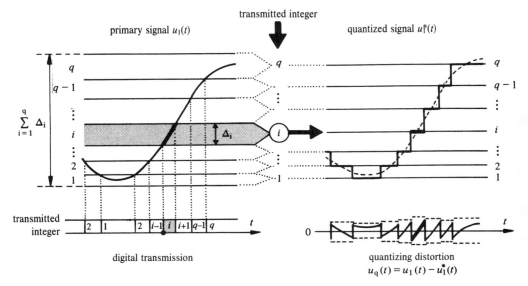

Fig. 7.3 Quantizing process.

integers. The sink, upon receiving this integer i, reconstructs the corresponding value of the signal as best as possible; lacking more precise information, it assumes that this value is found **in the middle of the interval** i, i.e., at the *quantizing step* i, from among the q possible steps (fig. 7.3).

7.2.3 Quantizing distortion and noise

The reconstructed signal $u_1^*(t)$ is only a discrete approximation of the original signal $u_1(t)$. It differs from this signal by a variable quantity called *quantizing distortion*, at most equal to half the interval Δ_i where the signal $u_1(t)$ is found:

$$u_q(t) = u_1(t) - u_1^*(t) \tag{7.1}$$

Quantizing distortion appears as noise superimposed on the signal, and is called *quantizing noise*. It is of random nature if $u_1(t)$ is random.

If the signal $u_1(t)$ is strictly equal to zero, the quantizing distortion is constant and approximately equals zero. The quantizing noise is, therefore, nonexistent. Unlike background noise, it appears only when the signal is present and surrounds it with a sort of veil, which is more objectionable than the effect of nonlinearities because it is nonharmonic. It must, therefore, be considered as *self-interference* in the sense defined in section 2.4.11.

DIGITAL MODULATION

To express the effect of quantizing distortion, we define a signal-to-quantizing noise ratio ξ_q between the power, designated by P_S, which the original signal has without the quantizing distortion and that of the quantizing distortion $u_q(t)$, designated by P_q.

$$\xi_q = \frac{P_S}{P_q} \tag{7.2}$$

7.2.4 Quantizing and sampling

In the case of figure 7.3, the moment of the transition from an integer i to the next $i + 1$ or $i - 1$ representing the primary signal $u_1(t)$, is determined by the signal $u_1(t)$ itself, or more precisely by the moment at which it crosses the limits of the interval i. We can, therefore, say that the signal samples itself at irregular moments, even random moments, if $u_1(t)$ is random.

More frequently, however, the quantizing is not carried out on the signal $u_1(t)$, which is continuous in time, but on equidistant samples taken at a frequency f_e. The integers representing the analog samples are, therefore, transmitted at regular intervals, with a period $T_e = 1/f_e$.

The result of the reconstruction of the sampled and quantized signal is a **staircase signal** $u_1^*(t)$ (figure 7.4) which differs from

- the case of figure 7.3 (quantizing without sampling) by the regular length T_e of its "steps" and by the fact that, from one sample to the next, it can jump several quantizing steps;
- the case of figure 4.17 ("sample-and-hold" without quantizing) by the loss of the exact value of the samples.

In this case, it is logical to define the quantizing distortion $u_q(t)$ by the difference between the staircase signal $u_2(t)$, which would have been obtained by sample-and-hold of the primary signal and the reconstructed quantized signal $u_1^*(t)$:

$$u_q(t) = u_2(t) - u_1^*(t) \tag{7.3}$$

This signal $u_q(t)$ can be considered to be the result of a sample-and-hold (during T_e) of the distortion due to the direct quantizing at q steps (without sampling) of the signal $u_1(t)$.

In fact, sampling and quantizing can be interchanged without changing either the reconstructed signal, or the quantizing distortion. As a result, it is possible to consider these two operations separately.

To comply with the sampling theorem, the frequency band of the original signal $u_1(t)$ must be limited by preliminary filtering to between 0 and $f_e/2$. In

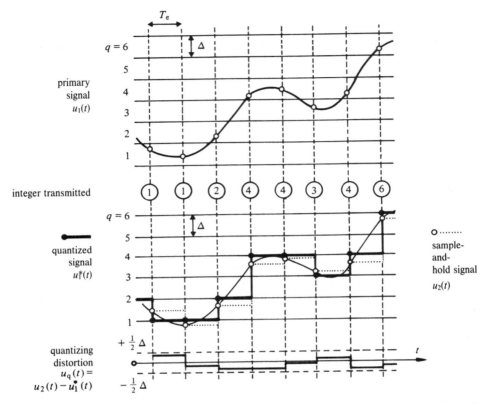

Fig. 7.4 Uniform quantizing (Δ = const.) of a sampled signal (sample-and-hold).

the same way, a low-pass filter at $f_e/2$ allows drawing a smoothed baseband signal from the quantized signal $u_1^*(t)$.

However, the spectral envelope of the staircase signal introduces **attenuation distortion** in $\sin \pi T_e f / \pi T_e f$ (sect. 4.3.7), which must be compensated for before obtaining the demodulated signal $u_1'(t)$, i.e., the approximately reconstituted primary signal (fig. 7.5).

The difference between $u_1(t)$ and $u_1'(t)$ represents the residual quantizing distortion (smoothed) $u_q'(t)$:

$$u_q'(t) = u_1(t) - u_1'(t) \qquad (7.4)$$

Because of the principle of superposition, this is the result of the (linear) operations of figure 7.5 on the distortion $u_q(t)$ according to (7.3).

DIGITAL MODULATION

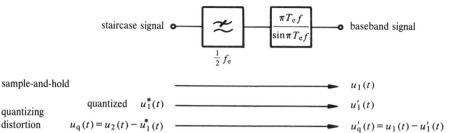

Fig. 7.5 Smoothing of a staircase signal.

7.3 UNIFORM QUANTIZING

7.3.1 Definition and characteristics of uniform quantizing

The quantizing operation is said to be *uniform* or ("linear") if the width Δ of all the quantizing intervals is the same, and, consequently, the quantizing steps are equidistant:

$$\Delta = \text{const.} \qquad \text{(uniform quantizing)} \qquad (7.5)$$

Figure 7.4 shows an example of uniform quantizing.

The *quantizing characteristic*, which ties the reconstructed quantized signal $u_1^*(t)$ to the primary signal $u_1(t)$ through the intermediary of q integers representing the steps and intervals, then has the form shown in figure 7.6.

Usually the quantizing domain is symmetrical with respect to $u_1 = 0$. It is limited to

$$\pm U_{1\max} = \pm \tfrac{1}{2} q\Delta \qquad (7.6)$$

or conversely,

$$\Delta = \frac{2U_{1\max}}{q} \qquad (7.7)$$

If the primary signal goes outside this domain, clipping occurs:

$$u_1^*(t) = \pm \tfrac{1}{2}(q-1)\Delta = \text{const.} \qquad \text{if } |u_1(t)| > U_{1\max} \qquad (7.8)$$

7.3.2 Quantizing distortion (uniform law)

Whatever the instantaneous value of the primary signal $|u_1(t)| < U_{1\max}$, the quantizing distortion $u_q(t)$ has an amplitude limited to $\pm\Delta/2$.

$$|u_q(t)| \leqq \tfrac{1}{2}\Delta \qquad (7.9)$$

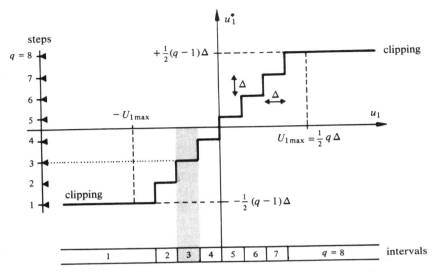

Fig. 7.6 Characteristics of uniform quantizing. Example: $q = 8$.

It has an approximately "sawtooth" form (fig. 7.3), which becomes even more triangular when the quantizing intervals are narrower with respect to the signal variations, i.e., when q is large.

7.3.3 Quantizing noise spectrum

The rigorous calculation of the spectrum quantizing distortion $u_q(t)$ in the general case of figure 7.3 is difficult [23] and of limited practical interest. An example is given in vol. VI, sect. 10.3.10.

On the other hand, for the frequent case of quantizing coupled with sampling, the form of this spectrum can be quickly evaluated by means of the autocorrelation function $\varphi_q(t)$ of the distortion $u_q(t)$. According to figure 7.4, $u_q(t)$ is a staircase voltage whose successive values are random with a zero mean. If the signal $u_1(t)$ is random, it can be assumed, further, that the successive values of $u_q(t)$ are practically not correlated. The autocorrelation function $\varphi_q(t)$, thus is a triangular pulse of length $2T_e$ (fig. 7.7). Its Fourier transform gives the (bilateral) power spectral density $\Phi_q(f)$ of the quantizing noise (vol. VI, sect. 5.3):

$$\varphi_q(t) \circ \!\!\xrightarrow{\;F\;}\!\!\bullet\; \Phi_q(f) = \frac{\Delta^2 T_e}{12} \left(\frac{\sin \pi T_e f}{\pi T_e f}\right)^2 \qquad (7.10)$$

DIGITAL MODULATION

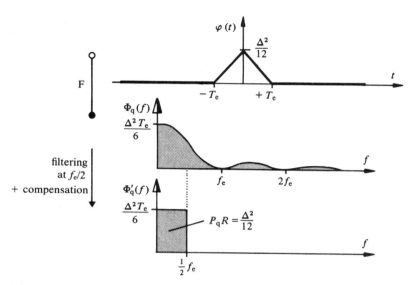

Fig. 7.7 Autocorrelation function $\varphi_q(t)$ and power spectral density (unilateral) $\Phi_q(f)$ of the quantizing noise.

After low-pass filtering at $f_e/2$ and compensating for the distortion in $\sin \pi T_e f / \pi T_e f$ according to figure 7.5, the power spectral density (bilateral) $\Phi_q(f)$ of the residual quantizing distortion $u'_q(t)$ is constant between 0 and $f_e/2$ (baseband).

The *quantizing noise* after demodulation, therefore, is a **white noise** in this frequency range, as long as the above hypotheses are satisfied. Note that this noise, although white, is certainly not Gaussian, since its instantaneous values are limited to between $-\Delta/2$ and $+\Delta/2$.

7.3.4 Calculation of the quantizing signal-to-noise ratio (uniform law)

The power P_q (over a resistance R) of the quantizing distortion depends on its probability density $p(u_q)$ according to the expression:

$$P_q = \frac{1}{R} \int_{-\Delta/2}^{+\Delta/2} u_q^2 \, p(u_q) \, du_q \qquad (7.11)$$

with

$$\int_{-\Delta/2}^{+\Delta/2} p(u_q) \, du_q = 1 \qquad (7.12)$$

If $q \gg 2$ (very fine quantizing), it can be assumed that the signal $u_1(t)$ has a uniform probability distribution inside a quantizing interval (but possibly different from one interval to the next). The result is that all the distortion values between $+\Delta/2$ and $-\Delta/2$ are approximately equally probable. Thus, the probability density $p(u_q)$, according to (7.12), is equal to

$$p(u_q) \cong \text{cste} = \frac{1}{\Delta} \tag{7.13}$$

Thus,

$$P_q = \frac{1}{R} \int_{-\Delta/2}^{+\Delta/2} u_q^2 \, p(u_q) \, du_q \cong \frac{1}{R} \cdot \frac{1}{\Delta} \int_{-\Delta/2}^{+\Delta/2} u_q^2 \, du_q = \frac{1}{R} \cdot \frac{\Delta^2}{12} \tag{7.14}$$

The power of the primary signal depends on its statistics, hence, on its probability density $p(u_1)$

$$P_S = \frac{1}{R} \int_{-\hat{U}_1}^{+\hat{U}_1} u_1^2 \, p(u_1) \, du_1 = \frac{1}{R} \, U_{1\text{eff}}^2 \tag{7.15}$$

The signal-to-quantizing noise ratio ξ_q defined by (7.2) has an overall value (for $q \gg 2$) of

$$\xi_q = \frac{P_S}{P_q} = 12 \, \frac{U_{1\text{eff}}^2}{\Delta^2} \tag{7.16}$$

7.3.5 Case of a sinusoidal signal

With the following hypotheses:
- sinusoidal signal with amplitude given by

$$\hat{U}_1 = \sqrt{2} \, U_{1\text{eff}} \tag{7.17}$$

- very fine quantizing ($q \gg 2$)

equation (7.16) becomes, with (7.7) (fig. 7.10):

$$\xi_q = 12 \, \frac{\hat{U}_1^2}{2\Delta^2} = \frac{3}{2} \left(\frac{\hat{U}_1}{U_{1\text{max}}} \right)^2 q^2 \tag{7.18}$$

In particular, if the sinusoidal signal covers the entire quantizing domain ($\hat{U}_1 = U_{1\text{max}}$), we have

$$\xi_q = \tfrac{3}{2} q^2 \tag{7.19}$$

The reconstructed signal $u_1^*(t)$ is affected by a (nonharmonic) distortion,

… DIGITAL MODULATION …

which, by analogy with relation (6.3), can be expressed by a quantizing distortion factor d_q

$$d_q = \sqrt{\frac{P_q}{P_S}} = \sqrt{\frac{1}{\xi_q}} \cong \frac{0.82}{q} \qquad (7.20)$$

For a sinusoidal signal of maximum amplitude, this distortion factor is given as a function of q in table 7.8.

Table 7.8 Distortion factor due to the uniform quantizing of a sinusoidal signal.

q	8	16	32	64	128	256
d_q	10.2%	5.1%	2.55%	1.28%	0.64%	0.32%

7.3.6 Instantaneous signal-to-quantizing noise ratio

By relating the instantaneous power $p_S(t)$ of a primary signal $u_1(t)$ to the mean power P_q (constant and independent of u_1) of the quantizing distortion, we can define an instantaneous signal-to-quantizing noise ratio $\xi_q(t)$:

$$\xi_q(t) = \frac{p_S(t)}{P_q} = 12 \frac{u_1^2(t)}{\Delta^2} \qquad (7.21)$$

With (7.7), we obtain

$$\xi_q(t) = 3 \left(\frac{u_1(t)}{U_{1\max}}\right)^2 q^2 \qquad (7.22)$$

or, as a logarithmic expression:

$$10 \lg \xi_q(t) = 20 \lg q + 20 \lg \frac{u_1(t)}{U_{1\max}} + 4.7 \qquad \text{dB} \qquad (7.23)$$

This ratio is expressed graphically in logarithmic form as a function of q and of the relative amplitude $u_1(t)/U_{1\max}$ in figure 7.9.

The global value of the signal-to-quantizing noise ratio ξ_q for any primary signal can be obtained by integrating $\xi_q(t)$, taking into account the probability distribution of the signal, since P_q is assumed to be independent of $u_1(t)$.

7.3.7 Remarks

Relations (7.18) and (7.22) and figure 7.9 bring up the following important remarks:

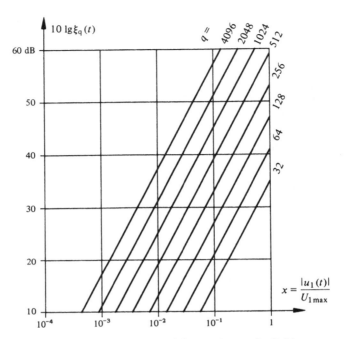

Fig. 7.9 Instantaneous signal-to-quantizing noise ratio $\xi_q(t)$.

- the signal-to-quantizing noise ratio ξ_q is proportional to the square of the number of intervals q. Doubling q increases $10 \lg \xi_q$ by 6 dB;
- the ratio ξ_q is proportional to the square of the relative amplitude of the signal (with respect to the limit $U_{1\max}$). Thus, it becomes very poor for weak signals, as the slope of 20 dB per decade of the family of lines shows in figure 7.9;
- the effect of *clipping* at $\pm U_{1\max}$ has been neglected. This effect depends on the statistics of $u_1(t)$, in particular on the probability that $|u_1(t)|$ exceeds $U_{1\max}$, for a given rms value $U_{1\mathrm{eff}}$. In the case of a sinusoidal signal, when $U_{1\mathrm{eff}} > U_{1\max}/\sqrt{2}$, the result is a harmonic distortion with a power P_d which, by virtue of (2.23) is added to the power of the quantizing distortion P_q. The resulting signal-to-"noise" ratio $\xi_{q\,\mathrm{res}}$ is, thus (fig. 7.10);

$$\xi_{q\,\mathrm{res}} = \frac{P_s}{P_q + P_d} \tag{7.24}$$

It is, nevertheless, necessary to emphasize that the subjective effect of the harmonic distortion due to clipping is very different from the (nonharmonic) distortion due to quantizing.

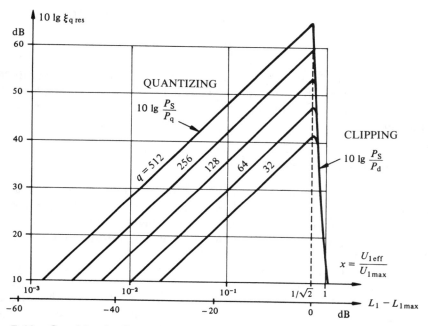

Fig. 7.10 Combined effect of the uniform quantizing and clipping on a sinusoidal signal.

7.4 NONUNIFORM QUANTIZING

7.4.1 Objectives

In many cases it is very undesirable for the ratio ξ_q to vary so strongly with the level of the signal. Ideally, we wish rather that it would stay constant, independently of $u_1(t)$.

It is, therefore, necessary to keep the *relative error* due to quantizing constant, whatever the amplitude of the signal, while the uniform quantizing introduces a maximum *absolute error* of $\pm 1/2\Delta$ in the entire quantizing range. The solution is to vary the quantizing step Δ as a function of u_1. A quantizing law in which the step Δ depends on the amplitude of the signal is called *nonuniform*.

7.4.2 Ideal nonuniform quantizing

From (7.21), if we set
$$\xi_q = \text{const.} \; \forall \; u_1(t) \tag{7.25}$$

then we must vary the quantizing step Δ proportionally to $|u_1(t)|$

$$\frac{\Delta(u_1)}{|u_1|} = \text{const.} \tag{7.26}$$

This condition is unfortunately unrealistic, since it gives rise to intervals that approach zero about the origin ($u_1 = 0$), which is incompatible with the necessity of covering the entire range of $-U_{1\,max}$ to $+U_{1\,max}$ with a *finite* number of intervals q.

7.4.3 Quantizing with companding

Nonuniform quantizing can be considered to be uniform quantizing preceded by *compression of the dynamic range* of the signal, which has the effect of favoring low amplitudes to the detriment of high amplitudes. The idea is the same as that presented in section 6.6.2 for analog transmission, the source of noise here being the quantizing distortion.

Instantaneous compression is applicable to a sampled transmission, as was mentioned in section 6.6.5.

The original dynamic range must be clearly re-established with respect to the demodulation by means of a strictly reciprocal *expansion* characteristic (fig. 7.11).

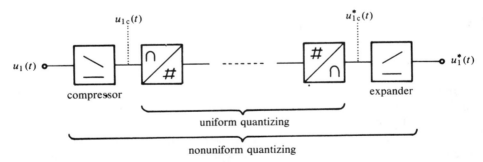

Fig. 7.11 Quantizing with companding.

The combination of three characteristics:

- compresssion: $u_{1C} = f_1(u_1)$
- uniform quantizing with q intervals of identical width Δ_0: $u^*_{1C} = f_2(u_{1C})$
- expansion: $u^*_1 = f_1^{-1}(u^*_{1C})$

gives a nonuniform quantizing characteristic $u^*_1 = f_3(u_1)$ according to figure 7.12.

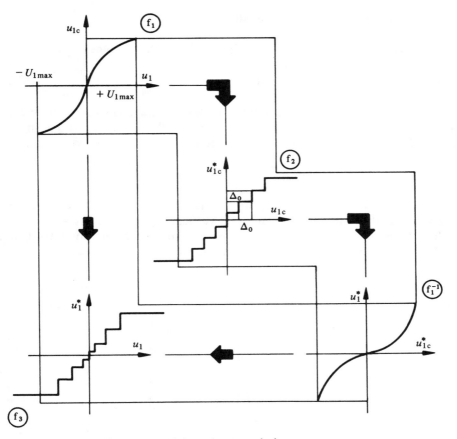

Fig. 7.12 Nonuniform quantizing characteristic.

The law of variation of the quantizing step Δ as a function of the instantaneous amplitude of the signal is given by the derivative of the compression characteristic:

$$\frac{\Delta(u_1)}{\Delta_0} = \frac{du_1}{du_{1C}} \tag{7.27}$$

Conversely, for a given quantizing characteristic $\Delta(u_1)$, the compression characteristic $u_{1C} = f_1(u_1)$ must satisfy the relation

$$du_{1C} = \frac{\Delta_0}{\Delta(u_1)} du_1 \tag{7.28}$$

7.4.4 Ideal compression characteristic

To satisfy the ideal condition (7.26), relation (7.28) becomes

$$\mathrm{d}u_{1C} = k\Delta_0 \frac{\mathrm{d}u_1}{|u_1|} \text{ with } k = \text{const.} \tag{7.29}$$

The solution to this diffferential equation is a *logarithmic compression* characteristic. Given the condition that the limit of clipping $U_{1\max}$ should not be affected by the compression, this characteristic takes the following form:

$$|u_{1C}| = U_{1\max} + k\Delta_0 \log \frac{|u_1|}{U_{1\max}} \tag{7.30}$$

As predicted in section 7.4.2, this characteristic, presented in figure 7.13, cannot be realized in practice due to its vertical asymptote at $u_1 = 0$.

The technical solutions for nonuniform quantizing must thus be satisfied with an approximation of this characteristic (sect. 7.5.3 and sect. 7.5.5).

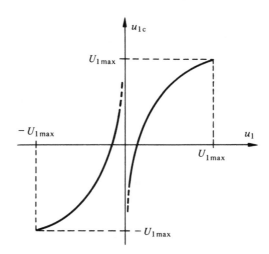

Fig. 7.13 Ideal compression characteristic.

7.5 PULSE CODE MODULATION

7.5.1 Definition

Pulse code modulation PCM is the combination of the three following operations [24]:

DIGITAL MODULATION

- sampling at frequency f_e
- quantizing, generally nonuniform in telecommunications, with q steps
- encoding of the q integers corresponding to the quantizing intervals, generally in binary form, i.e., with lb q binary digits ("bits").

The result of PCM modulation is a signal carrying digital information with a bit rate $\dot{D}$ corresponding to the number of samples taken per unit time, multiplied by the decision quantity lb q of each integer which represents them:

$$\dot{D} = f_e \, \text{lb} \, q \qquad \text{bit/s} \qquad (7.31)$$

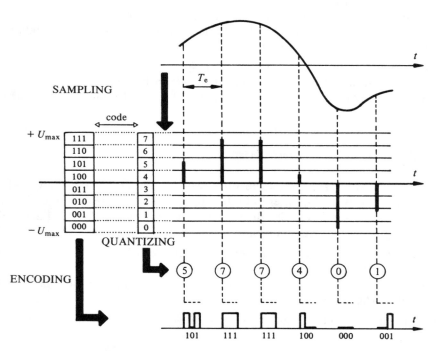

Fig. 7.14 Stages of PCM modulation.

7.5.2 Conditions for telephony

For the reasons mentioned in section 4.3.6, the sampling frequency for the telephone band from 300 to 3400 Hz has been set by international convention at

$$f_e = 8 \text{ kHz} \qquad (7.32)$$

Subjective tests have given rise to the demand for a signal-to-quantizing noise ratio of

$$10 \lg \xi_q \geq 35 \text{ dB} \qquad (7.33)$$

in such a way as to guarantee a good quality of speech transmission.

However, it is still necessary to consider:
- the statistical dispersion of the mean level of speech (sect. 1.7.1)
- the preponderance of small amplitudes in the statistical distribution of speech (fig. 1.9)
- the dispersion of the nominal level at the input of the PCM modulator, according to the attenuation of the path between the subscriber (more or less removed from one case to the next!) and this (shared!) modulator, across the switched network

Consequently, condition (7.33) must be respected not only for the maximum signal level, but also in the *range estimated at 40 dB* below this level.

In practical terms this requirement leads to the use of nonuniform quantizing.

7.5.3 A-law compression characteristic

According to the proposal of the European Conference of Posts and Telecommunications (CEPT), later adopted by the CCITT (Recommendation G.711), the ideal compression characteristic from figure 7.13 has been approached by the following compromise:
- *a logarithmic segment* for the relative amplitudes $x = |u_1|/U_{1\ max}$ located between $1/A$ and 1
- around the origin a *linear segment* for $x < 1/A$, tangential to the logarithmic segment (fig. 7.15)

Its expression in relative amplitude $y = |u_{1C}|/U_{1\ max}$, called the A-law, is

$$y = \frac{Ax}{1 + \ln A} \qquad \text{for } x \leq \frac{1}{A} \qquad (7.34)$$

$$y = \frac{1 + \ln Ax}{1 + \ln A} \qquad \text{for } \frac{1}{A} \leq x \leq 1 \qquad (7.35)$$

The slope of the linear segment at the origin is called the *compression ratio C*. It has been chosen to be equal to 16

$$C = \frac{A}{1 + \ln A} = 16 \qquad (7.36)$$

DIGITAL MODULATION

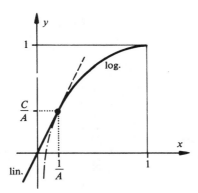

Fig. 7.15 A-law compression characteristic (first quadrant).

The result is the value of the parameter $A = 87.6$. With respect to a uniform quantizing step $\Delta_0 = 2U_{1\ max}/q$, from (7.10), the quantizing step is calculated by (7.27) as

$$\Delta(x) = \Delta_0 \frac{dx}{dy} \tag{7.37}$$

i.e., with (7.34), (7.35), and (7.36), respectively,

- linear part: $\quad \Delta(x) = \dfrac{1}{C} \Delta_0 = \text{const.} \quad \forall x \tag{7.38}$

- logarithmic part: $\Delta(x) = \dfrac{A}{C} \Delta_0 x \sim x \tag{7.39}$

7.5.4 Signal-to-quantizing noise ratio with A-law

By inserting the value of Δ from (7.38) or (7.39) into (7.22), we obtain the value of the instantaneous signal-to-quantizing noise ratio of (fig. 7.16):

- in the linear part ($x \leq 1/A$)

$$\xi_q(t) = 3q^2C^2x^2 \sim x^2 \tag{7.40}$$

With respect to uniform quantizing with the same number q of levels, the ratio ξ_q is thus increased by a factor C^2. Thus, we obtain the same quality for low amplitudes as with a uniform quantization $C = 16$ times finer, i.e., with 16 q levels.

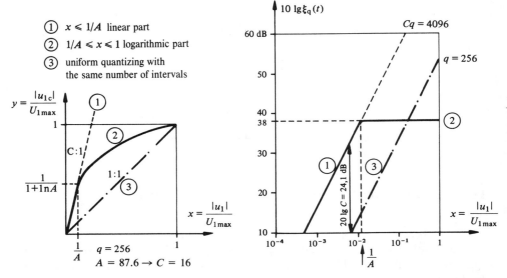

Fig. 7.16 Instantaneous signal-to-quantizing noise ratio using the A-law with $q = 256$ intervals (nonuniform).

- in the logarithmic part ($1/A \leq x \leq 1$)

$$\xi_q(t) = 3q^2 \frac{C^2}{A^2} = \text{const.} \quad \forall x \tag{7.41}$$

The objective of a ξ_q ratio independent of the amplitude is thus attained for $|x| > 1/A$. Nevertheless, for $x = 1$ (i.e., for $u_1 = U_{1\,\text{max}}$), ξ_q is decreased by a factor A^2/C^2 with respect to uniform quantizing with the same number q of levels.

To satisfy the condition (7.33) for $0.01 < x < 1$ (40 dB dynamic range) and obtain a number of steps compatible with binary coding, this number has been set (CCITT, Recommendation G.711):

$$q = 2^8 = 256 \text{ steps} \tag{7.42}$$

7.5.5 Comparison with the compression according to the μ-law

The approximation of the ideal logarithmic compression law (from 7.30) had given way, in the United States, to another solution called μ-law, with a mathematical expression:

$$y = \frac{\ln(1 + \mu x)}{\ln(1 + \mu)} \tag{7.43}$$

This compression characteristic (fig. 7.17) has an asymptotic behavior, which is

- linear, for $x \ll 1/\mu$
- logarithmic, for $1/\mu \ll x \leq 1$

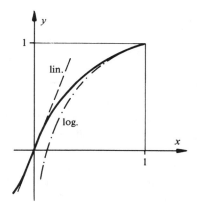

Fig. 7.17 μ-law compression characteristic.

For historical and economic reasons, the μ-law is also the subject of a CCITT recommendation (G.711) with $\mu = 255$. According to this law, the compression ratio (slope at the origin) is equal to

$$C = \frac{\mu}{\ln(1+\mu)} \cong 46 \tag{7.44}$$

The μ-law and A-law are in fact very similar. However, the difference in their compression ratios C at the origin, according to (7.36) and (7.44), gives rise to a better signal-to-quantizing noise ratio ξ_q with the μ-law for very weak signals. The difference depends on the statistics of the signal. It is presented in figure 7.18 in the case of speech signals modeled according to figure 1.9.

The difference between the two compression laws A and μ is nevertheless sufficiently important to make the compressor A incompatible with the expander μ and *vice versa*, and creates the risk of introducing nonlinear distortions and variations in level.

7.5.6 Coding of quantized samples

The quantizing procedure replaces the exact value of a sample with an integer that represents the interval in which this value is found. The transcrip-

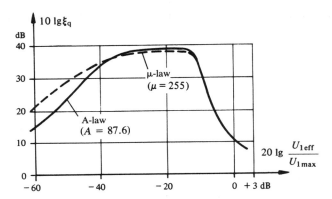

Fig. 7.18 Signal-to-quantizing noise ratio ξ_q with nonuniform quantizing according to the A-law or µ-law (signals with exponential statistics).

tion of the integer into a logical, generally binary, expression, called *PCM word*, constitutes *coding*, and the correspondence table between the q numbered intervals and their binary expression defines the *code*. The format of PCM word is in lb q bits, at least, i.e., 8 bits for $q = 256$.

Coding does not influence the modulation quality. The choice of a code is dictated by other considerations.

- technical advantages for the implementation of coding
- ease of implementing the decoding
- conditions relating to the digital transmission of coded samples: high clock rate content, canceling of the dc component, limitation of the frequency range occupied, *et cetera*

The third argument determines not only the code, but also the **mode** (from sect. 5.1.4). The choice of an adequate mode for PCM transmission is discussed in section 9.6 within the framework of digital systems.

7.5.7 Types of encoders

The operations of quantizing and coding are carried out with the same device, called the *encoder*, which brings about a certain number of **comparisons** between the analog value of the sample to be coded and the combinations of *standards* needed to obtain the best approximation. The binary word representing the value of the sample is derived from these combinations of standards by a rule that expresses the code.

DIGITAL MODULATION

According to the number of standards available and of necessary comparisons, there are three basic types of encoders, each of which can be used in multiple variations and combinations.

- serial encoder: a single standard is compared with the sample value up to q times (counting method);
- parallel encoder: the sample is compared in a single operation with a complete set of q standards (direct method);
- iterative encoder: the value of the sample is compared by successive approximations to different combinations of standards. A particularly interesting solution consists of effecting lb q comparisons with a set of lb q standards in binary progression (weighting method).

The number of successive comparisons necessary gives us an idea of the capability of the encoder to process high frequency signals. The most favorable from this point of view is the parallel encoder.

On the other hand, the physical complexity and the problems of tolerance increase with the number of standards (advantage of the serial encoder). The iterative encoder thus presents a good compromise between speed and cost.

In any case, the coding operation takes time. This is why it is preferable to proceed to a *sample-and-hold* before coding, so as to avoid any modification of the signal during this operation.

7.5.8 Choice of code

Among the $q!$ possible codes, the *simple binary code* (fig. 7.19 (a)) and its derivatives are the best adapted to serial coding. The iterative code and the process of decoding (reconstruction of the value of the sample by stacking standards) naturally requires a *weighted code* in which each binary element has a predetermined "weight."

For parallel coding, a code in which the contiguous intervals are represented by words differing only by a single binary element, such as *Gray's code* (fig. 7.19(d)), is particularly favorable, since imprecision in the single comparison with the q standards would only affect a single bit. This code, which is unweighted, however, is difficult to decode.

Because the signals to be quantized are bipolar (positive or negative, with zero mean), it is sufficient to code only the absolute value of the signal and to indicate the polarity by a supplementary element (sign bit). The code is thus *folded* (fig. 7.19(b)). It also has the advantage of not inverting all the bits as the signal passes zero.

In telephony, low amplitude signals are the most frequent (speech statistics (from figure 1.9), pauses, slack periods), giving rise to PCM words composed

256   TELECOMMUNICATION SYSTEMS

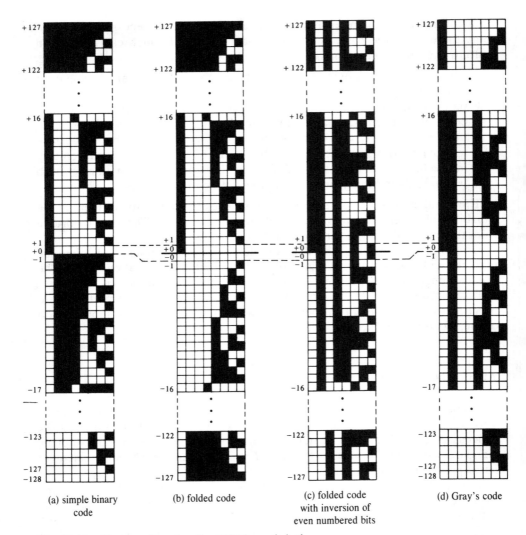

Fig. 7.19 Types of codes for PCM modulation.

almost exclusively of zeros, which would be very unfavorable for the clock rate content and, consequently, for the synchronization of the receiver and the intermediate regenerators. This is why the even numbered bits (in Europe), or all bits except the sign bit (in the USA and Japan) are systematically inverted in PCM words before being emitted (fig. 7.19(c)).

7.5.9 Nonuniform coding

The implementation of nonuniform quantizing with companding, according to section 7.4.3, and particularly with the A-law described in section 7.5.3, can be combined with the coding in three different ways:
- analog compression followed by uniform quantizing with an encoder at lb q = 8 bits (fig. 7.20(a));
- uniform quantizing with a number of levels Cq corresponding to the linear part of the compression characteristic, i.e., with lb (Cq) = 12 bits, followed by digital compression (transcoding) reducing the format of the PCM words to 8 bits (fig. 7.20(b));
- nonuniform coding at 8 bits with the digital compression characteristic incorporated into the encoder (fig. 7.20 (c)).

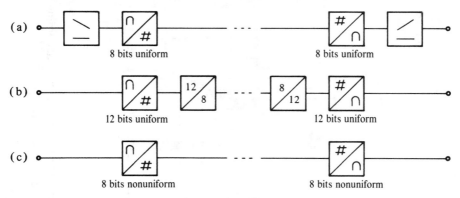

Fig. 7.20 Nonuniform quantizing and coding.

7.5.10 Approximation of digital A-law compression by a polygonal characteristic with 13 segments

The historical solution using analog compression (fig. 7.20(a)) poses problems of exact reciprocity of the characteristics of compression and expansion. It has now been replaced by a polygonal digital compression characteristic that is perfectly adapted to digital processing (fig. 7.21).

This characteristic consists of 13 linear segments, one of which, of slope 16, corresponding to the compression ratio C from (7.36), passes through the origin. The slope of these segments comprises a geometrical progression with a ratio of 1/2. Inside each segment, the quantizing is uniform at 16 intervals

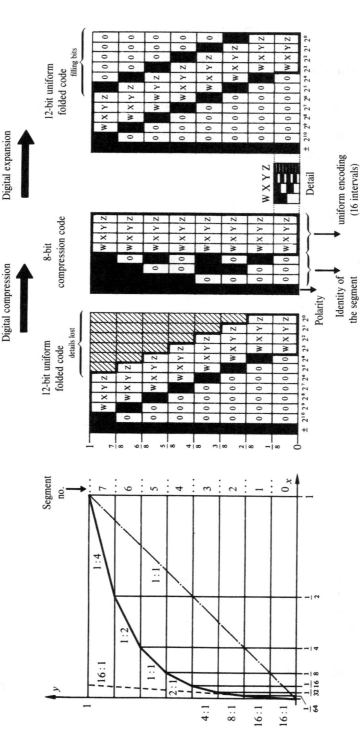

Fig. 7.21 Digital compression characteristic with 13 segments according to A-law, without inversion of the even numbered bits (only the positive values are presented).

(64 intervals in the central segment). The result is that the quantizing step Δ increases by a factor of 2 from one segment to the next as the absolute value of the primary signal increases.

In 8-bit PCM words coded in this way, the first bit indicates the polarity, the next three bits identify the segment (000 and 001 corresponding to the central segment), and the last four bits locate the quantizing step among the 16 within the segment. The weight attributed to these four bits during decoding depends on the segment.

A similar solution, with a 15-segment characteristic, has been defined to implement μ-law compression in digital form.

7.5.11 Signal-to-quantizing noise ratio for digital compression with 13 segments (A-law)

The polygonal approximation of A-law has repercussions for the instantaneous signal-to-quantizing noise ratio $\xi_q(t)$. For each segment, it is equal to the result of uniform quantizing with a number of intervals equal to the product of q times the slope of the segment (fig. 7.22).

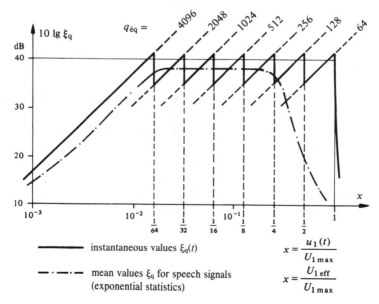

Fig. 7.22 Signal-to-quantizing noise ratio ξ_q ($q = 256$ intervals) with digital compression at 13 segments (A-law).

For random signals, such as telephone signals, the effect of discontinuities in slope practically disappears in the overall value ξ_q calculated by integration of $\xi_q(t)$, taking into account the statistics of $u_1(t)$.

7.5.12 Summary of PCM modulation parameters for telephony

By international convention (CCITT, Recommendation G 711), the following parameters have been set for the digital representation of telephone signals with PCM modulation:

- sampling frequency: $f_e = 8$ kHz
- nonuniform quantizing at $q = 256$ levels
- PCM word of lb $q = 8$ bits
- resulting bit rate (for one channel)

$$\dot{D} = f_e \text{ lb } q = 64 \text{ kbit/s} \tag{7.45}$$

- European coding law: digital compression according to A-law achieved by 13 segments (fig. 7.21), folded binary code with inversion of the even numbered bits (fig. 7.19 (c));
- American coding law: digital compresssion according to the μ-law achieved by 15 segments, folded binary code with inversion of all the bits, except the first (sign bit).

7.5.13 Application of PCM modulation to other signals

The application of PCM modulation, already in use for telephony, can also be considered for other analog information. However, international agreement has not yet been reached on all the parameters to be adopted.

Music (transmission):

- sampling frequency: $f_e = 32$ kHz
- uniform coding at 16 bits, $\dot{D} = 448$ kbit/s or nonuniform at 12 bits (compression with 5 segments), $\dot{D} = 384$ kbit/s, or nonuniform at 10 bits (A-law with 13 segments), $\dot{D} = 320$ kbit/s

Music (recording):

- sampling frequency: $f_e = 44.1$ kHz
- uniform coding at 16 bits, $\dot{D} \cong 0.7$ Mbit/s

Television (video signals):

- sampling frequency: $f_e = 13.3$ MHz
- uniform coding at 8 or 9 bits

DIGITAL MODULATION

7.6 MEASUREMENT OF QUANTIZING NOISE

7.6.1 Specifics of the problem

The quantizing noise is self-interference of the signal (sect. 7.2.3), i.e., the result of distortion, and thus cannot be measured except in the presence of the signal, in contrast to background noise, which can be measured in the absence of any signal.

In principle, relation (7.1) permits the measurement of the quantizing distortion $u_q(t)$ so as to arrange the two signals $u_1(t)$ and $u_1^*(t)$ exactly in phase. This does not occur in practice, given the delay introduced during signal processing, particularly by filtering.

In fact, only the power P_q of the distortion is of interest. Based on section 7.3.3, we will assume that its spectrum is uniform in the baseband.

7.6.2 Method of measurement

The principle consists of quantizing a *narrowband* primary test signal and measuring the spectral components of the quantizing distortion located *outside* the occupied signal band. By extrapolation, the total power P_q can be evaluated.

The test signal can be:

- a *sinusoidal* signal, the frequency of which is not a submultiple of the sampling frequency
- a *random* or *pseudo-random* signal with uniform and limited power spectral density (white noise)

In these two cases, the measurement results are subjected to statistical fluctuations and reflect the statistical distribution of the test signal itself as well.

Figure 7.23 gives a frequently used measuring arrangement. The power of the quantizing noise, which is assumed to be white, is measured in the band from 800 Hz to 3400 Hz, i.e. at a 2.6 kHz bandwidth, and must be multiplied by 3.1/2.6 so as to extrapolate it to the conventional telephone band of 300 Hz to 3400 Hz ($B_{tph} = 3.1$ kHz).

In apparent contradiction to definition (7.2), the power P_S of the signal is measured at the same point as the power P'_q of the quantizing distortion, i.e., at the output of the decoder (in the disjoint frequency bands). Thus, the possible differences in level between the input of the encoder and the output of the decoder are eliminated.

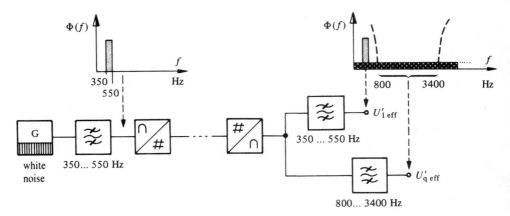

Fig. 7.23 Arrangement for measuring the quantizing noise.

In this way, the signal-to-quantizing noise ratio ξ_q is obtained by

$$10 \lg \xi_q = 10 \lg \frac{P'_S}{P'_q} = 20 \lg \frac{U'_{1\text{eff}}}{U'_{q\text{eff}}} - 10 \lg \frac{3.1}{2.6} \quad \text{dB} \tag{7.46}$$

7.7 EFFECTS OF TRANSMISSION ERRORS IN PCM

7.7.1 Context

As has been shown in section 5.4, the presence of noise and crosstalk in the transmission channel gives rise to irreversible regeneration errors with a probability of appearance ϵ, which can be evaluated if the following are known:
- the signal-to-noise ratio ξ'_R after transmission and equalization; or
- the rms value $U_{N\,\text{eff}}$ of the channel noise relative to the amplitude U'_{BR} of the received and equalized basic signal; and
- the statistical distribution (probability density) of the channel noise.

Because of regeneration errors accumulated in the course of transmission, the coded sample (PCM word) received is different from the digital value emitted. The more the interfered bit has a high weighting factor in the code used for the PCM modulation, the greater the difference between the demodulated "analog" value and the original analog value of sample.

If these errors are infrequent, the result is isolated crackling at the analog output (e.g., every 2.6 minutes on the average for a telephone channel of 64 kbit/s for $\epsilon = 10^{-7}$). For higher probabilities, these errors appear as additive noise at the output.

DIGITAL MODULATION

The power P'_{Ne} of this noise is added to that (P_q) of the quantizing distortion, to define a signal-to-noise ratio after demodulation ξ'_1, which is representative of the transmission quality of analog information, transmitted in digital form by PCM modulation in a noisy channel.

$$\xi'_1 = \frac{P'_{1S}}{P'_{1N}} = \frac{P'_{1S}}{P_q + P'_{Ne}} = \frac{1}{\delta} \cdot \xi_q \tag{7.47}$$

or

$$10 \lg \xi'_1 = 10 \lg \xi_q - 10 \lg \delta \tag{7.48}$$

where δ represents a *deterioration of the primary signal-to-noise ratio*, compared to the ideal case of error-free transmission in which the only noise is the quantizing noise. Clearly, δ is a function of the error probability ϵ, according to a law which in turn depends on

- the quantizing law
- the code used
- the statistical distribution of the errors over time

Finally, we can try to relate the deterioration δ of the signal-to-noise ratio after PCM demodulation to the transmission signal-to-noise ratio ξ'_R (after equalization), by the intermediary of ϵ, which depends on ξ'_R from (5.38) or (5.39) in the case of Gaussian noise.

7.7.2 Case of uniform quantizing

The calculation of δ can be done with the following simplifying hypotheses:
- uniform quantizing with $q = 2^b$ intervals ($q \gg 2$)
- simple binary code (fig. 7.19(c))
- ϵ sufficiently low to assume that there is at most only one error per coded sample
- each bit has the same probability ϵ of being wrongly received

The error signal $e(t)$ is defined by the difference between the reconstructed primary signal without error $u_1^*(t)$ and the primary signal effectively decoded $u_1^{*'}(t)$. It is a random variable, which can take the values indicated in table 7.24.

Its power P'_{Ne} is calculated by the statistical mean of $e^2(t)$:

$$P'_{Ne} = \frac{1}{R} \overline{e^2(t)} = \frac{\epsilon}{R} \left[4u_1^{*2} + \frac{U_{1max}^2}{q^2} \sum_{i=1}^{b-1} 2^{2i} \right]$$

$$= \frac{\epsilon}{R} \left[4u_1^{*2} + \frac{U_{1max}^2}{q^2} \cdot \frac{2^{2b} - 4}{3} \right] \cong \frac{\epsilon}{R} \left[4u_1^{*2} + \frac{U_{1max}^2}{3} \right] \tag{7.49}$$

Table 7.24 Value of the error according to the position of the wrongly received bit.

Weighting factor of this bit	b (sign bit)	$b-1$	...	i	...	1	no error
Order of the disturbed bit	$\pm$	2^{b-2}		2^{i-1}		2^0	
Corresponding value of $e(t)$	$2u_1^*(t)$	$\pm \frac{1}{2}U_{1\max}$		$\pm \frac{2^i}{q}U_{1\max}$		$\pm \frac{2}{q}U_{1\max}$	0
Corresponding probability	ϵ	ϵ		ϵ		ϵ	$1 - b \cdot \epsilon$

With (7.47), we obtain the deterioration δ:

$$\delta = \frac{\xi_q}{\xi_1'} = 1 + \frac{P'_{Ne}}{P_q} \qquad (7.50)$$

With (7.49) and taking P_q from (7.14), accounting for (7.7), it becomes

$$\delta = 1 + \epsilon \frac{4u_1^{*2} + \frac{1}{3}U_{1\max}^2}{\frac{U_{1\max}^2}{3q^2}} = 1 + q^2\epsilon \left(12 \frac{u_1^{*2}}{U_{1\max}^2} + 1\right) \qquad (7.51)$$

The ratio $u_1^*/U_{1\max}$ is a very good approximation (if $q \gg 2$) of the relative primary voltage $u_1/U_{1\max}$.

Figure 7.25 shows the degradation of the signal-to-noise ratio after demodulation ξ_1' as an instantaneous value, as a function of the error probability ϵ in the case where $q = 256$.

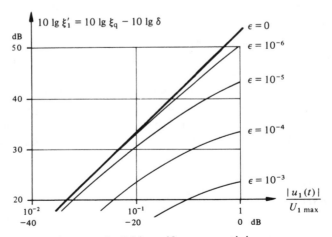

Fig. 7.25 Effects of errors in 8-bit uniform quantizing

7.7.3 Case of nonuniform quantizing

In nonuniform quantizing, the effect of regeneration errors is much more complex and depends strongly on the relative level of the primary signal and its statistics. By simulation, it is nevertheless possible to predict its effects. Figure 7.26 gives the results of such a simulation with exponential statistics for the primary signal and 8-bit nonuniform quantizing according to A-law.

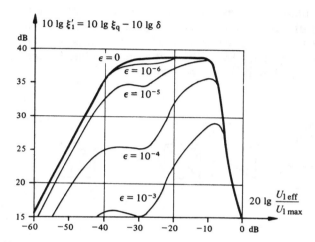

Fig. 7.26 Effects of errors in 8-bit nonuniform quantizing.

7.8 DIFFERENTIAL DIGITAL MODULATIONS

7.8.1 Principles

Differential digital modulation quantizes the *difference* between $u_1(t)$ and a value $v_1(t)$, estimated by *extrapolation* of the preceding values of $u_1(t)$, and not the instantaneous value of the signal as is the case by PCM modulation.

Extrapolation is not possible unless certain statistical characteristics of the signal $u_1(t)$ are known. A perfectly random signal of which nothing is known cannot be extrapolated. Differential modulation is more effective the more the signals represented are redundant and if the statistical laws which govern them are well known.

As with PCM modulation, differential digital modulation is almost always related to sampling at a fixed frequency f_e. It is in this form that they are discussed here.

7.8.2 Functions of the modulator and demodulator

In principle (fig. 7.27), the modulator quantizes and codes the difference between the sample $u_1(i)$ of the primary signal $u_1(t)$, taken at the instant iT_e, and a value $v_1(i)$ calculated from the preceding values of $u_1(t)$ and an adequate extrapolation algorithm:

$$v_1(i) = f[u_1(i-1), u_1(i-2), ...,] \quad (7.52)$$

The demodulator decodes the coded numbers it receives and reconstructs the quantized difference $u_1^*(i) - v_1^*(i)$, which is an approximation of the true difference $u_1(i) - v_1(i)$. Starting with $u_1^* - v_1^*$, it restores an approximation $u_1^*(i)$ of the primary signal by addition of an extrapolated value $v_1^*(i)$ calculated by the same algorithm as in the modulator.

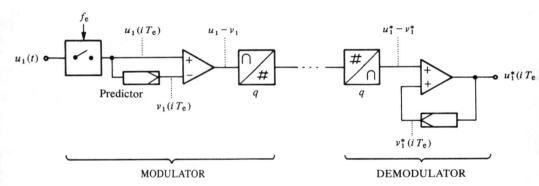

Fig. 7.27 Basic principle of differential digital modulation and demodulation.

The inconvenience of this first idea is that the extrapolated value v_1^* used in the demodulator is not exactly the same as the value v_1 that the modulator has subtracted from the signal u_1. It differs from it due to quantizing distortion. This inconvenience is avoided by extrapolating not the signal u_1 in the modulator, but a quantized signal u_1^*, obtained from the digital output of the modulator by the same circuit as is used in the demodulator (fig. 7.28).

The output of the demodulator from figure 7.27 or figure 7.28 is a quantized staircase signal $u_1^*(iT_e)$, which must still be filtered to give the demodulated signal $u_1'(t)$.

The numerous forms of differential digital modulation proposed in publications or patents are essentially distinguished by the type of *predictor* used and by the number q of quantizing steps. They are each more or less well adapted to the statistics of a particular signal, and, therefore, to a certain type

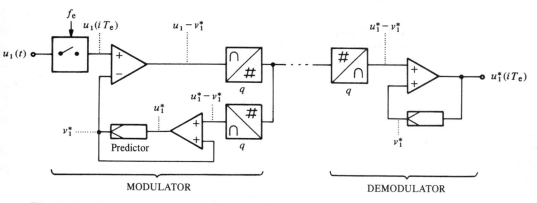

Fig. 7.28 Enhanced principle of a differential digital modulation.

of analog information. The objective is to try to reduce the bit rate $\dot{D}$ required, given by (7.31) as with PCM, while retaining satisfactory quality.

7.8.3 Delta modulation: procedure

In its basic form, *delta modulation* ΔM [25] is differential digital modulation characterized by

- *a zero-order extrapolation*, i.e., the estimated value v_1^* is simply equal to the quantized value u_1^* of the preceding sample;

$$v_1^*(i) = u_1^*(i - 1) \tag{7.53}$$

- quantizing of $u_1 - v_1^*$ at *a single bit* ($q = 2$) i.e., only the *sign* of the difference is considered.

Consequently, the two single parameters are:

- the *sampling frequency* f_e which at the same time gives the bit rate $\dot{D}$ at the output, because, with $q = 2$, (7.31) becomes

$$\dot{D} = f_e \tag{7.54}$$

- the quantizing step Δ, used in demodulation to reconstruct the difference $u_1^* - v_1^*$ from the sign of $u_1 - v_1^*$, transmitted in binary form

$$\text{if } u_1 > v_1^* \quad \text{bit transmitted: 1} \quad u_1^* - v_1^* = +\Delta \tag{7.55}$$

$$\text{if } u_1 \leq v_1^* \quad \text{bit transmitted: 0} \quad u_1^* - v_1^* = -\Delta \tag{7.56}$$

The reconstructed signal $u_1^*(t)$ is a staircase signal which can only vary by $+\Delta$ or $-\Delta$ at each sampling moment iT_e (fig. 7.29).

The quantizing distortion u_q is normally lower than Δ, i.e., less than the double of the limit of u_q in PCM modulation, according to (7.9):

$$|u_q(t)| = |u_1(t) - u_1^*(t)| \leq \Delta \qquad (7.57)$$

It is present even when $u_1(t)$ is noticeably constant (*granular noise*), or even null (*idle noise*). Thus, the distortion $u_q(t)$ is thus practically a square signal of amplitude Δ.

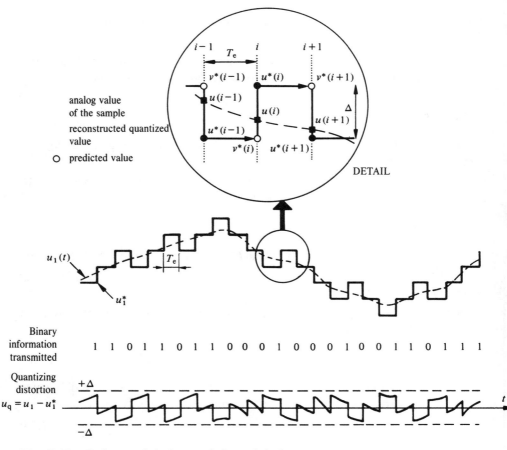

Fig. 7.29 Delta modulation and demodulation.

7.8.4 Slope overload in delta modulation

The ratio of the step value Δ to the sampling period T_e defines the steepest slope (positive or negative) that the reconstructed signal $u_1^*(t)$ can present. If the derivate of the primary original signal $u_1(t)$ exceeds this ratio, $u_1^*(t)$ can no longer follow $u_1(t)$. A *slope overload* appears (fig. 7.30), and is indicated by an increase in the quantizing distortion. The overload limit is given by

$$\left|\frac{du_1}{dt}\right|_{max} = \frac{\Delta}{T_e} = f_e \Delta \qquad (7.58)$$

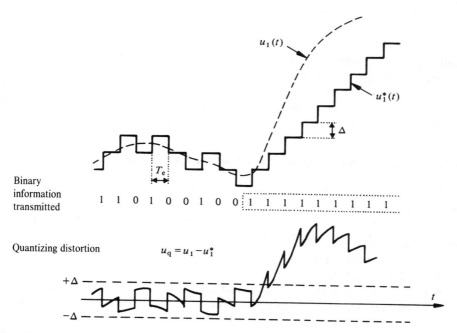

Fig. 7.30 Slope overload.

For a primary sinusoidal signal $u_1(t) = U_1 \sin 2\pi f_1 t$, relation (7.58) imposes a limit on the amplitude U_1 as a function of the frequency f_1 (fig. 7.31).

$$U_{1max} = \frac{f_e \Delta}{2\pi f_1} \qquad (7.59)$$

The only way to reconcile a low quantizing distortion (small Δ) with a sufficient amplitude (without slope overload), even at $f_{1\ max}$, is to choose a

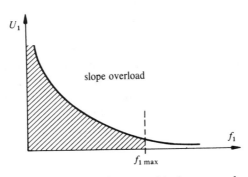

Fig. 7.31 Tolerable range (shaded) to avoid slope overload.

high sampling frequency, in fact, 5 to 10 times higher than the minimum given by the sampling theorem (4.16).

As a result of (7.59), delta modulation is particularly suitable for primary signals with a spectrum which decreases toward the high frequencies. This is notably the case for telephone signals (sect. 1.7.1).

7.8.5 Signal-to-quantizing noise ratio with delta modulation

The calculation of the signal-to-quantizing noise ratio ξ_q in ΔM is clearly more complex than with PCM and depends in greater measure on the statistics of the primary signal. Nevertheless, a rough evaluation can be made with the following simplifying hypotheses:

- primary signal with limited spectrum between 0 and $f_{1\,max}$
- no slope overload
- the quantizing distortion u_q is considered to be a random square signal of frequency f_e, of maximum amplitude $\pm \Delta$ and with a probability density assumed to be uniform between these extreme values, i.e.,

$$p(u_q) = \frac{1}{2\Delta} \tag{7.60}$$

- as with PCM (sect. 7.3.3), it is assumed that the spectrum of u_q is uniform at least up to f_e, which has been confirmed by measurements and simulations

The total power P_q (on a resistance R) of the quantizing distortion is thus

$$P_q = \frac{1}{R} \int_{-\Delta}^{+\Delta} u_q^2 \cdot p(u_q) \cdot du_q = \frac{1}{2R\Delta} \cdot \frac{2\Delta^3}{3} = \frac{1}{R} \frac{\Delta^2}{3} \tag{7.61}$$

DIGITAL MODULATION

Here, contrary to the case of PCM, because $f_e \gg f_{1\,max}$, only one part P'_q of this power really interferes with the demodulated baseband signal (low-pass filtering at $f_{1\,max}$). It can be estimated by

$$P'_q \cong \frac{f_{1max}}{f_e} \cdot P_q = \frac{1}{R} \cdot \frac{f_{1max}}{f_e} \cdot \frac{\Delta^2}{3} \qquad (7.62)$$

The signal-to-quantizing noise ratio ξ_q still depends on the primary signal statistics. In the particular case of a sinusoidal primary signal $u_1(t) = U_1 \sin 2\pi f_1 t$, of which the power is P_S, the ratio ξ_q becomes

$$\xi_q = \frac{P_S}{P_q} \cong \frac{3}{2} \cdot \frac{U_1^2}{\Delta^2} \cdot \frac{f_e}{f_{1max}} \qquad (7.63)$$

ξ_q increases with the square of the amplitude U_1 (fig. 7.32) and attains a maximum at the slope overload limit according to (7.59). This maximum depends on the frequency f_1 (fig. 7.33), and in effect in this case (7.63) becomes

$$\xi_{qmax} \cong \frac{3}{2} \frac{U_{1max}^2}{\Delta^2} \frac{f_e}{f_{1max}} = \frac{3}{8\pi^2} \left(\frac{f_e}{f_{1max}}\right)^3 \left(\frac{f_{1max}}{f_1}\right)^2 \qquad (7.64)$$

Beyond this limit, ξ_q decreases very rapidly due to the strong distortion introduced by the slope overload. Figure 7.34 illustrates the form of ξ_q in the case of a signal with uniform spectral density limited to $f_{1\,max}$.

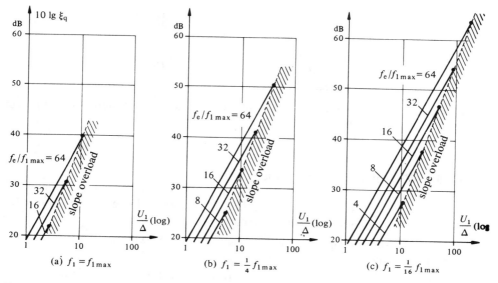

Fig. 7.32 Signal-to-quantizing noise ratio in delta modulation (primary sinusoidal signal of frequency f_1).

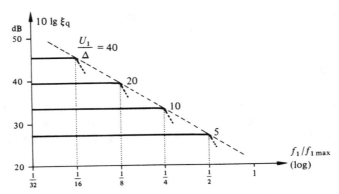

Fig. 7.33 Variation of the ratio ξ_q as a function of the frequency (effect of slope overload). Case in which $f_e = 16 f_{1\,\text{max}}$.

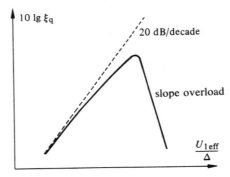

Fig. 7.34 Effect of slope overload on the ratio ξ_q (primary signal with uniform spectral density, limited to $f_{1\,\text{max}}$).

7.8.6 Differential PCM modulation

Differential PCM modulation, or DPCM, in principle very similar to delta modulation, is characterized in its simplest form by

- a *zero-order extrapolation* according to (7.53), as with ΔM
- quantizing of $u_1 - v_1^*$ at *q intervals* (in ΔM, $q = 2$).

The result is a decision rate $\dot{D}$ according to (7.31). Because of finer quantizing of the differences $u_1 - v_1^*$, the sampling frequency f_e can be chosen lower than with delta modulation. On the other hand, nonuniform quantizing

DIGITAL MODULATION

permits us to take into account the probabilty density of the difference $u_1 - v_1^*$ in the case of a particular signal u_1.

Consequently, the performance of a DPCM modulation is greatly dependent on the statistical properties of the signal, which render comparison with other digital modulations difficult.

7.8.7 Delta-sigma modulation

To compensate for the unfavorable effect of differential modulation on high frequencies, (e.g., figs. 7.31 and 7.33) we could imagine multiplying the spectrum of the primary signal, which is assumed to be uniform, by $1/f$ before proceeding to ΔM modulation, and then re-establishing the initial spectrum by multiplying by f at demodulation. This comes down to integrating the analog signal $u_1(t)$ at emission and deriving it with respect to time at reception, or, if the signal is already sampled, effecting the algebraic sum of these samples $u_1(iT_e)$ at the input of the ΔM modulator. The modulation thus defined is called *delta-sigma modulation* $\Delta\Sigma M$. Three possible and equivalent forms of implementation are given in figure 7.35. The two cumulative sums, which are identical, from figure 7.35(b), can be combined into one after the subtraction element (fig. 7.35(c)).

At the output of the decoder, we directly obtain a signal u_1^* quantized at q intervals, which corresponds to the primary signal $u_1(t)$, but has a very rich spectrum at high frequencies, especially if q is low ($\Delta\Sigma M$ is most often used with $q = 2$). Low-pass filtering is thus indispensable for extracting a replica $u_1'(t)$ of the primary signal.

In its simplest form, $\Delta\Sigma M$ uses quantizing with one bit ($q = 2$) and a predictor of zero order, according to a law similar to (7.53). Under these condtions, a sinusoidal signal modulated in $\Delta\Sigma M$ takes the form illustrated in figure 7.36.

Slope overload, typical of ΔM modulation, in $\Delta\Sigma M$ concerns the sum of the signal samples and not the samples themselves. For a sinusoidal signal $u_1(t) = U_1 \sin 2\pi f_1 t$, the variation of this sum is maximum when $u_1(t) = \pm U_1$. Its slope is thus equal to $U_1 f_e$, independent of f_1. The limiting condition for avoiding slope overload thus becomes

$$U_1 f_e \leq f_e \Delta \tag{7.65}$$

It is, therefore, necessary to choose the quantizing step Δ to be at least equal to the maximum amplitude of the primary signal, whatever the frequency f_1. The power spectral density of the quantizing distortion in $\Delta\Sigma M$ cannot be assumed to be uniform as with ΔM, but increases proportionally to f^2 because

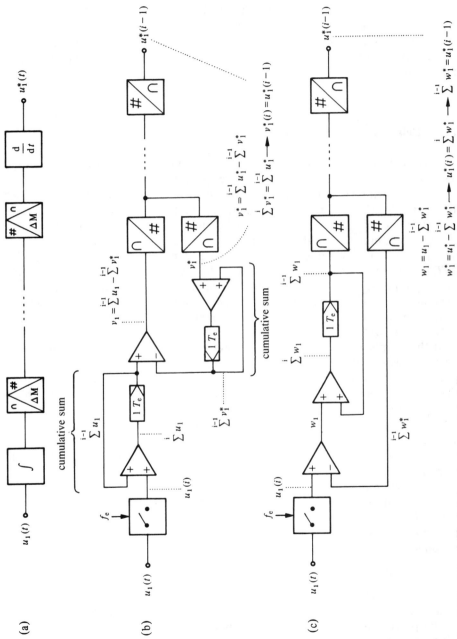

Fig. 7.35 Principle of delta-sigma modulation.

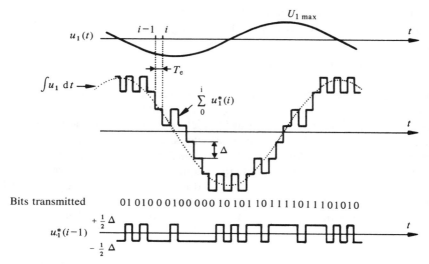

Fig. 7.36 $\Delta\Sigma$M (sinusoidal primary signal).

of the multiplication by f of the original signal spectrum (derivation with respect to time) after ΔM demodulation.

The great advantage of $\Delta\Sigma$M with respect to ΔM is that its signal-to-quantizing noise ratio ξ_q, proportional as in (7.64) at f_e^3, does not depend on the frequency of the primary signal.

7.9 ADAPTIVE DIGITAL MODULATION

7.9.1 Adaptive companding

The nonuniform quantizing presented in section 7.4 makes use of *instantaneous* companding, in the sense that the quantizing step Δ takes different but fixed values, within the range $-U_{1\,max}\ldots+U_{1\,max}$ and that the PCM code word depends only on the sample considered. Each PCM word contains its own compression information (segment identity) in the three bits following the sign bit (sect. 7.5.10).

In *adaptive companding quantizing*, the variable quantizing step Δ is derived from the *preceding samples* according to an appropriate algorithm. This idea is equally applicable to PCM modulation as to differential digital modulations such as ΔM and DPCM, in order to improve their performance for known signals.

7.9.2 Adaptive delta modulation

One way of avoiding the slope overload characteristic of ΔM (fig. 7.30) consists of progressively increasing the quantizing step Δ of the decoder when a series of identical bits (e.g., 111 or 000) lets us assume that the reconstructed signal $u_1^*(t)$ scarcely follows the primary signal $u_1(t)$. According to the number of preceding bits considered and the law of Δ variation as a function of the different patterns of these bits, several adaptation algorithms are possible. Not all are stable, i.e., the quantizing step Δ does not always converge toward its minimum value Δ_0 when the signal $u_1(t)$ stops varying. The step response (fig. 7.37) is a good criterion of stability for such an *adaptive delta modulation* AΔM.

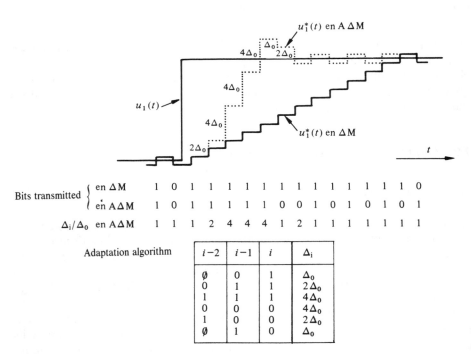

Fig. 7.37 Example of the step response compared for ΔM and AΔM.

7.10 COMPARISON AND APPLICATIONS OF DIGITAL MODULATIONS

7.10.1 Comparison criteria

For a given primary signal (specified by its power spectral density and the statistical distribution of its amplitudes) we compare

- the bit rate $\dot{D}$ necessary for similar transmission quality
- conversely, the quality obtained for an identical bit rate
- the sensitivity to noise in the course of transmission (regeneration errors)
- the complexity of the modulation equipment

These criteria bring up the following remarks:

- the notion of transmission quality, particularly in this case, is largely subjective. In effect, the nature of quantizing distortion can be very different depending on the type of modulation (quantizing noise, clipping, slope overload, granular noise, *et cetera*) and the only estimate of the signal-to-quantizing noise ratio is insufficient to account for the effect perceived by the receiver;
- the comparison is even more limited to a given type of signal or information the more the modulation is oriented toward redundancy reduction.

7.10.2 Comparison between PCM, ΔM, and DPCM for telephony

On the basis of relations (7.22) and (7.31) and with a sampling frequency f_e fixed at 8 kHz, it has been established that the quantizing signal-to noise ratio in PCM increases by 6 dB when the number of intervals q is doubled, i.e., when the bit rate $\dot{D}$ is increased by 8 kbit/s.

In ΔM, the best ξ_q that can be obtained, which is right at the limit of the slope overload, according to (7.64), depends on $f_e^3 = \dot{D}^3$. Doubling $\dot{D}$ increases $\xi_{q\,max}$ by 9 dB.

DPCM modulation is a compromise between the simplicity of ΔM and the generality of PCM modulation. For certain signals, its performance is superior to that of the other two modulations. For example, for telephone signals (fig. 7.38), DPCM approaches ΔM while q approaches 2, but tests and simulations have shown that it gives a ξ_q that is consistently better than PCM.

The case of adaptive modulation allows still less objective comparison. In any case, on the level of subjective quality, excellent results can be obtained with rates that are significantly lower than the usual 64 kbit/s in PCM.

The effect of regeneration errors on PCM transmission is analyzed in section 7.7. The inconvenience of differential and adaptive modulations is that they

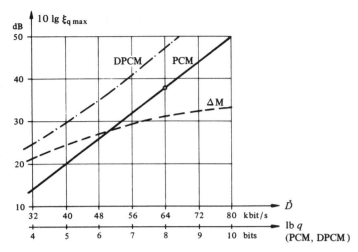

Fig. 7.38 Objective comparison of the maximum signal-to-quantizing noise ratios (telephone signals).

are more sensitive to these errors, which have effects that propagate to the following samples.

7.10.3 Applications of digital modulation

Due to direct quantizing of samples, each independently of the others, PCM modulation is the most universal. Its use is pervasive and generalized in the public telecommunications network (chapter 9), as much for balanced pairs, coaxial cables or optical fibers, as for radio transmissions (microwave links and satellite links). It also has numerous applications outside the domain of telecommunications (measurement techniques, recording, signal processing, *et cetera*).

ΔM modulation, attractive because of the extreme simplicity of its modulators and demodulators and as a result of its word format reduced to a single bit, can only compete qualitatively with PCM at the same rate $\dot{D}$ for rates lower than approximately 50 kbit/s (fig. 7.38). It is used in telephony in special applications where we try for a reduced rate without requiring high quality (military transmissions). The possibility of digitally converting ΔM modulation into PCM and *vice versa* has given rise to proposals for its use in the extreme periphery of the future all-digital network (subscriber lines, local network).

Numerous more elaborate forms of differential digital modulation (DPCM, AΔM, *et cetera*) allow performance to be significantly improved, though at the cost of more complex modulation equipment.

$\Delta\Sigma M$ modulation is sometimes used for telemetry by virtue of its 1-bit format and its transparency to continuous (dc) voltages (in contrast to ΔM). $\Delta\Sigma M$ can also be converted to PCM, even more easily than ΔM.

The transmission of images readily uses differential digital modulation, or even adaptive modulation, to take advantage of, on the one hand, the uniform ranges often presented in an image. On the other hand, in the case of moving images, we would wish to exploit the very frequently nearly complete similarity of two consecutive images. Diverse forms of DPCM modulation are envisaged for digital television and visiophony, likewise for fixed black-and-white images (facsimile, telecopying). These techniques are strictly speaking more closely related to coding (redundancy reduction) than to modulation.

Finally, we must note that the form in which digital modulations allow representation of analog information lends itself particularly well to all sorts of purely digital processing, such as, for example:

- ciphering and deciphering by combination of coded samples with a secret key
- redundancy reduction (source coding)
- adaptation to a highly noisy channel and recognition of a signal drowned in noise (channel coding, notably indispensable for space communications)
- switching, i.e., sorting and selective routing of coded samples belonging to different communications (time division multiplex)
- short-term storage of coded samples (delay, change of time slot)
- long-term semi-permanent storage (e.g., automatic spoken response)
- time division multiplexing
- digital filtering
- transcoding

Chapter 8

Analog Modulation

8.1 SINUSOIDAL CARRIER MODULATIONS

8.1.1 Types and notations

Analog modulations, such as those presented overall in section 4.2.5, are all based on an auxiliary periodic signal, called the *carrier*. For an important class of signals, the carrier is sinusoidal, and takes the following form

$$u_p(t) = U_p \cos \omega_p t = U_p \cos 2\pi f_p t = \text{Re}[\underline{u}_p(t)] \tag{8.1}$$

or, in complex notation (instantaneous complex value, from sect. 8.3.2)

$$\underline{u}_p(t) = U_p \exp(j\omega_p t) \tag{8.2}$$

The secondary signal $u_2(t)$, resulting from the modulation of the carrier by the primary signal $u_1(t)$, is not sinusoidal. Nevertheless, it can generally be written

$$u_2(t) = U_2(t) \cos [\varphi_2(t)] \tag{8.3}$$

or, in complex notation

$$\underline{u}_2(t) = U_2(t) \exp [j\varphi_2(t)] \tag{8.4}$$

The quantity $\varphi_2(t)$ is called the *instantaneous phase* of $u_2(t)$. Its derivative with respect to the time defines the *instantaneous angular frequency* $\omega_2(t)$

$$\omega_2(t) = 2\pi f_2(t) = \frac{d\varphi_2(t)}{dt} \tag{8.5}$$

The different types of modulation consist of varying one of the parameters of $u_2(t)$ proportionally to the signal $u_1(t)$, according to table 8.1. These varia-

Table 8.1 Types of analog modulation

Type	Symbol	$U_2(t)$	$f_2(t)$	$\varphi_2(t)$	law
Amplitude modulation	AM	$U_p + \Delta U(t)$	f_p	$\omega_p t$	$\Delta U(t) \sim u_1(t)$
Frequency modulation	FM	U_p	$f_p + \Delta f(t)$	$\omega_p t + \Delta \varphi(t)$	$\Delta f(t) \sim u_1(t)$
Phase modulation	ΦM	U_p	$f_p + \Delta f(t)$	$\omega_p t + \Delta \varphi(t)$	$\Delta \varphi(t) \sim u_1(t)$

tions $\Delta U(t)$, $\Delta f(t)$ or $\Delta\varphi(t)$ are called *instantneous deviations* of amplitude, frequency or phase, respectively.

Deviation ΔU, Δf or $\Delta\varphi$ designates the maximum value of the instantaneous deviation.

8.1.2 Phasor representation

The complex notation (8.2) or (8.4) lends itself to graphic representation in the form of *phasors* $\underline{U}$ or complex rotating vectors (vol. I, sect. 8.3.3). It must be specified that here *the phasors are represented as a peak value and not as an rms value*.

In diagrams of phasors of analog modulations with sinusoidal carriers, only the *relative angular speed* of the phasors, with respect to the angular speed ω_p of the carrier phasor is represented. It is therefore necessary to imagine the entire diagram rotating at speed ω_p. This general rotation will be specified in the diagrams by a white curvilinear arrow.

In AM modulation, the phasor $\underline{U}$ remains colinear with $\underline{U}_p$, whereas in angular modulation (FM or ΦM), it keeps the same modulus U_p but its phase varies (fig. 8.2).

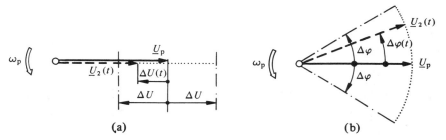

Fig. 8.2 Representation of analog modulations by phasors: (a) AM amplitude modulation; (b) angular modulation FM or ΦM.

8.2 AMPLITUDE MODULATION (AM)

8.2.1 Principles

In *amplitude modulation*, the amplitude of the carrier is a linear function of the primary signal $u_1(t)$. The expresssion of the secondary signal is thus

$$u_2(t) = [U_p + \Delta U(t)] \cos \varphi_p t \tag{8.6}$$

with the instantaneous amplitude deviation

$$\Delta U(t) = \alpha u_1(t) \tag{8.7}$$

where α is a constant factor which depends on the modulator.

The signal $U_p + \Delta U(t)$ constitutes the *envelope* of the modulated signal (fig. 8.3).

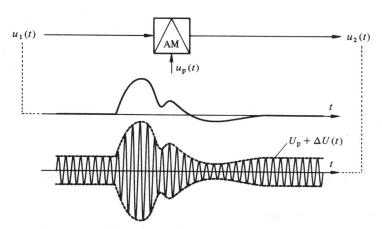

Fig. 8.3 Amplitude modulation.

8.2.2 Spectrum of the AM modulated signal

With (8.1), relation (8.6) can be written

$$u_2(t) = u_p(t) + \Delta U(t) \frac{u_p(t)}{U_p} \tag{8.8}$$

As a product in the time domain is transformed into a convolutional product in the frequency domain, the Fourier transform of (8.8) is

$$U_2(f) = U_p(f) + \Delta U(f) * \frac{U_p(f)}{U_p} \tag{8.9}$$

with

$$\Delta U(f) = \alpha U_1(f) \tag{8.10}$$

This operation is expressed graphically in figure 8.4.

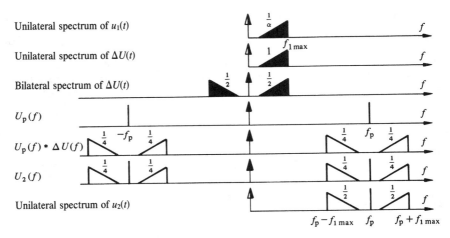

Fig. 8.4 Construction of the spectrum of the secondary signal in AM by convolution.

Consequently, the (unilateral) spectrum of the modulated signal is composed of

- a line U_p at the frequency f_p of the carrier
- two **sidebands**, one in a direct position, the other in reverse position, on both sides of f_p, and each corresponding to half of the (unilateral) spectrum of $\Delta U(t)$. The latter is itself similar (except for the factor α) to the spectrum of the primary (baseband) signal.

It occupies a bandwidth double that of the primary signal.

$$B_{2AM} = 2B_1 \tag{8.11}$$

8.2.3 Modulation factor

When the primary signal is *sinusoidal* $u_1(t) = U_1 \cos \omega_1 t$, the secondary signal from (8.6) becomes

$$u_2(t) = U_p \left(1 + \frac{\Delta U}{U_p} \cos \omega_1 t\right) \cos \omega_p t \tag{8.12}$$

The ratio of the (maximum) amplitude deviation ΔU to the amplitude U_p of the unmodulated carrier is called the *modulation factor m*

$$m = \frac{\Delta U}{U_p} \tag{8.13}$$

When the the modulation factor of an AM transmission is given, it is always understood to mean the value of $m = m_{nom}$ for the nominal signal, in the sense of section 2.4.14.

8.2.4 Spectrum of the AM signal modulated by a sinusoidal signal

In the case of a *primary sinusoidal signal* of frequency f_1, the two lateral bands are reduced to two sidelines at frequencies $f_p \pm f_1$ and with an identical amplitude equal to $\Delta U/2$, i.e., with (8.13), $mU_p/2$. The phasor $\underline{U}_2$ representing $u_2(t)$ is decomposed into three phasors: that of the carrier and those corresponding to the two sidelines, each rotating in one direction, at the relative speed $\pm\omega_1$ (fig. 8.5).

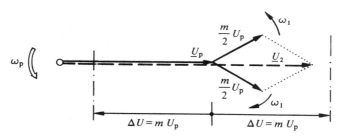

Fig. 8.5 AM phasor diagram.

If P_p is the power of the unmodulated carrier (of amplitude U_p), the power P_{2S} of the secondary signal is thus

$$P_{2S} = P_p + \frac{m^2}{4} P_p + \frac{m^2}{4} P_p = P_p \left(1 + \frac{m^2}{2}\right) \tag{8.14}$$

Let us note that the power P_{2S} depends on the modulation factor m. The power P_p of the carrier contains no information but is useful for demodulation. Thus while $m < 1$ (usual case), only less than a third of the secondary power is found in the sidebands.

8.2.5 Demodulation by envelope detection

AM modulation presents the advantage of allowing a very simple demodulation by *envelope detection*, i.e., by rectification (half- or full-wave) of the secondary signal, and then filtering (fig. 8.6). However, this procedure is only

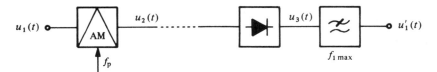

Fig. 8.6 Demodulation by envelope detection.

applicable if the envelope $U_p + \Delta U(t)$ is always positive. This is the case if $|\Delta U(t)| < U_p$, i.e., in sinusoidal modulation, if $m < 1$.

In the case of a full-wave rectification, the rectified signal is

$$u_3(t) = \sqrt{u_2^2(t)} = |u_2(t)| \qquad (8.15)$$

With (8.6), it becomes

$$u_3(t) = \sqrt{[U_p + \Delta U(t)]^2}\sqrt{\cos^2\omega_p t} \qquad (8.16)$$

If $U_p > |\Delta U(t)|$, i.e., if the envelope is always positive, (8.16) can be written

$$u_3(t) = [U_p + \Delta U(t)]\sqrt{\cos^2\omega_p t} \qquad (8.17)$$

The development in series of $\sqrt{\cos^2\omega_p t}$ is

$$\sqrt{\cos^2\omega_p t} = \frac{4}{\pi}\left(\frac{1}{2} + \frac{1}{3}\cos 2\omega_p t - \frac{1}{15}\cos 4\omega_p t + \frac{1}{35}\cos 6\omega_p t ...\right) \qquad (8.18)$$

The spectrum of $U_3(t)$ is the result of the convolution of the Fourier transform of the two terms of product (8.17), and is thus composed of lines at frequencies $0, 2f_p, 4f_p, 6f_p...$, and of sidebands on each side of these frequencies (fig. 8.7). A low-pass filtering at $f_{1\ max}$ allows the isolation of the baseband and the reconstruction of a signal $u_1'(t)$ similar to $\Delta U(t)$ except for the dc component, thus similar to the primary signal $u_1(t)$.

If a half-wave rectification is used, a supplementary line appears at f_p with two sidebands. Filtering is then found to be somewhat complicated.

If the envelope does not always remain positive, i.e., if $m > 1$, demodulation by detection restores a signal which is not similar to the envelope, but to its absolute value. It is therefore necessary to use another demodulation procedure (sect. 8.3.3).

8.2.6 Applications of AM modulation

AM modulation is typically that of long-, medium-, and short-wave radio broadcasting transmitters. Its characteristics are highly suitable for this type of application, with the following advantages:

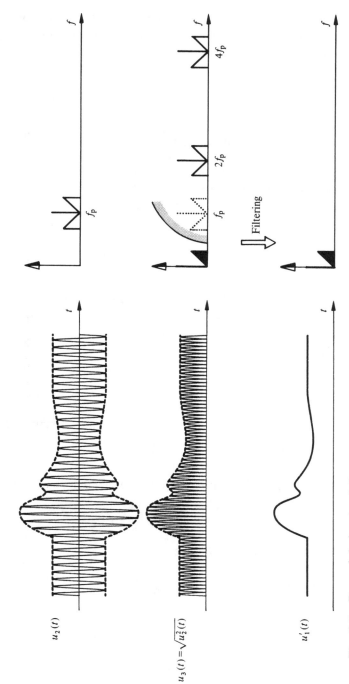

Fig. 8.7 Demodulation by full-wave rectification (stippled area: supplementary components if half-wave rectification is used).

- extreme simplicity of the detection demodulator, at a point such that reception has been historically possible even without electronic amplification methods (lead detector!).
- moderate bandwidth.

8.2.7 Envelope delay

In the course of transmission, the channel introduces a phase shift which depends on the frequency $b(\omega)$. If $\omega_1 \ll \omega_p$, the phase shift of the two sidebands with respect to the carrier is of the same value $\omega_1 db/d\omega$ and of opposite sign. The result is a modification of the phasor diagram (fig. 8.8).

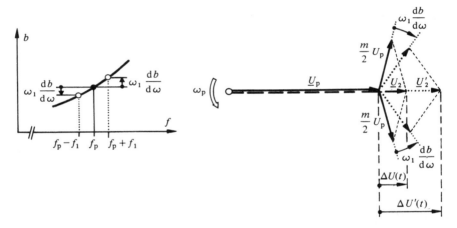

Fig. 8.8 Effect of channel phase change.

From figure 8.8 proceeds the amplitude variation $\Delta U(t)$ before transmission

$$\Delta U(t) = \frac{m}{2} U_p [\cos(+\omega_1 t) + \cos(-\omega_1 t)] = mU_p \cos \omega_1 t \tag{8.19}$$

and its value $\Delta U'(t)$ after phase change

$$\Delta U'(t) = \frac{m}{2} U_p \left[\cos\left(\omega_1 t - \frac{db}{d\omega}\omega_1\right) + \cos\left(-\omega_1 t + \frac{db}{d\omega}\omega_1\right) \right] \tag{8.20}$$

which can be written

$$\Delta U'(t) = mU_p \cos\left[\omega_1 \left(t - \frac{db}{d\omega}\right)\right] \tag{8.21}$$

With respect to the primary signal $U_1 \cos \omega_1 t$, the signal $\Delta U'(t)$, extracted from $u'_2(t)$ by envelope detection, is thus *delayed* by a time equal to $db/d\omega$ and called the *group delay* t_g. If the phase shift b is not proportional to the angular frequency ω, the propagation time ω of the envelope $t_g = db/d\omega$ if different than that of the carrier (phase propagation time $t_\varphi = b/\omega$). If, furthermore, $b(\omega)$ is not linear with respect to ω, t_g is not constant and the envelope undergoes distortions.

8.2.8 Effect of linear AM distortions

If the transmission channel presents linear distortions (sect. 2.4.7), the spectral components of $u_2(t)$ undergo different attenuations and group delays. The consequences are visible in the phasor diagram after transmission (phasor $\underline{U}'_2$) in the case of a sinusoidal modulation (fig. 8.9). The attenuation distortions have the effect of transforming the locus of the phasor extremity $\underline{U}'_2$ into an ellipse, and the phase distortions orient this location obliquely with respect to $\underline{U}_p$.

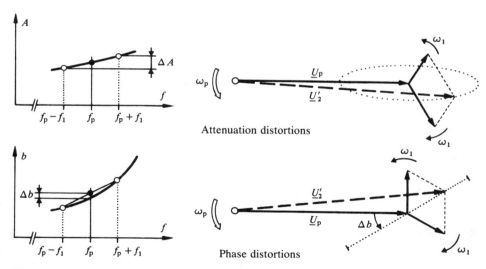

Fig. 8.9 Effect of AM linear distortions.

It has been established that the received secondary signal $u'_2(t)$ is modulated not only in amplitude but also in phase. Demodulation by envelope detection, which gives the modulus of the phasor $\underline{U}'_2$, thus restores a signal $u'_1(t)$ which is not strictly similar to $u_1(t)$ but which is affected by **nonlinear distortions**.

8.2.9 Effect of nonlinear distortions in AM. Cross modulation

Nonlinearities in the transmission channel give rise to intermodulation products (sect. 2.4.10). Those of the third order inevitably fall back into the original secondary band and after demodulation produce nonlinear and nonharmonic distortions in the baseband.

A major extra inconvenience appears when several AM transmissions share the same channel with different carriers (frequency division multiplex): certain third-order intermodulation products of the type $f_x - f_y + f_z$ between a carrier (f_x), its sidelines (f_y) and a second carrier (f_z), simulate a modulation of the second carrier with the primary signal of the first. The consequence is an extremely undesirable *intelligible crosstalk* (fig. 8.10). This phenomenon is called *cross modulation*. It is known in audio broadcasting by the name of the "Luxembourg effect". In this case, poorly elucidated nonlinearities of the ionosphere are at the origin of a transposition of the modulation of a particularly powerful emitter (the first with which this phenomenon was observed is that of Radio-Luxembourg at long-waves) toward other lower-power emitters.

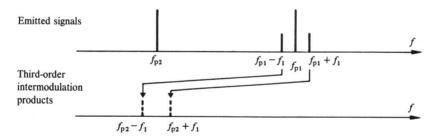

Fig. 8.10 AM cross modulation.

8.3 SUPPRESSED CARRIER AMPLITUDE MODULATION (DSBSC)

8.3.1 Principles

The carrier $u_p(t)$ is an auxiliary signal, necessary for AM modulation, but which is apparently not indispensable for transmission. The idea of *double sideband suppressed carrier DBSC modulation* consists of transmitting only the sidebands and reconstituting the carrier at reception.

The secondary signal $u_2(t)$ can thus be considered to be the following product:

$$u_2(t) = \Delta U(t) \cos \omega_p t \tag{8.22}$$

where $\Delta U(t)$ represents the envelope of the modulated signal, proportional, as in (8.7), to the primary signal. Contrary to AM modulation, $u_2(t)$ is zero if $u_1(t)$ is zero (fig. 8.11).

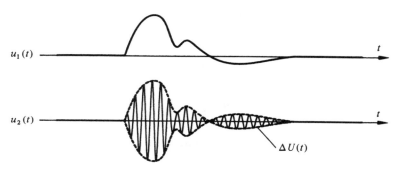

Fig. 8.11 DSBSC modulation.

Relation (8.22) shows that DSBSC modulation consists of effecting the product of a signal $\Delta U(t)$, proportional to the primary signal $u_1(t)$, with a normalized signal $u_p(t)/U_p$. This is why the DSBSC modulator is often called a *signal multiplier* (vol. VI, section 8.3).

8.3.2 Spectrum of the DSBSC modulated signal

The spectrum of $u_2(t)$ is the result of the convolution of $\Delta U(f)$ with the Fourier transform of $\cos \omega_p t$. With respect to the AM spectrum of figure 8.4, the line at f_p has disappeared (fig. 8.12).

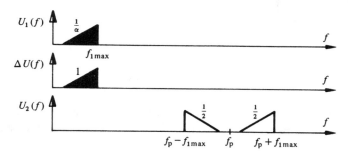

Fig. 8.12 Spectrum of the DSBSC secondary signal.

The secondary bandwidth B_2 remains equal to double the primary bandwidth B_1 from (8.11).

In sinusoidal modulation, the phasor diagram is reduced to two phasors representing the two sidelines at $f_p \pm f_1$ (fig. 8.13). The modulation factor no longer is applicable here. The amplitude of these lines is, as with AM, $\Delta U/2$, i.e., half of the envelope amplitude.

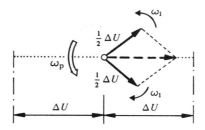

Fig. 8.13 DSBSC phasor diagram.

8.3.3 Demodulation by inverse modulation

The principle of demodulation by envelope detection according to section 8.2.5 is not applicable in DSBSC, as the result would not be the envelope signal $\Delta U(t)$, but its absolute value. The solution consists of submitting the secondary signal $u_2(t)$ to a new DSBSC modulation with a carrier $u'_p(t)$ restored in the receiver (fig. 8.14).

$$u'_p(t) = U_p \cos(\omega_p t + \varphi) \tag{8.23}$$

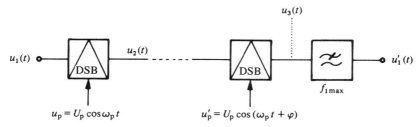

Fig. 8.14 Demodulation by inverse modulation.

φ expresses the phase error at the restoration of the carrier. If the frequency f'_p of $u'_p(t)$ is not identical to f_p, φ also depends on the time according to the relation

$$\varphi(t) = (\omega'_p - \omega_p)t \tag{8.24}$$

The signal $u_3(t)$ obtained after this second modulation is equal to

$$u_3(t) = u_2(t) \cos(\omega_p t + \varphi) = \Delta U(t) \cos \omega_p t \cos(\omega_p t + \varphi) \quad (8.25)$$

according to (8.22).

By breaking down the trigonometric product into a sum, we obtain

$$u_3(t) = \tfrac{1}{2}\Delta U(t)[\cos \varphi + \cos(2\omega_p t + \varphi)] \quad (8.26)$$

The second term of the sum gives rise to a DSBSC modulation of a carrier at $2f_p$ with the signal $\Delta U(f)/2$. These components can be eliminated by low-pass filtering in order to keep only the baseband signal (fig. 8.15).

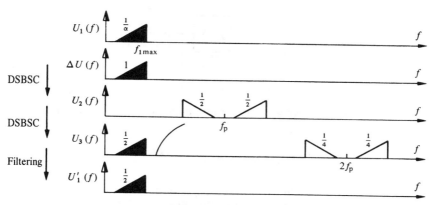

Fig. 8.15 Representation of inverse modulation in frequency domain (for $\varphi = 0$).

We therefore obtain

$$u_1'(t) = \tfrac{1}{2} \cos \varphi \cdot \Delta U(t) = \tfrac{1}{2}\alpha \cos \varphi \cdot U_1 \cos \omega_1 t \quad (8.27)$$

The consequences of this relation are significant:

- the phase error φ has repercussions for the amplitude of the demodulated signal. If $\varphi = \pi/2$, then $u_1'(t)$ is identically zero!
- in order for the demodulation to be optimal, it is necessary that the carrier be restored exactly *in frequency and in phase* ($\varphi = 0$). The demodulation would thus be *isochronous* or coherent;
- if the frequency f_p' of the restored carrier is slightly different from f_p, the amplitude of $u_1'(t)$ fluctuates between $\Delta U/2$ and 0 with a frequency equal to the difference $f_p - f_p'$, in agreement with relation (8.24).

Demodulation by inverse modulation is also practical in AM, particularly when $m > 1$. The isochronous restoration of the carrier then presents no difficulty, because the carrier is transmitted with the signal. For example, a phase-locked loop PLL allows the operation of this restoration (fig. 8.16).

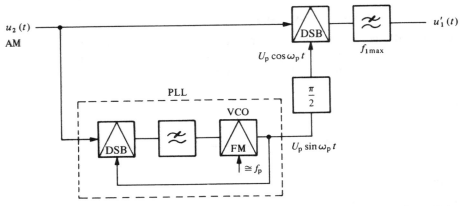

Fig. 8.16 Isochronous extraction of the carrier at reception (AM demodulation by inverse modulation).

8.3.4 Applications of DSBSC modulation

The necessity for isochronous reconstruction of the carrier at reception makes DSBSC modulation unsuitable for direct application in transmission. Nevertheless, it constitutes an indispensable intermediate stage in the implementation of single sideband modulation.

8.4 SINGLE SIDEBAND MODULATION (SSB)

8.4.1 Principles

The two sidebands of the spectrum of an AM or DSBSC modulation in fact carry the same information—that of the primary signal. One can therefore envisage transmitting only one. The secondary bandwidth B_2 is decreased by half and becomes equal to that of the primary signal (B_1).

$$B_{2\text{SSB}} = B_1 \tag{8.28}$$

Furthermore, if the carrier is not transmitted, the transmitted power is optimized in the sense that it corresponds only to the useful signal contained in one sideband. This procedure is called *single sideband modulation SSB*.

The most immediate and widely-used implementation of an SSB modulation consists of conserving one of the two sidebands at the output of a DSBSC modulator and eliminating the other by a filter. In principle, a low-pass filter (or respectively a high-pass filter) would suffice to isolate the lower sideband (or respectively the higher sideband). For reasons which will be specified in the context of carrier systems (section 10.1), it is generally a band-pass filter which performs this function (fig. 8.17). This filter must be even more selective when the primary signal contains lower frequencies.

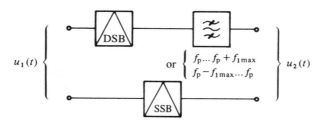

Fig. 8.17 SSB modulator.

Other principles of SSB modulation have been proposed, to avoid the use of a very selective filter (Hartley's modulator [27], Weaver's modulator [28]). These modulators eliminate one of the sidebands by the addition of two DSBSC signals which have sidebands in opposite phases. Although they are interesting from a theoretical viewpoint, their practical implementation is even more delicate than the principle of figure 8.17.

8.4.2 Spectrum of the SSB modulated signal

Figures 8.12 and 8.17 immediately give rise to the secondary signal spectrum with SSB (fig. 8.18).

It has been established that the unilateral spectrum of the primary signal $U_1(f)$ is translated by f_p (upper sideband) and, the case permitting, reversed by symmetry around f_p (lower sideband), hereby giving the spectrum $U_2(f)$ of the secondary signal.

$$U_2(f) = \alpha U_1(f - f_p) \text{ (upper sideband)} \tag{8.29}$$

$$U_2(f) = \alpha U_1(f_p - f) \text{ (lower sideband)} \tag{8.30}$$

In the particular case of a primary sinusoidal signal $U_1 \cos \omega_1 t$, the (unilateral) spectrum of the secondary signal has a single line at $f_p + f_1$ (or $f_p - f_1$). The modulated signal $u_2(t)$ is thus also sinusoidal, but of different frequency.

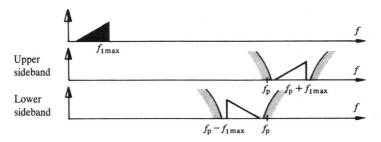

Fig. 8.18 Spectrum of the SSB secondary signal.

The phasor diagram is reduced to a single phasor rotating at angular speed $\omega_p + \omega_1$ or $\omega_p - \omega_1$.

Thus, although SSB modulation is derived from DSBSC modulation, it is not a matter of amplitude modulation (the signal envelope has no resemblance to the primary signal) but rather a *frequency translation* i.e., spectrum translation.

8.4.3 Demodulation of an SSB signal

As with DSBSC, demodulation requires an inverse modulation with the aid of a carrier restored with a possible phase error according to (8.23) and (8.24) (fig. 8.19).

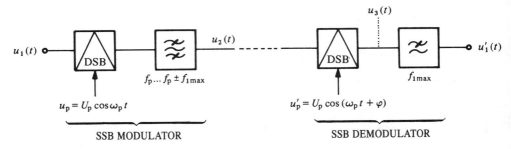

Fig. 8.19 SSB demodulation.

After the second modulation, the signal $u_3(t)$ is, according to (8.22):

$$u_3(t) = u_2(t) \cos(\omega_p t + \varphi) \qquad (8.31)$$

In the case of a primary sinusoidal signal $U_1 \cos \omega_1 t$, the secondary signal is

$$u_2(t) = \Delta U \cos[(\omega_p \pm \omega_1)t] \qquad (8.32)$$

ANALOG MODULATION

where ΔU is an amplitude proportional to U_1. The $+$ sign corresponds to the conservation of the upper sideband and the $-$ sign to that of the lower sideband. With (8.32), (8.31) becomes

$$u_3(t) = \Delta U \cos[(\omega_p \pm \omega_1)t] \cos(\omega_p t + \varphi)$$
$$= \tfrac{1}{2}\Delta U [\cos(\omega_1 t + \varphi) + \cos(2\omega_p t \pm \omega_1 t + \varphi)] \qquad (8.33)$$

A low-pass filter allows the elimination of the component at $2f_p \pm f_1$ and keeping only the baseband term $u'_1(t)$

$$u'_1(t) = \tfrac{1}{2}\Delta U \cos(\omega_1 t + \varphi) \qquad (8.34)$$

This result must be compared with that obtained by the same demodulation procedure in the case of DSBSC modulation from (8.27). It has been found that

- in DSBSC modulation, the *amplitude* of the signal $u'_1(t)$ depends directly on $\cos\varphi$
- in SSB modulation, the *phase* of $u'_1(t)$ depends on φ, but not its amplitude.

In SSB modulation, if the carrier is not exactly restored at reception, it produces the following effects:

- a frequency difference (asynchronism) $\Delta f_p = f'_p - f_p$ leads to a shift equal to Δf_p of all the frequencies contained in the primary signal. In effect, with (8.24), (8.34) becomes

$$u'_1(t) = \tfrac{1}{2}\Delta U \cos[(\omega_1 + \Delta\omega_p)t] \qquad (8.35)$$

- a phase difference (anisochronism) φ between the two carriers of the same frequency $f'_p = f_p$ (synchronous) causes a phase shift equal to φ of all the components of $u_1(t)$.

Thus, the isochronism of the restored carrier is only necessary if a conformal transmission (without phase distortion) is required.

8.4.4 Applications of SSB modulation

In *telephony*, the insensitivity of hearing to phase differences between spectral components (sect. 1.7.2) allows us to dispense with the conformity requirement and to accept an anisochronous restoration of the carrier at reception. SSB modulation is thus ideal as regards

- its secondary bandwidth $B_2 = B_1$
- the optimal useful power transmitted
- the relative simplicity of the modulation and demodulation equipments.

This is the basis of analog with frequency division multiplex transmission systems called *carrier systems* (section 10.1), which still constitute the essentials of telephone transmission systems at medium and long distances. It is also used as a preliminary stage for the construction of a frequency division multiplex of telephone channels, before a radio transmission (microwave links, satellite links) by means of another modulation.

8.4.5 Effect of linear distortions in SSB

Since SSB modulation is in fact a frequency translation, all the linear (attenuation and phase) distortions undergone by the secondary signal are found translated into the baseband after demodulation (fig. 8.20).

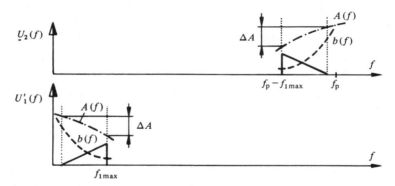

Fig. 8.20 Effect of linear distortion with SSB: frequency response in attenuation; frequency response in phase.

8.4.6 Effect of nonlinear distortions with SSB

By referring to section 2.4.10 and in particular to figure 2.12, it is apparent that the third-order intermodulation products are the cause of two types of interference of the demodulated signal (fig. 8.21):

- *self-interference* of the signal in the form of nonlinear and nonharmonic distortions (intermodulation noise);
- interference of the two neighboring channels (if frequency division multiplexing is used) due to the extensions of the intermodulation products over a bandwidth which is triple the original bandwidth. Therefore, there is *crosstalk* generation, fortunately unintelligible (equivalent to a noise) between these channels.

ANALOG MODULATION

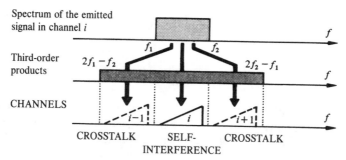

Fig. 8.21 Effect of third order intermodulation products.

8.4.7 SSB modulation with a carrier

If, in the case of a primary sinusoidal signal, we transmit the secondary signal $u_2(t)$ from (8.32) with its carrier $u_p(t)$, we obtain

$$u_2(t) = U_p \cos \omega_p t + \Delta U \cos [(\omega_p \pm \omega_1)t] \tag{8.36}$$

Under certain conditions, the signal can be demodulated by envelope detection. In effect, the envelope detection (sect. 8.2.4) extracts the primary demodulated signal $u_1'(t)$ from the envelope of $u_2(t)$, i.e., from the modulus of the phasor $\underline{U}_2$, which can be calculated by applying the cosine theorem to the phasor diagram (fig. 8.22):

$$U_2^2 = U_p^2 + (\Delta U)^2 - 2U_p \Delta U \cos(\pi - \omega_1 t) \tag{8.37}$$

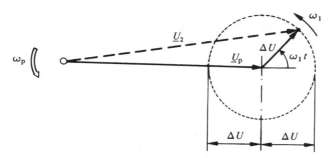

Fig. 8.22 Phasor diagram with SSB with a carrier.

From which we get the demodulated primary signal

$$u_1'(t) \sim U_2 = U_p \sqrt{1 + \left(\frac{\Delta U}{U_p}\right)^2 + 2 \frac{\Delta U}{U_p} \cos \omega_1 t} \tag{8.38}$$

If
$$\Delta U \ll U_p \tag{8.39}$$
then
$$u_1'(t) \sim U_2 \cong U_p + \Delta U \cos \omega_1 t \tag{8.40}$$

Thus, if the amplitude of the carrier is much greater than that of the single sideline, the SSB demodulation by envelope detection yields a signal $u_1'(t)$, approximately similar to the primary original signal.

8.5 VESTIGIAL SIDEBAND MODULATION (VSB)

8.5.1 The case of television transmission

The case of television signal transmission is special in the sense that
- the bandwidth of the video signal (baseband) is significant: approximately 5 MHz
- the video signal contains very low frequencies
- the transmission must be conformal (the video signal expresses by its form the nuances of an image line).

AM modulation is unfavorable because of its double secondary band. The filtering of a single sideband in SSB proves to be difficult due to the very low frequencies of video signals. On the other hand, the absence of a carrier would require an isochronous demodulation to implement the conformity. Thus, none of the modulations AM, DSBSC, SSB (even with a carrier) appears to be suitable. A compromise is necessary.

8.5.2 Principles of vestigial sideband modulation

This compromise is realized by *vestigial sideband modulation VSB* which in principle is an SSB modulation with a carrier, but in which one of the sidebands is only partially eliminated. Filtering is thus facilitated.

The block diagram of the VSB modulator is similar to figure 8.17 except that it contains an AM modulator instead of a DSBSC modulator and that from $f_p - f_0$ to $f_p + f_0$ the filter presents a slope with central symmetry around the frequency f_p, to which it attenuates the modulated AM signal by half (Nyquist filter). The result is presented in figure 8.23.

For $f_1 > f_0$, the signal is SSB modulated. However, for $f_1 < f_0$ it is equal to an AM signal in which the two sidebands are different, but which have a

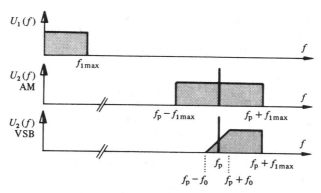

Fig. 8.23 Spectrum of a VSB-modulated signal.

constant sum, equal to ΔU. The corresponding phasor diagram shows that the locus of the phasor extremity $\underline{U}_2$ is a circle for $f_1 \geq f_0$ and ellipses of the same major axis for $f_1 < f_0$.

8.5.3 VSB demodulation

As with SSB modulation with a carrier (sect. 8.4.7), and even more when the ellipse of figure 8.24 is narrower ($f_1 \ll f_0$), envelope detection, applied to

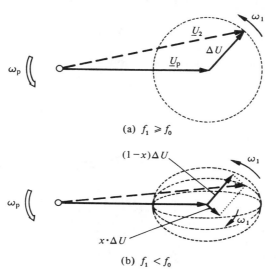

Fig. 8.24 Phasor diagram with VSB modulation.

the VSB modulated signal $u_2(t)$, restores a signal $u'_1(t)$ approximately equal to $U_p + \Delta U(t)$, i.e., similar to $u_1(t)$, as long as $\Delta U \ll U_p$. The VSB transmission is thus conformal without requiring an isochronous demodulation.

8.5.4 Applications of VSB modulation

In television, the video signal occupies in the baseband a range from almost 0 Hz (non inclusive) to 5 MHz. With VSB modulation, it can be transmitted in a secondary band of approximately 6 MHz (sect. 10.7.6). The Nyquist filter edge resulting from the assembly of the emission and reception filters extends from 1.25 MHz below to 1.25 MHz above the carrier.

Another application of VSB modulation is found in data transmission at 48 kbit/s in a band of 44 kHz corresponding to 11 analog telephone channels (section 11.4).

8.6 FREQUENCY MODULATION (FM)

8.6.1 Principles

In *frequency modulation* the amplitude U_2 of the secondary signal is constant and equal to that of the carrier U_p, but the instantaneous frequency $f_2(t)$, defined according to (8.5), varies proportionally to the primary signal $u_1(t)$ (fig. 8.25).

$$f_2(t) = f_p + \Delta f(t) \tag{8.41}$$

with the instantaneous frequency deviation

$$\Delta f(t) = \nu u_1(t) \tag{8.42}$$

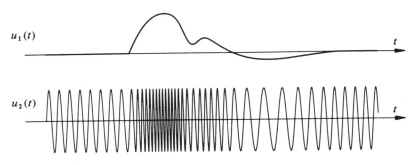

Fig. 8.25 Frequency modulation.

where v is a constant coefficient which depends on the modulator (dimension: Hz/V).

By virtue of (8.5), the instantaneous phase $\varphi_2(t)$ varies linearly with the integral of the primary signal, in effect:

$$\varphi_2(t) = 2\pi \int f_2(t)dt = \omega_p t + \Delta\varphi(t) \tag{8.43}$$

and

$$\Delta\varphi(t) = 2\pi \int \Delta f(t)dt = 2\pi v \int u_1(t)dt \tag{8.44}$$

Thus the secondary signal takes the form

$$u_2(t) = U_p \cos[\varphi_2(t)] = U_p \cos[\omega_p t + 2\pi v \int u_1(t)dt] \tag{8.45}$$

An important note comes up: contrary to the definition of AM, DSBSC, SSB, and VSB modulations which establish a linear relation between $u_1(t)$ and $u_2(t)$, relation (8.45) relates $u_2(t)$ to $u_1(t)$ by the intermediary of a trigonometric function. **It is therefore not linear.** Consequently, **the principle of superposition is not valid for FM**: the signal $u_2(t)$ modulated by the sum of two primary signals is not equal to the sum of the two secondary signals resulting from the modulation by each of these signals separately!

8.6.2 Modulation index: definition

When the primary signal is sinusoidal $u_1(t) = U_1 \cos \omega_1 t$, the instantaneous frequency of the secondary signal becomes

$$f_2(t) = f_p + \Delta f \cos \omega_1 t \tag{8.46}$$

where Δf is the *frequency deviation*, proportional to the amplitude U_1. The instantaneous frequency thus varies between the limits $f_p - \Delta f$ and $f_p + \Delta f$.

The instantaneous phase is, therefore, according to (8.43)

$$\varphi_2(t) = 2\pi \left(f_p t + \frac{\Delta f}{\omega_1} \sin \omega_1 t \right) \tag{8.47}$$

or

$$\varphi_2(t) = \omega_p t + \frac{\Delta f}{f_1} \sin \omega_1 t \tag{8.48}$$

or still

$$\varphi_2(t) = \omega_p t + \Delta\varphi \sin \omega_1 t \tag{8.49}$$

The *phase deviation* $\Delta\varphi$ is an important characteristic value of the frequency modulation by a sinusoidal signal. It is more often called the *modulation index* δ.

$$\delta = \frac{\Delta f}{f_1} = \Delta\varphi \tag{8.50}$$

The modulation index δ defined above is quite different from the modulation factor m of AM modulation, from (8.13). It varies both with the amplitude U_1 of the primary signal (since Δf is proportional to it) and with its frequency f_1.

8.6.3 Spectrum of the FM signal modulated by a sinusoidal signal

By transcribing (8.45) using the complex notation (8.4) and introducing the expression $\varphi_2(t)$ from (8.49), we obtain

$$\underline{u}_2(t) = U_p \exp[j(\omega_p t + \Delta\varphi \sin \omega_1 t)] \tag{8.51}$$

thus, with (8.2) and (8.50)

$$\underline{u}_2(t) = \underline{u}_p(t) \exp(j\delta \sin \omega_1 t) \tag{8.52}$$

It can be shown mathematically (for example, in [30]) that

$$\exp(j\delta \sin x) = \sum_{n=-\infty}^{+\infty} J_n(\delta) \exp(jnx) \tag{8.53}$$

where $J_n(\delta)$ is the *Bessel function* of the first kind, of nth order and of argument δ (fig. 8.26).

Thus, (8.52) becomes

$$\underline{u}_2(t) = U_p \sum_{n=-\infty}^{+\infty} J_n(\delta) \exp[j(\omega_p + n\omega_1)t] \tag{8.54}$$

The spectrum of the secondary FM signal modulated by a sinusoidal signal of frequency f_1 is derived from this expression. In particular, it can be concluded that
- it is a spectrum composed of lines at $f_p + nf_1$, with n being an integer between $-\infty$ and $+\infty$
- the amplitude of the line at $f_p + nf_1$ is given by the product of U_p (amplitude of the non-modulated carrier) by the value of the Bessel function of nth order and of argument equal to the modulation index δ

$$U(f_p + nf_1) = U_p|J_n(\delta)| \tag{8.55}$$

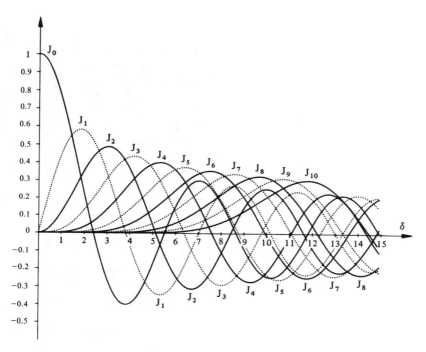

Fig. 8.26 Bessel functions of the first kind for $n \geq 0$.

Figure 8.27 represents in rough perspective the values that these lines take (for $n > 0$) with respect to U_p, the parameter of "depth" being δ.

From some properties of the Bessel functions, the following spectrum characteristics result from:

- $J_{-n}(\delta) = (-1)^n J_n(\delta)$ (n integer) \hfill (8.56)

  The two sidebands at $f_p \pm nf_1$ have the same amplitude. The *spectrum is symmetrical* with respect to f_p

- $\lim_{n \to \infty} J_n(\delta) = 0$ \hfill (8.57)

  Although the number of lines and consequently the bandwidth of the secondary signal are ***theoretically infinite***, the amplitude of the lateral lines further from f_p decreases. This decrease comes later as δ rises

- $\sum_{n=-\infty}^{+\infty} J_n^2(\delta) = 1$ \hfill (8.58)

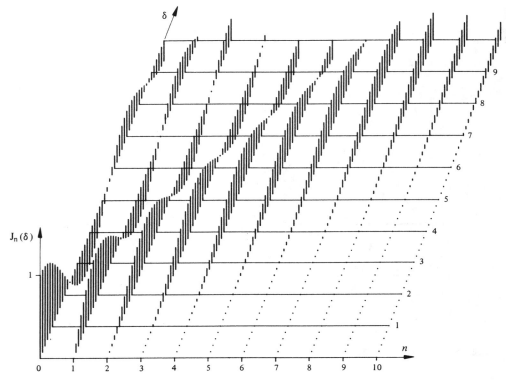

Fig. 8.27 Amplitude of the sidelines (FM with sinusoidal primary signal).

The power P_{2S} of the secondary signal, equivalent to the sum of the powers of the spectral lines, becomes with (8.55)

$$P_{2S} = \frac{1}{R}\frac{U_p^2}{2}\sum_{n=-\infty}^{+\infty} J_n^2(\delta) = \frac{1}{R}\frac{U_p^2}{2} \qquad (8.59)$$

which corresponds to the fact that $u_2(t)$ is a signal with constant amplitude equal to U_p. Its power depends neither on Δf nor on δ. However, the distribution of this power between the line at f_p and the side lines is very different according to the value of δ.

Figure 8.28 shows the form that the spectrum takes when δ is varied by modifying either the amplitude U_1 of the primary signal (and thus the frequency deviation Δf), or its frequency f_1.

Relations (8.54) and (8.56) permit the representation of the *phasor diagram* corresponding to a sinusoidal FM modulation (fig. 8.29). The geometric locus

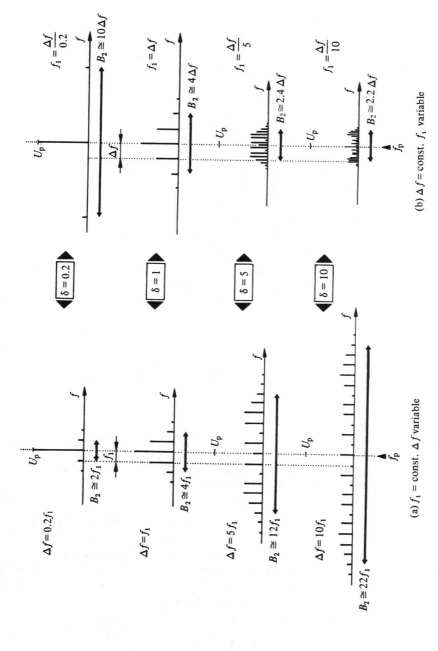

Fig. 8.28 Variations of the unilateral FM spectrum with Δf or f_1.

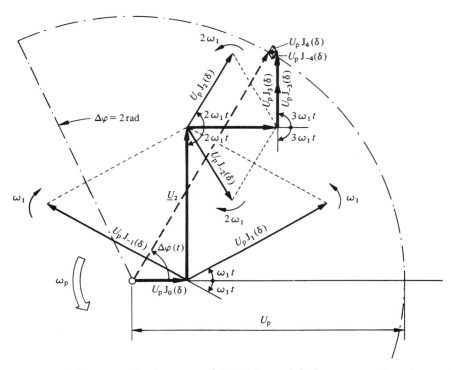

Fig. 8.29 Diagram of phasors with FM modulation, example: $\delta = 2$, $\omega_1 t = \pi/6$.

of the extremity of the phasor $\underline{U}_2$ is an arc of a circle of radius U_p and angle $2\Delta\varphi = 2\delta$. It is approached by the sum of pairs of phasors with a two-by-two resultant which is sometimes in phase (even n) and sometimes in quadrature (odd n) with $\underline{U}_p$, conforming to relation (8.56).

8.6.4 Limitation of the bandwidth. Carson's rule (sinusoidal modulation)

It is unthinkable to devote an infinite bandwidth to the transmission of an FM modulated signal. The limitation, technically and economically necessary, of the secondary band leads to ceasing to transmit the furthest extended sidelines of the carrier. The consequences of this are:
- a reduction of the transmitted power
- an approximation of the amplitude and phase of the phasor $\underline{U}_2$ (fig. 8.29), which leads to demodulation distortions.

It is generally assumed that distortions resulting from a bandpass filtering which lets at least 98% of the power of the secondary signal pass are tolerable. In taking into account the values of the Bessel function, it has been established that it is enough to conserve only the lines with an amplitude exceeding $0.1\ U_p$. This is the case for a number of lines empirically equal to the integer part of $\delta + 1$. From this finding *Carson's rule* is derived (fig. 8.30):

- a signal frequency modulated (FM) by a sinusoidal primary signal of frequency f_1 occupies a bandwidth which can be estimated at

$$B_{2FM} \cong 2(\delta + 1)f_1 \qquad (8.60)$$

or, in using (8.50)

$$B_{2FM} \cong 2(f_1 + \Delta f) \qquad (8.61)$$

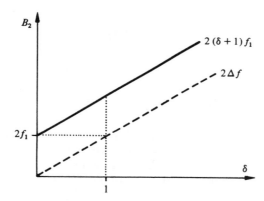

Fig. 8.30 Carson's rule for the estimation of the secondary bandwidth.

This approximation will hereafter be assumed to be sufficient for relative calculations of the FM modulation.

Carson's rule brings up the following remarks, confirmed by figure 8.28:

- if $\delta \ll 1$, i.e., if $\Delta f \ll f_1$, ("narrowband" FM), B_2 approaches $2f_1$ and the *spectrum is reduced to the carrier with two sidelines* of amplitude $U_p J_1(\delta) \cong \delta U_p/2$ at $f_p \pm f_1$. It therefore has the same form as in AM modulation, but is distinguished from it by the relative phase of the sidelines with respect to the carrier (fig. 8.31). Paradoxically, although the instantaneous frequency $f_2(t)$ is strictly limited between $f_p - \Delta f$ and $f_p + \Delta f$, the spectrum of $u_2(t)$ extends well beyond these limits, up to $f_p \pm f_1$;
- if $\delta \gg 1$, i.e., if $\Delta f \gg f_1$, ("wideband" FM) B_2 extends toward $2\Delta f$ and the spectrum contains a large number (approximately 2δ) of lines.

310   TELECOMMUNICATION SYSTEMS

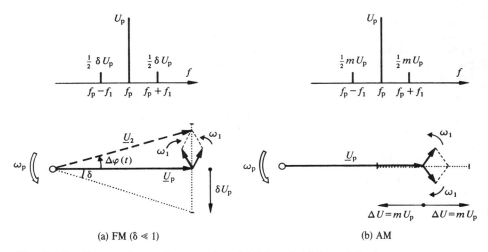

Fig. 8.31 Comparison of narrowband FM and AM modulations.

8.6.5 Measurement of the frequency deviation

A direct measurement of the frequency deviation Δf in sinusoidal FM modulation is not possible. In fact the instantaneous frequency has no physical reality. One can, however, derive Δf from the modulation index δ which itself is identifiable by the measuring the amplitude of the spectral lines referred to that (U_p) of the unmodulated carrier.

The simplest course is to measure selectively the line at f_p, whose amplitude is given according to (8.55) by $U_p J_0(\delta)$, and to vary the frequency f_1 of the primary signal while maintaining its amplitude (thus Δf) constant. For certain values of f_1, thus of δ according to (8.50), the line at f_p is cancelled, conforming to the zeros of the Bessel function J_0 (table 8.32).

This measurement, repeated at several zeros to eliminate ambiguity, allows a very precise determination of Δf.

8.6.6 Case of any primary signal

The preceding sections (8.6.5 to 8.6.2) concern only the case of a sinusoidal primary signal. If the signal $u_1(t)$ is any signal, or even a random signal, of r.m.s. value $U_{1\text{ eff}}$, we have, from (8.45)

$$u_2(t) = U_p \cos\left[\omega_p t + \Delta\varphi(t)\right] \tag{8.62}$$

ANALOG MODULATION

Table 8.32 First zeros of the zero-order Bessel function J_0.

$J_0(\delta) = 0$ for $\delta =$	2.405	5.520	8.654	11.792	14.931	...

where $\Delta\varphi(t)$, according to (8.44), is proportional to the integral of $u_1(t)$. The importance of the phase deviation $\Delta\varphi(t)$, expressed by its r.m.s. value $\Delta\varphi_{\text{eff}}$, is a criterion permitting the definition of two extreme cases:

- $\Delta\varphi_{\text{eff}} \ll 1$ rad: *"narrowband" FM modulation*. The secondary phasor $\underline{U}_2$ thus oscillates around $\underline{U}_p$, but without notably separating from it.
- $\Delta\varphi_{\text{eff}} \gg 1$ rad: *"wideband" FM modulation*. The phasor $\underline{U}_2$ can thus effect several turns on both sides of the phasor $\underline{U}_p$.

It must be noted that the commonly used terms "narrowband" and "wideband" have only a relative meaning for one of the procedures compared with the other.

8.6.7 Spectrum of a "narrowband" FM signal modulated by any signal

The analogy between the AM spectrum and the "narrowband" FM spectrum illustrated in figure 8.31 is valid for each component (sinusoidal at f_i) of the signal $U_1(t)$. The result is that, as far as one can assimilate the angle $\Delta\varphi(t)$ to its tangent, the principle of superposition is despite everything applicable and the spectrum of the "narrowband" FM signal $u_2(t)$ contains pairs of lines at $f_p \pm f_i$ each corresponding to a component of the primary signal. It thus possesses, as does the AM spectrum, two sidebands, one direct, the other inverse, each similar to the spectrum of the primary signal (fig. 8.33). Nevertheless, the carrier at f_p is dephased by $\pi/2$ with respect to that of the same signal modulated in AM.

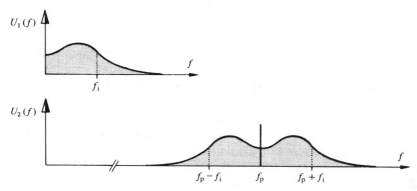

Fig. 8.33 Secondary "narrowband" FM spectrum.

8.6.8 Spectrum of a "wideband" FM signal modulated by any signal

This case is completely different from the preceeding case in that the instantaneous phase deviation $\Delta\varphi(t)$ takes very high values—for example, several times 2π. Consequently, relation (8.62) which relates $u_2(t)$ to $\Delta\varphi(t)$ is certainly not linear. The spectrum of $u_2(t)$ thus cannot be deduced simply from the spectrum of $u_1(t)$.

By virtue of (8.50), if $\Delta\varphi$ is very large, it means that the frequency deviation Δf must be much larger than all the frequencies f_1 contained in the primary signal. The result is that the instantaneous frequency $f_2(t)$ of the modulated signal varies greatly, but *very slowly*, as it follows the variations of $u_1(t)$.

Then and then only can it be stated that each range of values of the primary signal between u_1 and $u_1 + du_1$ corresponds in a quasi-stationary manner to a secondary signal whose power is located in the frequency domain between $f_p + \Delta f(u_1)$ and $f_p + \Delta f(u + du_1)$. This appears with a probability given by the probability density $p(u_1)$ of the signal $u_1(t)$ (fig. 8.34).

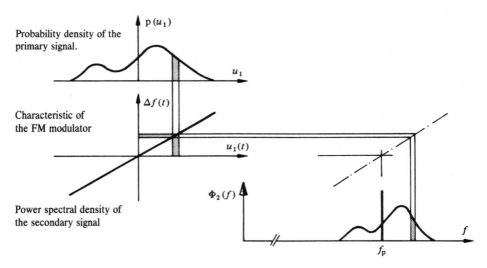

Fig. 8.34 Envelope of the FM spectrum with large phase deviation.

This empirical statement in fact expresses a theoretical result known as the *adiabatic theorem* or *Woodward's theorem*. A rigorous proof can be found in [4]. It is stated as follows:

In an FM modulation with large phase deviation ("wideband" FM), the envelope of the spectrum or of the power spectral density of the secondary signal has the same form as the probability density of the primary signal.

ANALOG MODULATION

This theorem allows us to state that

- the secondary spectrum of an FM signal modulated by a primary signal with Gaussian distribution approaches a Gaussian power spectral density when the r.m.s. phase deviation $\Delta\varphi_{\text{eff}}$ becomes very large. The r.m.s. value of the secondary bandwidth thus becomes $B_{2\text{eff}} \cong 2\Delta f_{\text{eff}} = 2\nu U_{1\,\text{eff}}$
- if $u_1(t)$ is a signal with uniform statistics between two limits $U_{1\,\text{min}}$ and $U_{1\,\text{max}}$ (for example, a triangular or sawtooth signal), the power frequency distribution becomes approximately uniform between $f_p + \nu U_{1\,\text{min}}$ and $f_p + \nu U_{1\,\text{max}}$
- if the statistical distribution of $u_1(t)$ is not symmetrical with respect to 0, the spectrum presents the same dissymmetry with respect to f_p.

8.6.9 Spectrum of an FM signal modulated by a bounded spectrum signal

If the signal is frequency modulated with an r.m.s. phase deviation $\Delta\varphi_{\text{eff}}$ by a primary signal with a spectrum limited to $f_{1\,\text{max}}$, nothing much can be assumed about the secondary spectrum.

For the nominal level of the primary signal (corresponding to $U_{1\,\text{nom}}$) the frequency deviation Δf_{nom} is the same whatever $f_1 < f_{1\,\text{max}}$. However, the phase deviation $\Delta\varphi_{\text{nom}}$ varies with f_1 (figure 8.35)

$$\Delta\varphi_{\text{nom}} = \frac{\Delta f_{\text{nom}}}{f_1} \tag{8.63}$$

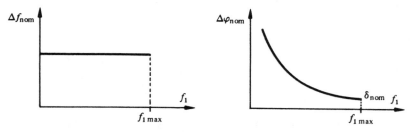

Fig. 8.35 Nominal frequency and phase deviations in FM modulation.

and takes a minimum value which is designated the *nominal modulation index*

$$\delta_{\text{nom}} = (\Delta\varphi_{\text{nom}})_{\text{min}} = \frac{\Delta f_{\text{nom}}}{f_{1\,\text{max}}} \quad \text{(in FM)} \tag{8.64}$$

The empirical Carson's rule (8.60) applied to $f_{1\,\text{max}}$ gives then a good approximation of the nominal secondary bandwidth

$$B_{2\text{nom FM}} \cong 2(\delta_{\text{nom}} + 1)f_{1\text{max}} \tag{8.65}$$

or, with (8.64), and by analogy with (8.61)

$$B_{2\text{nom}} \cong 2(f_{1\text{max}} + \Delta f_{\text{nom}}) \tag{8.66}$$

This important result contains the limiting cases of the two preceding sections. In effect

- if $\delta_{\text{nom}} \ll 1$ ("narrowband" FM modulation):

$B_{2\text{nom}}$ approaches $2f_{1\,\text{max}}$

- if $\delta_{\text{nom}} \gg 1$ ("wideband" FM modulation):

$B_{2\text{nom}}$ approaches $2\Delta f_{\text{nom}}$

8.6.10 Demodulation of an FM signal

In principle, any quadripole with a transfer function $H(f)$ which varies linearly with the frequency permits the extraction of a signal $u_1'(t)$ from $u_2(t)$, proportional to the instantaneous variations $\Delta f(t)$ of the instantaneous frequency $f_2(t)$. This is particularly the case if $H(f)$ is proportional to f, i.e., if the quadripole effects a mathematical *derivation* of the received signal $u_2'(t)$. A signal $u_3(t)$ is thus obtained, such that, with (8.45):

$$u_3(t) \sim \frac{du_2'(t)}{dt} = -U_p' 2\pi[f_p + vu_1(t)]\sin[\varphi_2(t)] \tag{8.67}$$

The signal $u_3(t)$ is at the same time

- amplitude modulated: in effect its amplitude is proportional to $U_p'|f + vu_1(t)|$ and thus varies linearly with $u_1(t)$.
- frequency modulated, since $\varphi_2(t)$ linearly depends on the integral of $u_1(t)$ according to (8.45).

An *envelope detection* (sect. 8.2.5) applied to $u_3(t)$ permits the restoration of the instantaneous amplitude, independently of the instantaneous frequency of $u_3(t)$.

In any case, the mathematical derivation of relation (8.67) presupposes that the amplitude U_p' of the received secondary signal is constant. Frequently this is not the case following attenuation variations or attenuation distortions in the channel or excessive limitation in the secondary band. The amplitude of the derived signal $u_3(t)$ then depends on the amplitude variations of the received signal. This is why *it is indispensable to clip $u'_2(t)$ strictly* before deriving it (fig. 8.36). The zero crossings (and thus the instantaneous frequency) of the signal are not affected by this clipping.

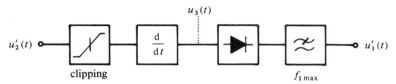

Fig. 8.36 Principle of FM demodulation by derivation.

Another form of FM demodulation, which is more direct, consists of extracting the instantaneous frequency from the secondary signal with the aid of a *phase-locked loop PLL* whose principle is described in [35] and [36]. This demodulator (fig. 8.37), also called a *coherent demodulator*, is based on a phase comparison between the signal $u_2(t)$ preliminarily clipped, and a voltage-controlled oscillator VCO modulated in frequency by the demodulated primary signal $u'_1(t)$, obtained at the output of the phase comparator.

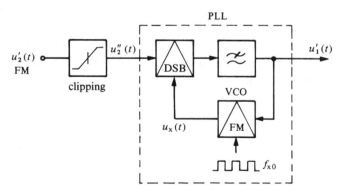

Fig. 8.37 Coherent FM demodulation by a phase-locked loop (PLL).

The phase comparator can be realized by a DSBSC modulator multiplying the two square-wave signals $u''_2(t)$ and $u_x(t)/U_x$, followed by a filtering to eliminate the double-frequency component. Relation (8.27) shows that its output voltage $u'_1(t)$ depends on the cosine of the phase difference between $u''_2(t)$ and

$u_x(t)$. If things are arranged so that this difference remains very close to $\pi/2$, we would have

$$u_1'(t) \sim \cos[\varphi_2(t) - \varphi_x(t)] = \sin\left[\varphi_2(t) - \varphi_x(t) - \frac{\pi}{2}\right]$$

$$\cong \varphi_2(t) - \varphi_x(t) - \frac{\pi}{2} \tag{8.68}$$

The instantaneous frequencies are

$$f_2(t) = f_p + vu_1(t) = \frac{1}{2\pi} \frac{d\varphi_2(t)}{dt} \tag{8.69}$$

$$f_x(t) = f_{x0} + v_x u_1'(t) = \frac{1}{2\pi} \frac{d\varphi_x(t)}{dt} \tag{8.70}$$

with the equilibrium condition of the loop

$$f_2(t) = f_x(t) \tag{8.71}$$

The output signal is thus

$$u_1'(t) = \frac{f_p - f_{x0}}{v_x} + \frac{v}{v_x} u_1(t) \tag{8.72}$$

i.e., similar, except for a dc component, to the primary original signal $u_1(t)$.

Coherent FM demodulation by PLL [3] has become economical and currently used due to its implementation in integrated circuit form. Compared to demodulation by derivation, it has the advantage of better insensitivity to the threshold effect (sect. 8.8.11).

8.6.11 Applications of FM modulation

The principal properties of angular modulations (FM and ΦM) are
- independence of the output level after demodulation with respect to the level of the secondary signal received by the demodulator (as a consequence of the clipping at the input);
- good noise immunity if the modulation index is high (sect. 8.8.9);
- secondary bandwidth always larger than in AM modulation and even larger when δ is high.

These properties make them suitable for applications in telecommunications for cases in which
- the attenuation between emitter and receiver is submitted to fluctuations, for example, in microwave links (variations of atmospheric conditions)

or for mobile radiocommunication, notably in an urban milieu (reflections, loss of direct visibility between emitter and receiver);
- a particularly good quality is desired despite unfavorable conditions, for example, in the case of satellite links (enormous attenuation), of microwave links (noise, interferences), audio radiobroadcasting of high quality (ultra short-wave, sound TV) or mobile radiotelephony in a highly noisy medium;
- the necessary bandwidth is available at the same time.

Consequently, FM modulation is reserved for transmission applications at high radio frequencies, where it presents substantial advantages compared to amplitude modulation. It is in fact very often combined with a phase modulation ΦM (pre-emphasis, sections 10.7.3 and 12.4.3).

8.6.12 Effect of linear distortions in FM

If the attenuation of the channel depends on the frequency (attenuation distortions), the amplitude of the received FM signal varies with the instantaneous frequency. The result is a superimposed amplitude modulation which can be eliminated by the clipping preceding the demodulation.

Nevertheless, by referring to figure 8.29, it has been established that, if two lines of the same pair no longer have the same amplitude, not only the amplitude but also the phase of the secondary phasor $\underline{U}_2$ is affected (very weakly, in fact). The consequence of this is *nonlinear distortion* on the demodulated signal $u'_1(t)$.

Phase distortions in the channel produce the same effect, but in a much more noticeable manner, by turning the pairs of lines of figure 8.29, which makes their sum two by two not exactly in phase or in quadrature with $\underline{U}_p$. In effect, in the range $f_p - \Delta f_{nom} \ldots f_p + \Delta f_{nom}$, if the group delay is not constant but varies approximately linearly with respect to f (fig. 8.38), the nominal amplitude signal $u_1(t)$ undergoes a delay proportional to its instantaneous value

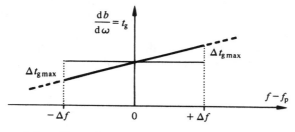

Fig. 8.38 Group delay distortion.

(fig. 8.39). This distortion of the demodulated primary signal can be roughly expressed by the addition of a second harmonic of relative amplitude d_2 (fig. 8.39).

For a sinusoidal signal $u_1(t)$ of frequency f_1 and a maximum group delay distortion of $\pm \Delta t_{g\,max}$, we have, according to figure 8.39,

$$d_2 \sin(2\omega_1 \Delta t_{gmax}) = 1 - \cos(\omega_1 \Delta t_{gmax}) \tag{8.73}$$

so that, if $\omega_1 \Delta t_{g\,max}$ is very small:

$$d_2 = \frac{1 - \cos(\omega_1 \Delta t_{gmax})}{\sin(2\omega_1 \Delta t_{gmax})} \cong \frac{\frac{1}{2}\omega_1^2 (\Delta t_{gmax})^2}{2\omega_1 \Delta t_{gmax}} = \frac{\pi}{2} f_1 \Delta t_{gmax} \tag{8.74}$$

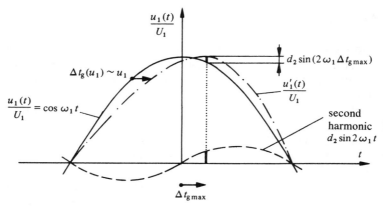

Fig. 8.39 Estimation of the second harmonic.

d_2 expresses the second order distortion factor after demodulation. It has been established that it is not only proportional to the amplitude of the primary signal (by the intermediary of $\Delta t_{g\,max}$), but also proportional to the frequency of the primary signal.

If the primary signal is that of a wideband frequency division multiplex (for example, multichannel microwave links, chapter 12), the nonlinear distortions on the demodulated signal are particularly undesirable (causing unintelligible crosstalk, section 8.4.6). Very strict requirements on phase linearity must be posed for the transmission of the secondary FM modulated signal. They constitute one of the essential difficulties of FM transmission.

8.7 PHASE MODULATION (ΦM)

8.7.1 Principle

In *phase modulation* ΦM, it is the instantaneous phase $\varphi_2(t)$ which varies linearly with the primary signal $u_1(t)$

$$\varphi_2(t) = \omega_p t + \Delta\varphi(t) \tag{8.75}$$

the instantaneous phase deviation $\Delta\varphi(t)$ being proportional (coefficient of proportionality β in rad/V) to $u_1(t)$

$$\Delta\varphi(t) = \beta u_1(t) \tag{8.76}$$

It follows from (8.5) that the instantaneous frequency deviation $\Delta f(t)$ is thus proportional to the derivative of the primary signal

$$\Delta f(t) = \frac{1}{2\pi} \frac{\mathrm{d}\Delta\varphi(t)}{\mathrm{d}t} = \frac{\beta}{2\pi} \frac{\mathrm{d}u_1(t)}{\mathrm{d}t} \tag{8.77}$$

8.7.2 Relationships and differences between ΦM and FM

Relations (8.44) and (8.77) show that the ΦM and FM modulations are distinguished only by a preliminary derivation or integration of the primary signal (fig. 8.40).

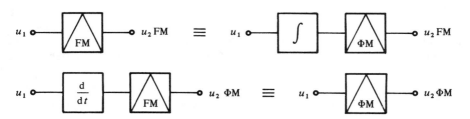

Fig. 8.40 Equivalences of FM and ΦM modulators.

This also means that an FM signal can be demodulated with a ΦM demodulator on the condition that the reconstructed primary signal is derived afterwards. Inversely an FM demodulator followed by an integrator allows the perfect demodulation of a ΦM signal.

In the case of a modulation by a *sinusoidal primary signal* of given frequency, these preliminary operations consist simply of changing the phase of the primary signal of $\pm \pi/2$. The definition of the *modulation index* from (8.50) is not affected. In the same way, the spectrum of the ΦM signal with sinusoidal modulation is *identical to the FM spectrum* with the same modulation index, as presented in section 8.6.3.

However, in the case of any primary signal, in particular with a bounded spectrum ($f_1 < f_{1\,max}$), important differences appear. In effect, while the nominal frequency deviation is constant with FM for the entire primary band, Δf_{nom} increases linearly with f_1 with ΦM. This is thus the phase deviation $\Delta\varphi_{nom}$ which remains constant (compare fig. 8.41 with figure 8.35).

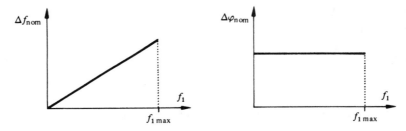

Fig. 8.41 Nominal frequency and phase deviations with ΦM modulation.

The definition of the *nominal modulation index* from (8.64) remains valid, with the following convention:

$$\delta_{nom} = \Delta\varphi_{nom} = \frac{\Delta f_{nom}(f_1)}{f_1} \qquad \text{(in ΦM)} \qquad (8.78)$$

the same is true for Carson's rule (8.65).

The principal difference between ΦM and FM concerns their behavior in the presence of noise (sect. 8.8.10).

8.7.3 Phase modulation by Armstrong's method

The remarks made in sections 8.6.4 (fig. 8.31) and 8.6.7 concerning the analogy between the AM spectrum and the "narrowband" FM spectrum (i.e., ΦM at low phase deviation) lead to the following idea: the addition of a carrier phase-shifted by $\pi/2$ to a DSBSC signal permits better simulation of an angular modulation as the amplitude U_p of the carrier increases with respect to the amplitude deviation of the DSBSC signal. In effect, the phasor diagram (fig. 8.42) shows that

$$\tan[\Delta\varphi(t)] = \frac{\Delta U(t)}{U_p} \sim \frac{u_1(t)}{U_p} \qquad (8.79)$$

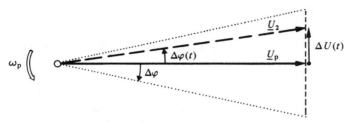

Fig. 8.42 Phasor diagram in Armstrong's modulator.

If $\Delta U(t) \ll U_p$, this relation becomes

$$\tan[\Delta\varphi(t)] \cong \Delta\varphi(t) \sim u_1(t) \tag{8.80}$$

It is thus a matter of phase modulation.

It can also be stated that the general relation (8.62), when developed, becomes

$$u_2(t) = U_p \cos[\omega_p t + \Delta\varphi(t)] = U_p[\cos \omega_p t \cos \Delta\varphi(t) - \sin \omega_p t \sin \Delta\varphi(t)] \tag{8.81}$$

If $\Delta\phi(t) \ll 1$, then $\cos\Delta\varphi(t) \cong 1$ and $\sin\Delta\varphi(t) \cong \Delta\varphi(t)$. We therefore obtain

$$u_2(t) = U_p \cos \omega_p t - U_p \Delta\varphi(t) \sin \omega_p t \cong U_p \cos\varphi_p(t) - \Delta U(t) \sin \varphi_p(t) \tag{8.82}$$

This relation demonstrates that ΦM modulation with a small deviation comes to the sum of a DSBSC modulated signal and of the carrier in quadrature.

This type of phase modulator (fig. 8.43) is frequently used as a preliminary stage, followed by a frequency multiplication destined to increase the frequency of the carrier and at the same time increase the modulation index up to the desired values.

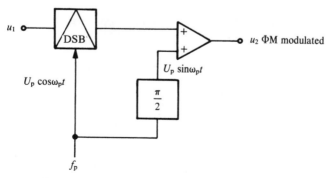

Fig. 8.43 Phase modulation by Armstrong's method.

8.8 EFFECT OF NOISE

8.8.1 Interference of a sinusoidal signal of small amplitude on an unmodulated carrier

If an unmodulated sinusoidal carrier $u_p = U_p \cos\omega_p t$ is combined by addition with a sinusoidal interfering signal $u_N = U_N \cos\omega_N t$, the resulting phasor $\underline{U}'_{2N}$ is simultaneously modulated in amplitude and in phase (fig. 8.44).

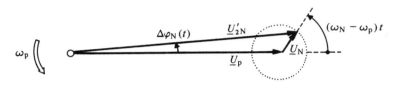

Fig. 8.44 Addition of a sinusoidal interfering signal.

The demodulator therefore receives a signal $u'_{2N}(t)$ such that

$$u'_{2N}(t) = u_p(t) + u_N(t) = U_p \left(\cos \omega_p t + \frac{U_N}{U_p} \cos \omega_N t \right) \tag{8.83}$$

By the cosine theorem, the amplitude of the phasor $\underline{U}'_{2N}$ is calculated

$$U'_{2N} = U_p \sqrt{1 + \left(\frac{U_N}{U_p}\right)^2 + 2\frac{U_N}{U_p} \cos(\omega_N - \omega_p)t} \tag{8.84}$$

If $U_N \ll U_p$, we have

$$U'_{2N} \cong U_p \left[1 + \frac{U_N}{U_p} \cos(\omega_N - \omega_p)t \right] \tag{8.85}$$

The instantaneous amplitude deviation is thus approximately

$$\Delta U_N(t) \cong U_N \cos(\omega_N - \omega_p)t \tag{8.86}$$

The instantaneous phase deviation $\Delta\varphi_N(t)$ is calculated by

$$\tan \Delta\varphi_N(t) = \frac{U_N \sin(\omega_N - \omega_p)t}{U_N \cos(\omega_N - \omega_p)t + U_p} \tag{8.87}$$

If $U_N \ll U_p$, this relation becomes

$$\tan \Delta\varphi_N(t) \cong \Delta\varphi_N(t) \cong \frac{U_N}{U_p} \sin(\omega_N - \omega_p)t \tag{8.88}$$

The instantaneous frequency deviation $\Delta f_N(t)$ is thus

$$\Delta f_N(t) = \frac{1}{2\pi} \frac{d\Delta\varphi_N(t)}{dt} \cong \frac{U_N}{U_p}(f_N - f_p)\cos(\omega_N - \omega_p)t \qquad (8.89)$$

8.8.2 Effect of a small sinusoidal interference on AM modulation

The AM demodulator (envelope detection) extracts from signal $u_2'(t)$ a signal $u_1'(t)$ proportional to the amplitude deviation $\Delta U'(t)$ of $u_2'(t)$. Disturbing this secondary signal by a small sinusoidal interference ($U_N \ll U_p$), the parasitic amplitude deviation is given by (8.86).

The following remarks can be made about the resulting primary parasitic signal $u'_{1N}(t)$:

- it is sinusoidal;
- its amplitude U'_{1N} is proportional to that of the interference U_N;
- its frequency f_{1N} is equal to the difference $|f_N - f_p|$ between those of the carrier and the interference (fig. 8.45);
- any sinusoidal interference of frequency f_N included between $f_p - f_{1\,max}$ and $f_p - f_{1\,min}$ on the one hand, between $f_p + f_{1\,min}$ and $f_p + f_{1\,max}$ on the other hand, gives rise to a parasitic signal in the baseband ($f_{1\,min} < f_{1N} < f_{1\,max}$). These ranges are represented in figure 8.46.

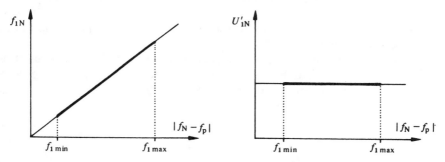

Fig. 8.45 Characteristics of the AM or ΦM demodulated parasitic signal.

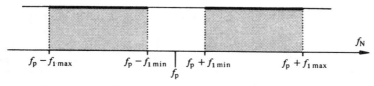

Fig. 8.46 Frequency ranges capable of interfering with the baseband after demodulation (AM, FM, ΦM).

8.8.3 Effect of sinusoidal interference on SSB modulation

In the SSB demodulator, the interference $u_N(t) = U_N \cos \omega_N t$ is frequency translated, like any other signal, into a parasitic signal $u'_{1N}(t)$ with a frequency $|f_N - f_p|$.

The same remarks as with AM modulation are valid, except that only the interferences located in the single secondary sideband (upper or lower) give rise to a parasitic signal at the output of the demodulator. The others can be eliminated by filtering *before* demodulation.

8.8.4 Effect of a small sinusoidal inteference on FM modulation

The process of FM demodulation restores a signal $u'_1(t)$ proportional to the instantaneous frequency deviation $\Delta f(t)$ of the received signal. The parasitic deviation is given by (8.89) when a sinusoidal interference $U_N \ll U_p$ is added to the carrier. The amplitude Δf_N of this parasitic deviation is thus

$$\Delta f_N \cong \frac{U_N}{U_p} |f_N - f_p| \qquad (8.90)$$

It is important to state (fig. 8.47) that, under these conditions:

- the parasitic demodulated signal is sinusoidal with a frequency $f_{1N} = |f_N - f_p|$
- its amplitude is proportional to the *ratio* between U_N and U_p *and to the frequency difference* $|f_N - f_p|$
- although the frequency range occupied by the FM modulated signal (Carson's rule) can be much broader, only the interferences appearing in the frequency ranges represented at fig. 8.46 (independently on δ) give rise to signals in the baseband ($f_{1\,\text{min}} < f_{1N} < f_{1\,\text{max}}$).

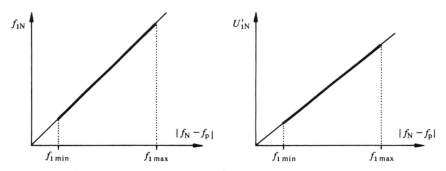

Fig. 8.47 Characteristics of the FM demodulated parasitic signal.

ANALOG MODULATION

The advantages of FM modulation proceed from this dependence of U'_{1N} with respect to $|f_N - f_p|$.

8.8.5 Effect of a small sinusoidal interference on ΦM modulation

The ΦM demodulator gives a parasitic signal $u'_{1N}(t)$ proportional to $\Delta\varphi_N(t)$ at its output, given by (8.88) when $U_N \ll U_p$. The amplitude of this parasitic phase deviation is thus

$$\Delta\varphi_N \cong \frac{U_N}{U_p} = \frac{U_{Neff}}{U_{peff}} \tag{8.91}$$

The characteristics of the demodulated parasitic signal are the same as with AM (fig. 8.45), except that, as with FM, it is the *ratio* of U_N to U_p which determines the amplitude of $u'_{1N}(t)$.

8.8.6 Effect of white noise. Signal-to-noise ratio after demodulation

The case of a sinusoidal interference is very specfic. Most often the interference is random in character. In the frequent case of thermal noise, its power spectral density can be considered uniform. Each element df_N of the frequency range brings a contribution dP_N to the disturbing power and gives rise, after demodulation, to an elementary parasitic signal of power dP'_{1N}. The integral of these elementary contributions provides the power P'_{1N} of the noise at the output of the demodulator. Only the noise components of frequencies f_N such that the demodulated parasitic signal falls in the baseband must be taken into consideration for the integration. They are located in the ranges presented in figure 8.46.

The *signal-to-noise ratio after demodulation* ξ'_1 is given by the ratio between the power $P'_{1S\,nom}$ of the nominal demodulated signal and the power P'_{1N} of the noise at the output of the demodulator.

For a given modulation, the power of the demodulated signal is *proportional to the square of the deviation* of the characteristic parameter of this modulation (ΔU, Δf or $\Delta\varphi$), whether the origin of this deviation is a true signal (for example, the nominal signal) or an interference producing a parasitic signal.

The ratio ξ'_1 is therefore calculated by the *ratio of the squares of the nominal and parasitic deviations* (r.m.s. values).

The contributions of each element df_N are incoherent. They must therefore be *added in a quadratic manner* to evaluate the resulting parasitic deviation.

Figure 8.48 specifies the notations used in the case of a baseband located between $f_{1\,min}$ and $f_{1\,max}$.

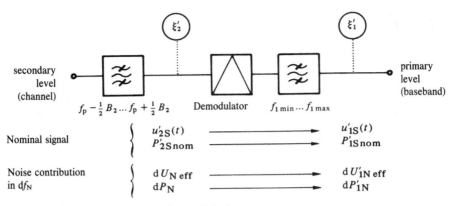

Fig. 8.48 Noise effect at demodulation.

The relationships between $P'_{2S\,nom}$ and $P'_{1S\,nom}$ on the one hand, and between dP_N and dP'_{1N} on the other, depend on the type of modulation and demodulator used.

For thermal (white) noise, there is, at the secondary level (input of the demodulator), according to (2.40)

$$dU_{Neff} = \sqrt{RdP_N} = \sqrt{RkTdf_N} \tag{8.93}$$

Only this case is considered in the following sections. Nevertheless, if the interference in the channel is not a white noise (i.e., its power spectral density is not uniform), it is necessary to take this into account in the integration of du'_{1N}.

8.8.7 Effect of thermal noise on AM modulation (envelope detection)

The characteristic parameter being the amplitude deviation ΔU, we have, in this case

$$\xi'_{1AM} = \frac{(\Delta U'_{nom\,eff})^2}{(\Delta U_{Neff})^2} = \frac{(\Delta U'_{nom\,eff})^2}{\int d(\Delta U_{Neff})^2} \tag{8.94}$$

with, according to (8.86) and (8.93)

$$d(\Delta U_{Neff})^2 = d(U_{Neff})^2 = RkTdf_N \tag{8.95}$$

The limits of integration for f_N are $f_p - f_{1\,max}$ and $f_p - f_{1\,min}$ on the one hand, $f_p + f_{1\,min}$ and $f_p + f_{1\,max}$ on the other hand (fig. 8.46).

$$(\Delta U_{Neff})^2 = R2kT(f_{1max} - f_{1min}) = R2kTB_1 \tag{8.96}$$

ANALOG MODULATION

Furthermore, $\Delta U'_{\text{nom}}$ represents the amplitude deviation (i.e., the amplitude of the envelope) of the received secondary signal. By introducing the (nominal) modulation factor m defined by (8.13) and the expression of the received secondary power P'_{2S} according to (8.14), we obtain

$$(\Delta U'_{\text{nom eff}})^2 = m^2 U'^2_{\text{peff}} = R \cdot m^2 P'_p = R \cdot \frac{m^2}{1 + \frac{m^2}{2}} P'_{2S\text{nom}} \tag{8.97}$$

Thus the signal-to-noise ratio after demodulation (8.94) becomes

$$\xi'_{1\text{AM}} = \frac{m^2}{m^2 + 2} \frac{P'_{2S\text{nom}}}{kTB_1} \tag{8.98}$$

With AM, this ratio is thus even greater when the modulation factor m is greater. Nevertheless, m must remain lower than 1 so that envelope detection is possible.

It can be shown that relation (8.98) is also valid in the case of an inverse isochronous demodulation.

8.8.8 Effect of thermal noise on SSB modulation

With SSB demodulation, the ratio between the input and the output voltages of the demodulator is the same, whether or not it concerns the nominal signal or an interference.

$$\frac{U'_{2\text{nom}}}{U'_{1\text{nom}}} = \frac{dU_{\text{Neff}}}{dU'_{1\text{Neff}}} \tag{8.99}$$

Relation (8.92) becomes

$$\xi'_{1\text{SSB}} = \frac{(U'_{1\text{nom eff}})^2}{\int d(U'_{1\text{Neff}})^2} = \frac{(U'_{2\text{nom eff}})^2}{\int d(U_{\text{Neff}})^2} \tag{8.100}$$

The limits of integration here are $f_p - f_{1\,\text{max}}$ and $f_p - f_{1\,\text{min}}$ (lower sideband) *or* $f_p + f_{1\,\text{min}}$ and $f_p + f_{1\,\text{max}}$ (upper sideband).

In the case of thermal noise, with (8.93), the ratio ξ'_1 becomes

$$\xi'_{1\text{SSB}} = \frac{P'_{2S\text{nom}}}{kTB_1} \tag{8.101}$$

Since, with SSB, $B_2 = B_1$, it is noted that the signal-to-noise ratio at the output of the demodulator ξ'_1 is the same as at the input. SSB modulation is the only one which has this property.

8.8.9 Effect of thermal noise on FM modulation

In this case it is the ratio of the squared r.m.s. values of the nominal and parasitic frequency deviations which determine ξ'_1. Since the nominal signal is sinusoidal, $\Delta f_{\text{nom eff}} = \Delta f_{\text{nom}}/\sqrt{2}$ and we have

$$\xi'_{1\text{FM}} = \frac{(\Delta f_{\text{nom eff}})^2}{(\Delta f_{\text{Neff}})^2} = \frac{\frac{1}{2}(\Delta f_{\text{nom}})^2}{\int d(\Delta f_{\text{Neff}})^2} \tag{8.102}$$

with, according to (8.89)

$$d(\Delta f_{\text{Neff}}) \cong \frac{1}{U'_p} |f_N - f_p| dU_{\text{Neff}} \tag{8.103}$$

For thermal noise, (8.93) allows us to write, with $P'_p = RU'^2_p/2$,

$$d(\Delta f_{\text{Neff}})^2 \cong \frac{1}{2P'_p}(f_N - f_p)^2 kT df_N \tag{8.104}$$

The limits of integration are the same as with AM (sect. 8.8.7), because all the components of noise located in the ranges represented in figure 8.46, and these only, give rise to contributions within the passing band of the filter at the output of the demodulator (fig. 8.49).

The result of integration is

$$(\Delta f_{\text{Neff}})^2 \cong \frac{1}{P'_p} \frac{kT}{3} (f^3_{1\text{max}} - f^3_{1\text{min}}) \tag{8.105}$$

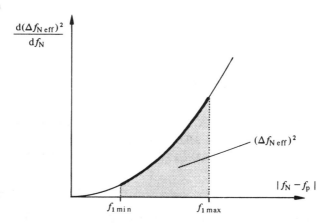

Fig. 8.49 Effect of white noise on FM as a function of the frequency difference between the noise band df_N and the carrier.

ANALOG MODULATION

It is noted that with FM, the power of the received secondary signal P'_{2S} is always equal to the power P'_p of the unmodulated carrier at the same point, from (8.59). The ratio ξ'_1 finally becomes

$$\xi'_{1FM} = \frac{3}{2} \frac{(\Delta f_{nom})^2}{f^3_{1max} - f^3_{1min}} \frac{P'_{2S}}{kT} \tag{8.106}$$

The following particular cases are interesting in practice:
- if $f_{1\,min} = 0$ and $f_{1\,max} = B_1$ (baseband starting from zero), relation (8.106) is written

$$\xi'_{1FM} = \frac{3}{2} \frac{(\Delta f_{nom})^2}{f^2_{1max}} \frac{P'_{2S}}{kTB_1} \tag{8.107}$$

With definition (8.65) of the nominal modulation index δ_{nom}, the ratio $\xi'_{1\,FM}$ thus becomes

$$\xi'_{1FM} = \frac{3}{2} \delta^2_{nom} \frac{P'_{2S}}{kTB_1} \tag{8.108}$$

- if $B_1 \ll f_{1\,max}$, (very narrow primary band, separated from zero), then

$$f^3_{1max} - f^3_{1min} = (f_{1max} - f_{1min})(f^2_{1min} + f_{1max}f_{1min} + f^2_{1min}) \tag{8.109}$$

$$\cong 3B_1 f^2_{1max}$$

and (8.106) becomes

$$\xi'_{1FM} \cong \frac{1}{2} \delta^2_{nom} \frac{P'_{2S}}{kTB_1} \tag{8.110}$$

In these two particular realistic cases (for example, FM radio broadcasting or higher channels of a multichannel microwave link), the *signal-to-noise ratio* after FM demodulation is proportional to the square of the modulation index δ. The price to be paid for the increase of ξ'_1 by increasing δ is the enlargement of the secondary band which proceeds from it (Carson's rule).

8.8.10 Effect of thermal noise on ΦM modulation

The signal-to-noise ratio after demodulation ξ'_1 proceeds from (8.92) with (8.91) and (8.93) by a calculation analogous to that of section 8.8.7.

$$\xi'_{1\Phi M} = \frac{(\Delta \varphi_{nom\,eff})^2}{(\Delta \varphi_{Neff})^2} = \frac{\frac{1}{2}(\Delta \varphi_{nom})^2}{RkT\int df_N} U'^2_p \tag{8.111}$$

The limits of integration are also the same as with AM (fig. 8.46); the result is, with $P'_{2S} = RU'^{2}_{p}/2$,

$$\xi'_{1\Phi M} = (\Delta\varphi_{nom})^2 \frac{P'_{2S}}{2kTB_1} \qquad (8.112)$$

Remembering that, from (8.78), the nominal phase deviation is identical to the nominal modulation index δ_{nom}, we finally obtain

$$\xi'_{1\Phi M} = \frac{1}{2} \delta^2_{nom} \frac{P'_{2S}}{kTB_1} \qquad (8.113)$$

For a primary signal with $f_{1\,min} = 0$ and for the same nominal modulation index, and thus for the same secondary bandwidth (according to Carson's rule), the phase modulation ΦM gives a signal-to-noise ratio after demodulation which is systematically 3 times worse (-4.7 dB) than the frequency modulation FM. Nevertheless, it presents the advantage of a uniform power spectral density for the resulting noise after demodulation. This noise is thus white, while in FM (fig. 8.49) its spectral density increases strongly (with f^2). Good use is made of this advantage by means of a pre-emphasis in multichannel microwave links (sect. 12.4.4) and for ultra-short-wave radiobroadcasting (sect. 10.7.3).

8.8.11 Threshold effect in FM and ΦM angular modulations

In the preceding sections, it has always been implicitly assumed that the interference $u_N(t)$ at the input of the demodulator was much lower than the amplitude U_p of the carrier. This justified the approximate values of the parasitic phase deviation $\Delta\varphi_N(t)$ from (8.88) and frequency deviation $\Delta f_N(t)$ according to (8.89).

When the r.m.s. value $U_{N\,eff}$ of the interference is not very small with respect to U_p, the probability that $u_N(t)$ takes comparable, or even higher values than U_p is no longer negligible. The result can be the situation presented in figure 8.50(b).

The resulting phasor $\underline{U}'_{2N}$ can thus make a complete turn around the carrier $\underline{U}_p$, which means that a very small variation of $\underline{U}_N$ (in amplitude or in phase) can lead to a very large variation of $\Delta\varphi_N(t)$,. As the FM demodulator reacts to the derivative of $\Delta\varphi_N(t)$, it produces a parasitic spike at the output.

This phenomenon gives rise to a *threshold effect*. Its probability is never zero, but increases sharply when the signal-to-noise ratio in the channel is degraded.

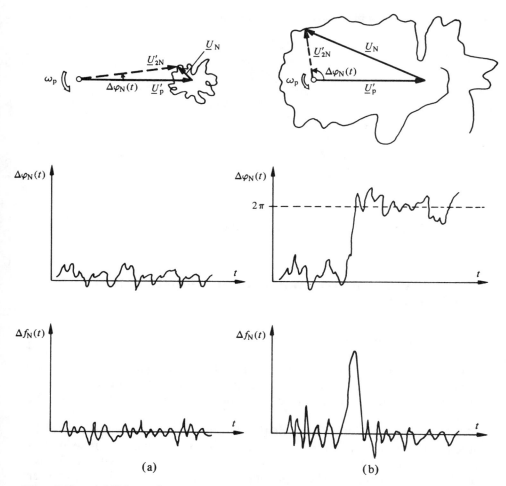

Fig. 8.50 Addition of a random interference. Effect on the instantaneous phase: (a) weak interference $U_{\text{Neff}} \ll U_p$; (b) significant interference $u_N(t) \cong U_p$.

It appears as a change in the statistical characteristics of the noise after demodulation and by a rapid lowering of the signal-to-noise ratio ξ_1'. The latter, instead of decreasing proportionally to the received secondary power P'_{2s}, as relation (8.106) would imply, is degraded much more rapidly (fig. 8.51) as P'_{2s} is decreased beyond a certain threshold, which is higher when δ_{nom} is larger.

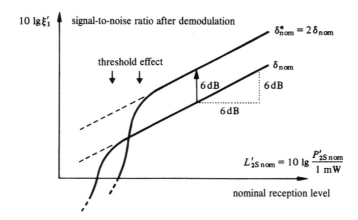

Fig. 8.51 Threshold effect in angular modulation.

8.9 SUMMARY AND COMPARISON OF SINUSOIDAL CARRIER MODULATIONS

8.9.1 Reference signal-to-noise ratio

In the presence of uniform power spectral density noise, such as thermal noise for example, the noise power found at the input of the demodulator is proportional to the (secondary) bandwidth B_2 occupied by the modulated signal.

To compare the performance of the different modulations applied to the same primary signal, it is appropriate to be cautious, because their secondary bandwidths are different. An objective comparison must start with the following identical parameters. These parameters express real conditions, as seen by the user of the link:

- bandwidth B_1 of the primary signal to be transmitted
- power P'_{2S} of the secondary signal available at the input of the demodulator, taking into account the emission power P_{2S} and the attenuation of the channel
- spectral density (assumed to be uniform) of the channel noise, for example kT in the case of thermal noise.

The properties of modulations will be compared

- *with the same primary signal* (characterized by its bandwidth B_1)

ANALOG MODULATION

- across the *same channel* (characterized by its attenuation and the power spectral density of the noise)
- with the *same nominal power* of the secondary signal.

These data allow the definition of a *reference signal-to-noise ratio* ξ_0

$$\xi_0 = \frac{P'_{2\text{Snom}}}{kTB_1} \tag{8.114}$$

It is necessary to emphasize that this ratio is arbitrary and without physical reality, in the sense that it introduces a **hypothetical** noise power kTB_1 which cannot be measured anywhere in the system.

8.9.2 Comparison of the performances of different modulations

Two criteria must be taken into consideration:

- the signal-to-noise ratio after demodulation ξ'_1 (criterion of transmission quality)
- the bandwidth B_2 of the modulated signal (criterion of spectral occupancy).

The comparison of the ratios ξ'_1 is done relative to the reference ratio ξ_0 defined above. It is noted that the expression of ξ_0 from (8.114) appears in each of the relations (8.98), (8.101), (8.110), and (8.113). We can therefore write

$$\xi'_{1\text{AM}} = \frac{m^2}{m^2 + 2} \xi_0 \tag{8.115}$$

$$\xi'_{1\text{SSB}} = \xi_0 \tag{8.116}$$

$$\xi'_{1\text{FM}} = \tfrac{3}{2} \delta^2 \xi_0 \qquad (\text{if } f_{1\max} = B_1) \tag{8.117}$$

$$\xi'_{1\Phi\text{M}} = \tfrac{1}{2} \delta^2 \xi_0 \tag{8.118}$$

Relation (8.116) incidentally gives a physical meaning to the definition of ξ_0: the reference signal-to-noise ratio ξ_0 is that which is obtained at the output of the demodulator of a transmission with SSB modulation.

The secondary bandwidths, in the case of a primary signal included between 0 and $f_{1\max}$, are given by relations (8.11), (8.28), and (8.65).

Vestigial sideband modulation VSB is located between the SSB and AM modulations, according to the relative importance of the remainder of the sideband and the carrier.

The comparison according to the two criteria (fig. 8.52) brings up the following remarks:
- AM modulation, despite its double secondary bandwidth, is always worse performing in the presence of noise than SSB modulation;
- the angular modulations FM and ΦM allow a significant gain with respect to ξ'_1 compared to SSB modulation, at the cost, however, of a considerable enlargement of the bandwidth. This gain is only positive when $\delta > \sqrt{2/3}$ with FM and $\delta > \sqrt{2}$ with ΦM.

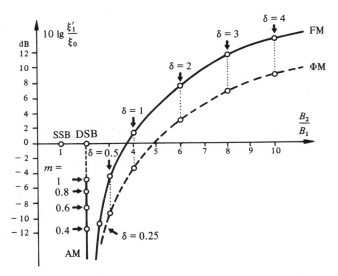

Fig. 8.52 Comparison of the secondary bandwidths B_2 and the signal-to-noise ratios after demodulation ξ'_1 for different analog modulations.

These comparisons show that it is not sufficient to increase the bandwidth B_2 to automatically improve ξ'_1 (with ξ_0 constant), or to tolerate a lower ratio ξ_0 (while keeping ξ'_1 constant). B_2 and ξ_0 are involved in the definition of the channel capacity C according to (4.12), that is representative of its quality. They are only interchangeable within the limits of the technical modulation procedures.

Figure 4.4 has already brought up this interchangeability. Here a system 1 (representative of modulations AM, SSB, VSB) which has a small bandwidth, but which requires a high ξ_0, is compared with a system 2 (representative of "wideband" FM and ΦM modulations) where ξ_0 can be reduced, provided that B_2 can be increased. This justifies the application of angular modulations to sound radio broadcasting with ultra-short-waves and to microwave links.

8.9.3 Comparison of the signal-to-noise ratios at the input and output of the demodulator

Before the input of the demodulator, a bandpass filter of width B_2 allows the definition of the frequency band occupied by the secondary signal and the elimination of the noise located outside this band (fig. 8.48).

In the presence of a white noise (for example, thermal noise) in the channel, the signal-to-noise ratio ξ'_2 at the input of the demodulator is thus determined by the power P'_{2Snom} of the received secondary signal, modulated by the nominal primary signal and by the bandwidth B_2, necessary for transmission.

$$\xi'_2 = \frac{P'_{2Snom}}{kTB_2} \tag{8.119}$$

The different types of demodulators behave in a very different way with regard to the signal-to-noise ratio ξ'_1 which they give at their output in the presence of the *same* signal-to-noise ratio ξ'_2 at their input. Let us note that this criterion of comparison between the modulations takes into account their different secondary bandwidth requirements. It thus leads to different results from those mentioned in the preceding paragraph.

For different modulations, the values of ξ'_2, ξ'_1 and their ratio are compared in table 8.53 and presented as a function of the characteristic parameters (nominal modulation factor m, nominal modulation index δ) in figures 8.54 and 8.55.

Concerning these comparisons, the following remarks can be made:

Table 8.53 Signal-to-noise ratios before and after demodulation.

	B_2	ξ'_2	ξ'_1	ξ'_1/ξ'_2
AM	$2B_1$	$\dfrac{1}{2}\xi_0$	$\dfrac{m^2}{m^2+2}\xi_0$	$\dfrac{2m^2}{m^2+2}$
SSB	B_1	ξ_0	ξ_0	1
FM ($f_{1\max}=B_1$)	$2(\delta+1)B_1$	$\dfrac{1}{2(\delta+1)}\xi_0$	$\dfrac{3}{2}\delta^2\xi_0$	$3\delta^2(\delta+1)$
ΦM ($f_{1\max}=B_1$)	$2(\delta+1)B_1$	$\dfrac{1}{2(\delta+1)}\xi_0$	$\dfrac{1}{2}\delta^2\xi_0$	$\delta^2(\delta+1)$

$$\xi_0 = \frac{P'_{2Snom}}{kTB_1}$$

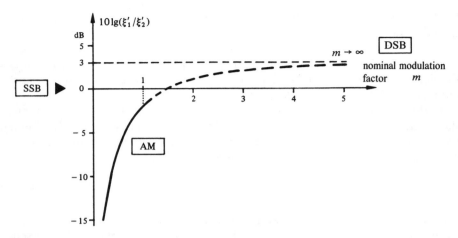

Fig. 8.54 AM demodulation.

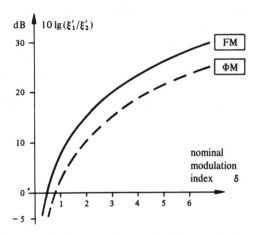

Fig. 8.55 FM/ΦM demodulation ($f_{1\ min} = 0$; $f_{1\ max} = B_1$).

- the AM demodulator (envelope detection) systematically reduces the signal-to-noise ratio the lower the power of the envelope compared to that of the carrier (low m);
- the FM demodulator increases ξ'_1 with respect to ξ'_2 as long as δ is higher than 0.475;
- the same goes for the ΦM demodulator if $\delta > 0.755$

ANALOG MODULATION

- when the modulation index is increased, the noise power P'_{1N} at the output of the demodulator remains constant while that of the signal P'_{1S} increases with δ^2, from which there is an increase of ξ'_1. This reasoning is only valid when the implicit hypothesis assuming that $u_N(t) \ll U_p$ is proven. The increase of δ leads to an increase in the secondary band and a simultaneous increase in the noise power at the input of the demodulator and thus of U_{Neff}, which may no longer be negligible vis-a-vis U_p (while $U'_{N1\,eff}$ remains constant). The threshold effect (sect. 8.8.11) finally strictly limits the increase of ξ'_1.

8.9.4 Summary of the principal properties of sinusoidal carrier modulations

The principal properties of the analog modulations with a sinusoidal carrier presented in sections 8.2 (AM), 8.3 (DSBSC), 8.4 (SSB), 8.6 (FM) and 8.7(ΦM), are collected in tables 8.56 to 8.60 and presented in the form of a synopsis.

8.10 ANALOG PULSE MODULATIONS

8.10.1 Pulse amplitude modulation PAM

In section 4.3.3, the sampling of a signal $u_1(t)$ by a series of periodic pulses of frequency f_e and duration τ has already been presented as a modulation of the amplitude of these pulses by the primary signal $u_1(t)$ (suppressed carrier pulse amplitude modulation).

If a dc component U_0 is added to the (alternative) signal $u_1(t)$ before it is sampled (fig. 8.61), or if a sequence of pulses is added to the sampled signal isochronously with the sampling function $e(t)$ and with an amplitude U_0, such that

$$U_0 > |u_1(t)| \tag{8.120}$$

the *pulse amplitude modulation PAM* (with carrier) is obtained. In this the amplitude of the modulated pulses $u_2(t)$ always remains positive.

The spectrum of the modulated PAM signal follows immediately from figure 4.12 or 4.13 (a) (if $\tau \ll T_e$) by the addition of lines at $f = 0$ and at the multiples of f_e (fig. 8.62).

In principle, demodulation is done simply by low-pass filtering at $f_{1\,max}$. In practice, this filtering is almost always preceded by a *hold* (sect. 4.3.7) which transforms the signal $u_2(t)$ into a *staircase signal*. In this way the primary signal is reconstructed without attenuation, but with a linear distortion according to a law in $\sin \pi T_e / \pi T_e f$ which must be compensated for (fig. 8.63).

Table 8.56 Amplitude modulation AM.

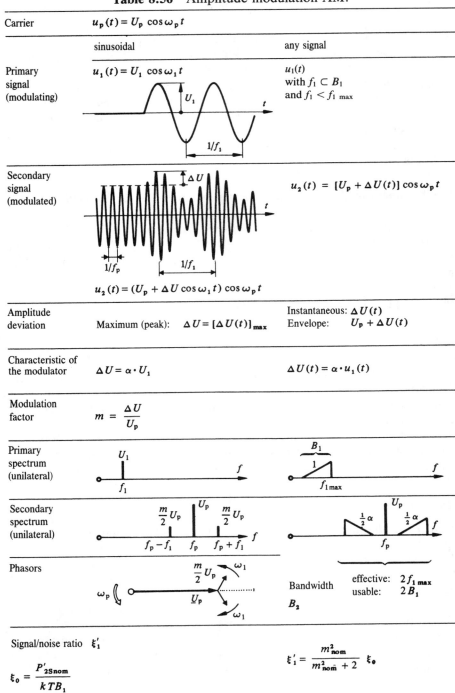

Table 8.57 Double sideband suppressed carrier modulation DSBSC.

Carrier	$u_p(t) = U_p \cos \omega_p t$	
	sinusoidal	any signal
Primary signal (modulating)	$u_1(t) = U_1 \cos \omega_1 t$	$u_1(t)$ with $f_1 \subset B_1$ and $f_1 < f_{1\,max}$
Secondary signal (modulated)	$u_2(t) = \Delta U \cos \omega_1 t \cos \omega_p t$	$u_2(t) = \Delta U(t) \cos \omega_p t$ $u_2(t) \sim u_1(t) \dfrac{u_p(t)}{U_p}$
Amplitude deviation	Maximum (peak): $\Delta U = [\Delta U(t)]_{max}$	Instantaneous: $\Delta U(t)$ Envelope: $\|\Delta U(t)\|$
Characteristic of the modulator	$\Delta U = \alpha U_1$	$\Delta U(t) = \alpha u_1(t)$ Multiplication of $u_p(t)/U_p$ by $u_1(t)$
Primary spectrum (unilateral)	U_1 at f_1	B_1 up to $f_{1\,max}$
Secondary spectrum (unilateral)	$\tfrac{1}{2}\Delta U$ at $f_p - f_1$, f_p, $f_p + f_1$	$\tfrac{1}{2}\alpha$ sidebands around f_p
Phasors	$\tfrac{1}{2}\Delta U$ with ω_1, ω_p, $\tfrac{1}{2}\Delta U$ with ω_1	Bandwidth B_2 effective: $2f_{1\,max}$ usable: $2B_1$
Signal/noise ratio ξ_1'		
$\xi_0 = \dfrac{P_{2S\,nom}'}{kTB_1}$		$\xi_1' = \xi_0$

Table 8.58 Single sideband modulation SSB.

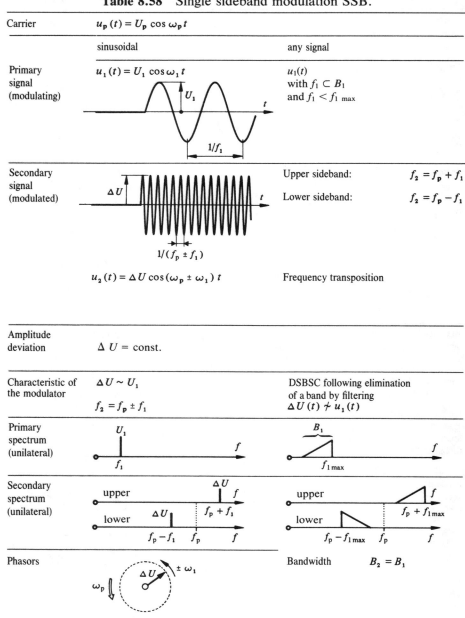

	sinusoidal	any signal
Carrier	$u_p(t) = U_p \cos \omega_p t$	
Primary signal (modulating)	$u_1(t) = U_1 \cos \omega_1 t$	$u_1(t)$ with $f_1 \subset B_1$ and $f_1 < f_{1\,max}$
Secondary signal (modulated)	$u_2(t) = \Delta U \cos(\omega_p \pm \omega_1) t$	Upper sideband: $f_2 = f_p + f_1$ Lower sideband: $f_2 = f_p - f_1$ Frequency transposition
Amplitude deviation	ΔU = const.	
Characteristic of the modulator	$\Delta U \sim U_1$ $f_2 = f_p \pm f_1$	DSBSC following elimination of a band by filtering $\Delta U(t) \not\sim u_1(t)$
Primary spectrum (unilateral)		
Secondary spectrum (unilateral)		
Phasors		Bandwidth $\quad B_2 = B_1$
Signal/noise ratio ξ_1' $\xi_0 = \dfrac{P_{2S\,nom}'}{kTB_1}$		$\xi_1' = \xi_0$

Table 8.59 Frequency modulation FM.

Carrier	$u_p(t) = U_p \cos \omega_p t$	
	sinusoidal	any signal
Primary signal (modulating)	$u_1(t) = U_1 \cos \omega_1 t$	$u_1(t)$ with $f_1 \subset B_1$ and $f_1 < f_{1\,max}$
Secondary signal (modulated)	$f_2(t) = f_p + \Delta f \cos \omega_1 t$ $\varphi_2(t) = \omega_p t + \dfrac{\Delta f}{f_1} \sin \omega_1 t$	$u_2(t) = U_p \cos[\varphi_2(t)]$ $f_2(t) = f_p + \Delta f(t)$ $\varphi_2(t) = 2\pi \int f_2(t)\,dt = \omega_p t + \Delta \varphi(t)$
Frequency and phase deviation	$\Delta f = [\Delta f(t)]_{max}$ $\Delta \varphi = [\Delta \varphi(t)]_{max} = \dfrac{\Delta f}{f_1}$	$\Delta f(t) = f_2(t) - f_p$ $\Delta \varphi(t) = \varphi_2(t) - \omega_p t = 2\pi \int \Delta f(t)\,dt$
Characteristic of the modulator	$\Delta f = \nu U_1$ $\Delta \varphi = \nu U_1 / f_1$	$\Delta f(t) = \nu u_1(t)$ $\Delta \varphi(t) = 2\pi \nu \int u_1(t)\,dt$
Modulation index	$\delta = \dfrac{\Delta f}{f_1} = \Delta \varphi$	$\delta = \Delta \varphi_{min} = \dfrac{\Delta f}{f_{1\,max}}$
Primary spectrum (unilateral)	U_1 at f_1	B_1 up to $f_{1\,max}$
Secondary spectrum (unilateral)	$U_n = U_p J_n(\delta)$ at $f_p \pm n f_1$, f_p	spectrum centered at f_p
Phasors	$\underline{U}_2$, $\Delta \varphi$, $\underline{U}_p$, ω_p	Bandwidth B_2 : theoretical $B_2 = \infty$ practical (Carson) $B_2 \cong 2(\delta + 1) f_{1\,max}$ $B_2 \cong 2(\Delta f + f_{1\,max})$
Signal/noise ratio ξ_1' $\xi_0 = \dfrac{P'_{2S\,nom}}{k T B_1}$		si $B_1 = f_{1\,max}$: $\xi_1' = \dfrac{3}{2} \delta^2_{nom} \xi_0$ si $B_1 \ll f_{1\,max}$: $\xi_1' = \dfrac{1}{2} \delta^2_{nom} \xi_0$

Table 8.60 Phase modulation ΦM.

Carrier	$u_p(t) = U_p \cos \omega_p t$	
	sinusoidal	any signal
Primary signal (modulating)	$u_1(t) = U_1 \cos \omega_1 t$	$u_1(t)$ with $f_1 \subset B_1$ and $f_1 < f_{1\,max}$
Secondary signal (modulated)	$\varphi_2(t) = \omega_p t + \Delta\varphi \cos \omega_1 t$ $f_2(t) = f_p - f_1 \Delta\varphi \sin \omega_1 t$	$u_2(t) = U_p \cos[\varphi_2(t)]$ $\varphi_2(t) = \omega_p t + \Delta\varphi(t)$ $f_2(t) = f_p + \Delta f(t) = \dfrac{1}{2\pi}\dfrac{d\varphi_2(t)}{dt}$
Phase and frequency deviation	$\Delta\varphi = [\Delta\varphi(t)]_{max}$ $\Delta f = [\Delta f(t)]_{max} = f_1 \Delta\varphi$	$\Delta\varphi(t) = \varphi_2(t) - \omega_p t$ $\Delta f(t) = f_2(t) - f_p = (d\varphi_2/2\pi dt) - f_p$
Characteristic of the modulator	$\Delta\varphi = \beta U_1$ $\Delta f = \beta f_1 U_1$	$\Delta\varphi(t) = \beta u_1(t)$ $\Delta f(t) = \dfrac{1}{2\pi}\beta\, du_1(t)/dt$
Modulation index	$\delta = \Delta\varphi = \dfrac{\Delta f}{f_1}$	$\delta = \Delta\varphi$
Primary spectrum (unilateral)	spectral line U_1 at f_1	spectrum in B_1 up to $f_{1\,max}$
Secondary spectrum (unilateral)	$U_n = U_p J_n(\delta)$ at $f_p \pm n f_1$, f_p	spectrum around f_p
Phasors	$\underline{U}_2$, $\Delta\varphi$, $\underline{U}_p$, ω_p	Bandwidth B_2: theoretical $B_2 = \infty$ practical (Carson) $B_2 \cong 2(\delta + 1)f_{1\,max}$ $B_2 \cong 2(\Delta f_{max} + f_{1\,max})$
Signal/noise ratio ξ_1' $\xi_0 = \dfrac{P_{2S\,nom}'}{kTB_1}$		$\xi_1' = \dfrac{1}{2}\delta_{nom}^2 \xi_0$

ANALOG MODULATION

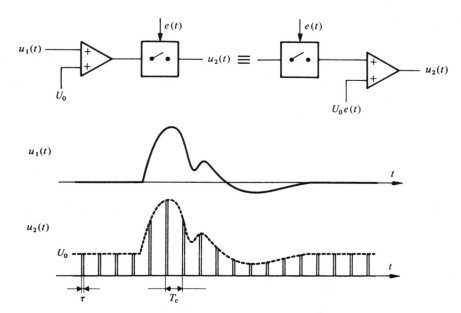

Fig. 8.61 Pulse amplitude modulation.

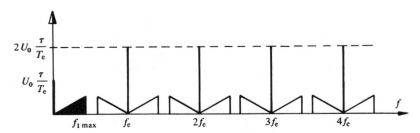

Fig. 8.62 Spectrum of a PAM signal modulated by a primary signal with a limited band $f_1 < f_{1\,max} < f_e/2$ (in the case in which $\tau < T_e$).

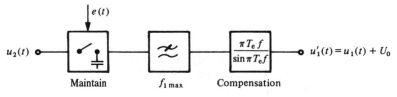

Fig. 8.63 Unmodulation of the signal PAM.

8.10.2 Pulse duration modulation PDM

Another form of pulse modulation consists of varying the pulse duration τ proportionally to the primary signal while keeping its amplitude A constant. Generally, one of the edges of each pulse is strictly periodic and the other is modulated in time.

$$\tau(iT_e) = \tau_0 + \Delta\tau(iT_e) \tag{8.121}$$

with

$$\Delta\tau(iT_e) \sim u_1(iT_e) \tag{8.122}$$

We thus obtain *pulse duration modulation PDM*.

This can be generated from a PAM modulation, followed by an amplitude-time conversion with the aid of a linear slope (fig. 8.64).

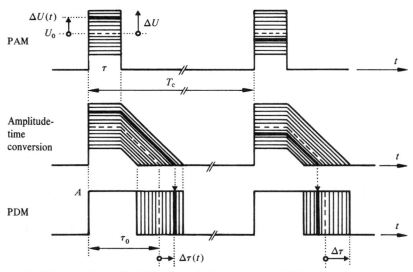

Fig. 8.64 Conversion of a PAM modulation into a PDM modulation.

For the demodulation of a PDM signal, an inverse process can be used, i.e., a linear time-amplitude conversion which takes the time modulation PDM to an amplitude modulation PAM, then a PAM demodulation by holding, filtering, and compensation, according to figure 8.63.

8.10.3 Pulse position modulation PPM

In PDM modulation according to figure 8.64, only the trailing edges of the pulses carry information. The duration of the unmodulated pulses represents an energy which is not strictly necessary for transmission. *Pulse position modulation PPM* optimizes transmission by emitting pulses of constant amplitude A, of constant and very short duration τ_0, at an instant shifted by $\Delta t(iT_e)$, with respect to the periodic sampling instants, with

$$\Delta t(iT_e) \sim u_1(iT_e) \tag{8.123}$$

We can easily go from a PDM modulation to PPM modulation by generating a short pulse of duration τ_0 at the moment of the trailing edge of the PDM-modulated pulse (fig. 8.65).

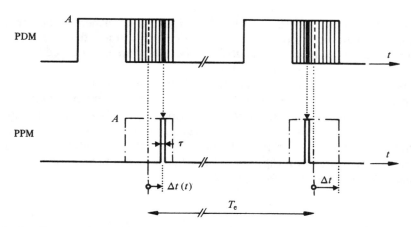

Fig. 8.65 Conversion of a PDM modulation into a PPM modulation.

In the same way, for demodulation, the appearance of PPM pulses can determine the trailing edge of pulses whose leading edge has been periodically triggered, at frequency f_e. The result is PDM-modulated pulses, which can be demodulated by conversion into PAM.

8.10.4 Pulse frequency modulation PFM

PPM modulation is in fact a phase modulation, the time deviation $\Delta t(t)$ being interpreted as a phase deviation

$$\Delta \varphi(t) = 2\pi \Delta t(t)/T_e \tag{8.124}$$

On the contrary, if the instantaneous frequency $f(t)$, i.e, the inverse of the interval $T(t)$ between two consecutive modulated pulses is a linear function of $u_1(t)$, we obtain a *pulse frequency modulation* PFM. The time deviation $\Delta t(t)$ with respect to the periodic moments iT_e (corresponding to the frequency of unmodulated pulses) is thus **inversely proportional to the integral** of the primary signal. Thus, it is not bounded. The PFM pulses are, therefore, more or less brought together, according to the polarity of the primary signal.

This property of PFM modulation precludes its use for the construction of a time division multiplex. However, with PDM and PPM, for a limited primary voltage, the modulated pulses remain inside periodic and bounded time domains. It is therefore possible to interpose other pulses between them.

8.10.5 Uniform sampling and natural sampling

The forms of PDM, PPM and PFM modulation described in the preceding sections are derived from *uniform sampling*, in the sense that the deviations of duration $\Delta\tau(iT_e)$ and position $\Delta t(iT_e)$ respectively are determined by samples $u_1(iT_e)$ taken at *strictly periodic* instants.

It is possible to imagine another form of time modulation (PDM or PPM) in which the deviation $\Delta\tau(iT_e)$ is determined by the moment in which a signal with linear slope, starting from zero at periodic instants iT_e, is equal to $u_1(t)$ (fig. 8.66). This procedure is called *natural sampling*. It allows the generation of PDM or PPM modulation with the help of a comparator without going through PAM modulation (fig. 8.67). The transmitted information corresponds to *nonequidistant samples* of $u_1(t)$.

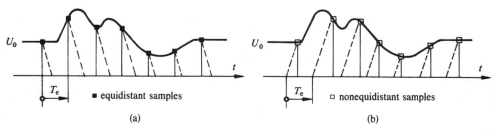

Fig. 8.66 Sampling: (a) uniform; (b) natural.

Aside from its simplicity of implementation, natural sampling presents some theoretical peculiarities regarding the spectrum of signals modulated in this way.

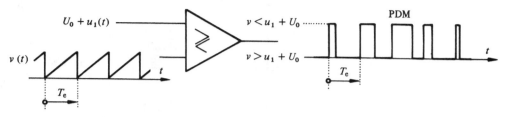

Fig. 8.67 Natural sampling.

8.10.6 Spectrum of PDM and PPM modulated signals

While interesting from the theoretical point of view [31], the spectrum of PDM or PPM modulated pulses is of no great practical interest. In any case it is necessary to have a very large secondary bandwidth, even infinite, if the pulse edges are vertical. These spectra have the following structure (fig. 8.68), when the primary signal is sinusoidal with frequency f_1:

- a *line at the origin* corresponding to the mean value of the unmodulated pulses $A\tau_0/T_e$;
- *lines at multiples* nf_e of the sampling frequency, of which the amplitude is practically identical and equal to $2A\tau_0/T_e$ (for $n \ll T_e/\tau_0$). Rigorously, the amplitude of the lines contains furthermore as a factor the zero-order Bessel function $J_0(2\pi nf_e \Delta\tau)$ with PDM and $J_0(2\pi nf_e \Delta t)$ with PPM, but these factors are practically very close to 1 for a low n;
- groups of *sidelines* at $nf_e \pm kf_1$ whose amplitude depends on the Bessel functions of order k and with an argument increasing with n. Therefore, these groups are broader the greater n is;
- the line at the origin ($n = 0$) is also accompanied by sidelines at kf_1. In *natural sampling*, all the lines for $k \pm 1$ have a zero amplitude; this is not the case with uniform sampling, which thus excludes the possibility of a direct demodulation of the baseband by low-pass filtering.

8.10.7 Time crosstalk in a PAM-modulated system

One of the principal motivations for pulse modulation is the construction of a time division multiplex of z channels, the samples belonging to the different channels being shifted in time and interleaved. If these pulses are perfectly rectangular, they do not influence each other. However, if they are deformed, for example by the presence of a parasitic capacitance on the conductor which carries them, an annoying interference from one time channel to its neighbor appears (fig. 8.69).

	Line amplitude at nf_e	Sideline amplitude nf_e+kf_1	$nf_e + kf_1$	Unilateral spectrum graph
	$n \neq 0$, n integer $\lessgtr 0$	n any integer $-\infty \ldots +\infty$		
	$k = 0$	$k \neq 0$, k integer $\lessgtr 0$		
PDM	$\dfrac{A}{\pi n} \cdot \sin(\pi n f_e \tau) \cdot J_0(2\pi n f_e \cdot \Delta\tau)$	$\dfrac{A}{2\pi} \cdot \dfrac{f_e}{f_x} \cdot J_k(2\pi \cdot f_x \cdot \Delta\tau)$		
PPM	$\dfrac{A}{\pi n} \cdot \sin(\pi n f_e \tau_0) \cdot J_0(2\pi n f_e \cdot \Delta t)$	$\dfrac{A}{\pi} \cdot \dfrac{f_e}{f_x} \cdot \sin[\pi(nf_e + kf_1)\tau] \cdot J_k(2\pi \cdot f_x \cdot \Delta t)$		

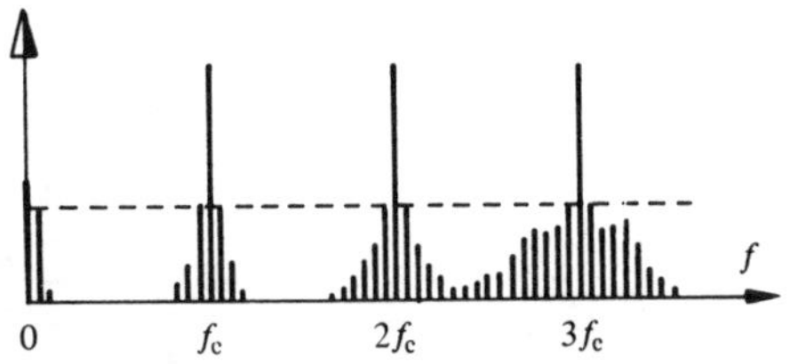

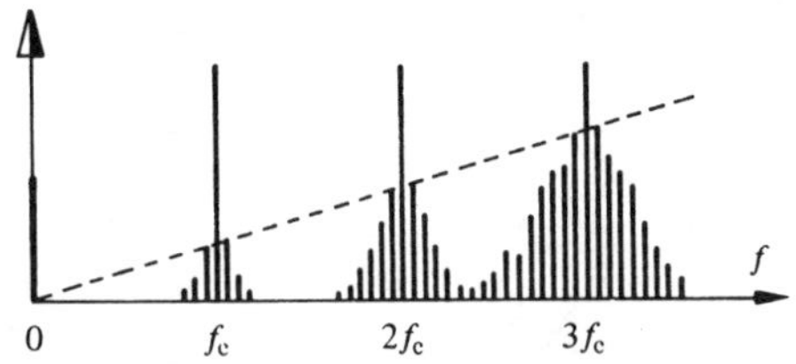

Uniform sampling: $f_x = nf_e + kf_1$
Natural sampling: $f_x = nf_e$

Fig. 8.68 Spectrum of PDM and PPM signals with sinusoidal modulation.

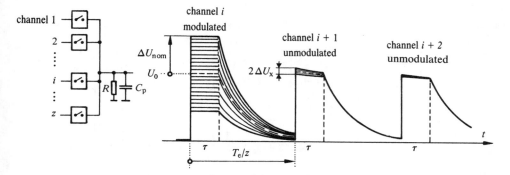

Fig. 8.69 Time crosstalk due to parasitic capacitance.

The result is a partial transfer of the modulation from channel i to the next time channel (channel $i + 1$), called *time crosstalk* (intelligible!).

The product of the load resistance R on the multiplex and of its parasitic capacitance C_p determines the time constant of the exponential decay of the pulses. We can therefore evaluate the parasitic amplitude deviation ΔU_x which affects the channel $i + 1$, as a function of the nominal deviation ΔU_{nom} of the channel i

$$\Delta U_x \cong \Delta U_{nom} \cdot \exp\left(-\frac{\frac{T_e}{z} - \tau}{RC_p}\right) \tag{8.125}$$

and the resulting crosstalk attenuation

$$A_x = 20 \lg \frac{\Delta U_{nom}}{\Delta U_x} \cong \frac{8.68}{RC_p}\left(\frac{T_e}{z} - \tau\right) \quad \text{dB} \tag{8.126}$$

It is therefore necessary to reduce the time constant RC_p as much as possible.

The deformation of PAM pulses due to a high-pass transfer characteristic (series coupling capacitor, coupling by transformer, etc.) also gives rise to time crosstalk (fig. 8.70). It will be assumed that the time constant T_0 representative of this coupling is much greater than τ, without being infinite.

The modulation residue after the pulse gives rise to a vertical displacement of *all* the following pulses (if $T_0 \gg T_e/z$), i.e., general crosstalk in all the time channels. We can estimate the parasitic amplitude deviation ΔU_x

$$\Delta U_x = \Delta U_{nom}\left[1 - \exp\left(-\frac{\tau}{T_0}\right)\right] \cong \Delta U_{nom}\frac{\tau}{T_0} \tag{8.127}$$

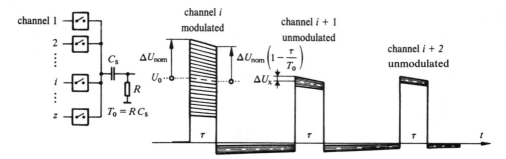

Fig. 8.70 Time crosstalk due to high-pass type of coupling.

The crosstalk attenuation proceeds from this:

$$A_x = 20 \lg \frac{\Delta U_{nom}}{\Delta U_x} \cong 20 \lg \frac{T_0}{\tau} \qquad \text{dB} \qquad (8.128)$$

There is therefore an interest in obtaining a time constant T_0 which is as high as possible.

8.10.8 Effect of noise on PPM

If noise $u_N(t)$ is superimposed onto the PPM modulated signal in the course of transmission, the pulses are vertically shifted. However, they do not undergo any modification regarding the *moment* (information carrying parameter) of their appearance. Nevertheless practically, the leading edge of the pulse serves as a criterion to determine the time deviation $\Delta t(t)$ with the aid of a fixed voltage threshold. The moment of threshold crossing is not affected by the interference $u_N(t)$ provided that the edge considered is **vertical**. The inevitable limitation of the secondary bandwidth to a value B_2 nevertheless involves a non-zero rise time t_m according to (4.2) and approximately equal to $0.4/B_2$. The situation is thus that of figure 8.71.

Thus a parasitic time deviation Δt_N appears at the height of the threshold

$$\Delta t_N(t) = t_m \frac{u_N(t)}{A} \qquad (8.129)$$

In the case of thermal noise collected in band B_2, we have

$$(\Delta t_{Neff})^2 = t_m^2 \frac{U_{Neff}^2}{A^2} = t_m^2 R \frac{kTB_2}{A^2} \qquad (8.130)$$

ANALOG MODULATION

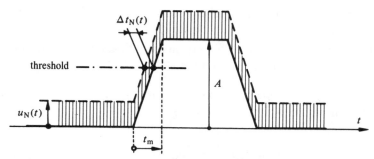

Fig. 8.71 Superimposition of interference on a pulse at a vertical edge.

The signal-to-noise ratio after demodulation ξ'_1 is given, in analog form as in section 8.8.6, by the ratio of the squares of the nominal and parasitic time deviations (r.m.s. values).

$$\xi'_1 = \frac{(\Delta t_{\text{nom eff}})^2}{(\Delta t_{\text{Neff}})^2} \tag{8.131}$$

Taking into account (8.129) and (4.2), we obtain

$$\xi'_{1\text{PPM}} \cong (\Delta t_{\text{nom eff}})^2 \frac{A^2}{RkTB_2} \frac{B_2^2}{(0.4)^2} \sim B_2 \tag{8.132}$$

In conclusion, the quality of a PPM transmission increases linearly with the bandwidth assigned to it.

8.10.9 Applications of analog pulse modulations

In telecommunications, PAM modulation is rarely used due to its sensitivity to interference (as with AM) and time crosstalk problems. However, it intervenes as an intermediate stage (sampling) in other pulse modulations (PDM, PPM or PCM).

Time modulations (PDM, PFM and PPM) present the following advantages:

- good noise immunity (at the price of a large secondary bandwidth);
- no amplitude linearity requirement: the pulses thus modulated can be used to modulate a transmitter by on-off keying (OOK, sect. 8.11.3).

Further, the PPM and PFM modulations lead to a signal with a constant mean power (independent of the modulating signal) which can be very small (if $\tau_0 \ll T_e$).

Consequently, the principal applications of these modulations in telecommunications are
- optical transmission (analog television on optical fibers) or infrared beams (remote control, cordless telephone station);
- telemetry
- in some cases, microwave links with time division multiplex of a small number of channels.

8.11 DISCRETE ANALOG MODULATIONS

8.11.1 Definition

Conforming to section 4.2.5, *discrete analog modulations* are designated here as procedures in which
- the primary signal carries binary or *m*-ary **digital information**, and due to this fact can take m discrete values;
- the carrier is a *sinusoidal* signal $u_p(t) = U_p \cos \omega_p t$;
- the secondary signal is obtained by the **discrete variation** of one of the parameters (amplitude, frequency, or phase) of the carrier according to the m possible values of the primary signal.

These modulations are analog in the sense that they do not modify the nature of the information carried by the signals before and after modulation. In this case, however, this information is digital. These procedures are just particular cases of sinusoidal carrier modulations in which the primary signal can be considered to be a signal with m discrete values, tied to a clock (clocked in the sense of section 5.1.3) or not. In reality, the primary signal often does not exist and it is the primary digital information which directly determines the parameters of the secondary signal.

8.11.2 Types

Depending on the parameter being modulated, one can distinguish the following types:
- *amplitude shift keying ASK*, with m values, of which a particular case is the *on-off keying OOK* with $m = 2$ amplitude values: 0 and U_p (fig. 8.72);
- *frequency shift keying FSK*, in which the frequency of the secondary signal takes m discrete values, for example for $m = 2$: $f_p - \Delta f$ and $f_p + \Delta f$ (fig. 8.72);

ANALOG MODULATION 353

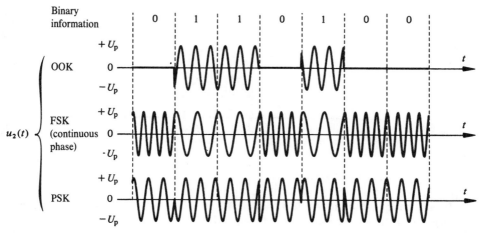

Fig. 8.72 Binary analog modulations.

- *phase shift keying PSK* whose phase takes m values, $2\pi/m$ apart, according to the values of the moments (fig. 8.72).

8.11.3 On-off keying OOK

The secondary signal has the expression

$$u_2(t) = a_k U_p \cos\omega_p t \qquad (8.133)$$

ere a_k takes the values 0 or 1.

By analogy with relation (8.22), it has been established that OOK modulation is equivalent to a DSBSC modulation by a *unipolar* binary signal $(0; +1)$, which is generally random.

Consequently, the power spectral density $\Phi_2(f)$ of the OOK signal contains

- a line at f_p (due to the dc component of the unipolar signal);
- two sidebands, symmetrical of each other at f_p, each corresponding to the power spectral density of the unipolar binary signal.

The bandwidth B_2 is theoretically infinite. An (inevitable) limitation of B_2 incurs the risk of intersymbol interference after demodulation. This danger can be avoided by a preliminary restriction of the primary band (nonrectangular basic signal), or by a soft limitation of the secondary band, including the condition of central symmetry of the filter slope with respect to the frequencies $f_p \pm \dot{M}/2$, imposed by Nyquist's expanded criterion (5.15)(fig. 8.73).

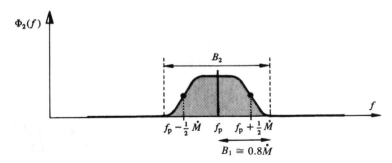

Fig. 8.73 OOK signal spectrum.

In any case, the bandwidth B_2 necessary in OOK is the double of that which is required in the baseband according to (4.9).

$$B_{2\text{OOK}} = 2B_1 \cong 1.6\dot{M} \qquad (8.134)$$

8.11.4 Frequency shift keying FSK

In the binary case ($m = 2$), the secondary signal is of the form

$$u_2(t) = U_p \cos[2\pi(f_p + a_k\Delta f)t] \qquad (8.135)$$

where a_k takes the values -1 or $+1$.

It is therefore a case of FM modulation by a *antipolar* primary binary signal. Due to the nonlinearity of the FM procedure, the secondary spectrum is very complex. It is simplified with the following hypothesis: if the FSK signal is generated by switching between two sinusoidal generators at $f_p + \Delta f$ and $f_p - \Delta f$, it can be considered to be the sum of two OOK signals having one or the other of these frequencies as a carrier and modulated, one by the original binary signal, and the other by its logical inverse (fig. 8.74).

At the switching instants, the relative phase of the two generators can be anything. The result is phase discontinuities for the FSK signal.

Since the Fourier transform is a linear operation, the spectrum of the FSK signal is obtained by the sum of the two OOK spectra (fig. 8.75).

From this calculation we get an expression for the secondary bandwidth

$$B_{2\text{FSK}} = 2B_1 + 2\Delta f \cong 1.6\dot{M} + 2\Delta f \qquad (8.136)$$

which is similar to that given by Carson's rule (8.66). Lines are presented at $f_p - \Delta f$ and $f_p + \Delta f$.

In reality, the FSK signal is rarely generated in this way, but rather by frequency variation of a voltage-controlled oscillator VCO, which leads to a

ANALOG MODULATION

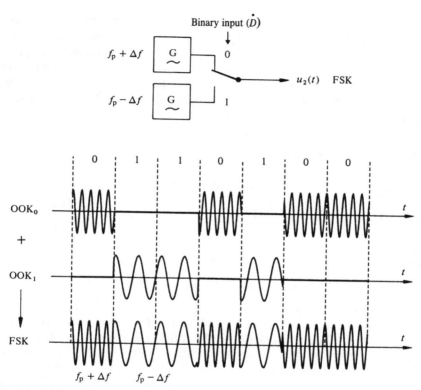

Fig. 8.74 FSK signal with discontinuous phase.

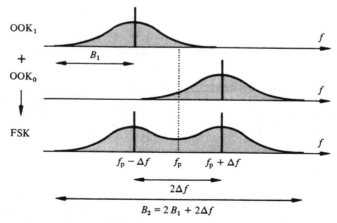

Fig. 8.75 Spectrum of an FSK signal with discontinuous phase.

secondary signal with *continuous phase*. This detail, apparently insignifiant, noticeably modifies the spectrum (fig. 8.76): the discrete lines at $f_p \pm \Delta f$ disappear, and the power spectral density $\Phi_2(f)$ presents maxima spaced approximately $2\Delta f$ apart. These are more accentuated the greater Δf is with respect to B_1 (thus with respect to the bit rate $\dot{D}$). The position of these maxima with respect to f_p depends again on the probability of the two binary values, conforming to the adiabatic theorem (sect. 8.6.8).

8.11.5 Phase shift keying (PSK)

The expression of an *m*-ary PSK modulation is

$$u_2(t) = U_p \cos\left(\omega_p t + a_k \frac{2\pi}{m}\right) \tag{8.137}$$

where $a_k = 0...m - 1$.
For $m = 2$, this expression is reduced to

$$u_2(t) = U_p \cos(\omega_p t + a_k \pi) = \pm U_p \cos\omega_p t \tag{8.138}$$

We can therefore consider this phase modulation to be a DSBSC modulation by a binary *antipolar* signal $(+1, -1)$. Its bandwidth and spectrum are identical to those of OOK modulation, with the slight difference that there is no line at f_p if the two binary values are equally probable. The bandwidth B_2 remains equal to $2B_1$, even if $m>2$, because a PSK modulation with m phases can be decomposed into the sum of two ASK modulations with carriers of the same frequency in quadrature.

8.11.6 Isochronous demodulation

Demodulation is said to be *isochronous* or *coherent* if the receiver arranges a carrier $u'_p(t)$ exactly in phase with the carrier corresponding to the signal $u'_2(t)$ which it receives. A demodulation by inverse modulation is thus possible (sect. 8.3.3). In the case of FSK ($m = 2$), this demodulation is done separately for the two characteristic frequencies $f_p \pm \Delta f$. The result of these operations, followed by a low pass filtering at $f_{1\,\text{max}} = B_1$ is shown in figure 8.77.

8.11.7 Incoherent demodulation

The isochronous restoration (without phase error) of the carrier at reception meets difficulties and requires the use of particular methods (transmission of a pilot, use of a phase-locked loop PLL). A simpler demodulation is often

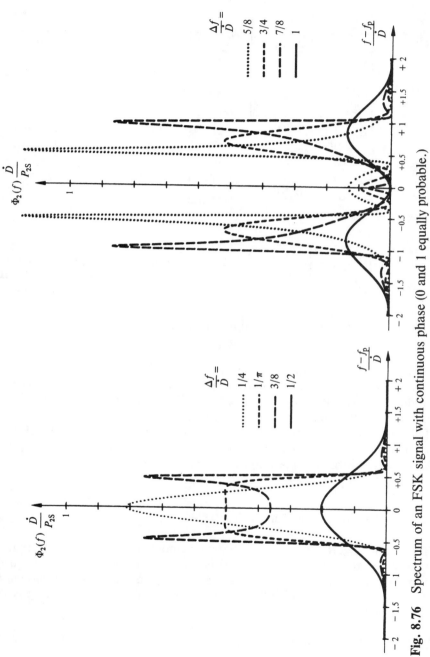

Fig. 8.76 Spectrum of an FSK signal with continuous phase (0 and 1 equally probable.)

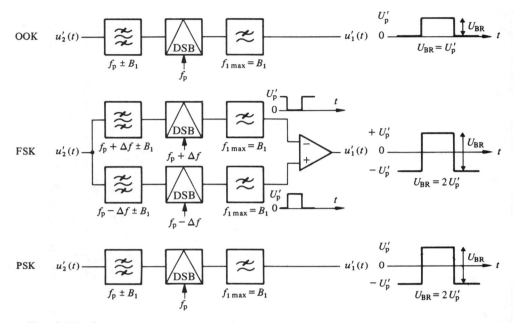

Fig. 8.77 Isochronous demodulation (binary case $m = 2$).

preferred, of the envelope detection type, for example. This in principle restores the same signal with OOK or FSK (in the latter case by separately detecting the two OOK signals at $f_p \pm \Delta f$.) The operation results in a slight degradation of the noise immunity with respect to isochronous demodulation.

This method fails, however, in the case of PSK since the envelope is constant and the carrier frequency unique. A stable reference phase is thus indispensable to the interpretation of the received symbols.

8.11.8 Differential phase shift keying DPSK

To permit nevertheless the incoherent demodulation of a PSK signal, we generally do not evaluate the *absolute* phase of a symbol with respect to the carrier, but only consider the *relative* phase shift of a symbol **with respect to the precedent** $i - 1$. This phase shift $\Delta\varphi_i$ thus takes a value among the m possible values at each symbol.

$$\Delta\varphi_i = \varphi_i - \varphi_{i-1} = a_k \frac{2\pi}{m} \tag{8.139}$$

As with any differential system, the interpretation errors at reception have the tendency to propagate from one symbol to the following symbols, which leads to a higher error probability than with an isochronous demodulation.

To limit the number of bits affected by an interpretation error of $2\pi/m$ at demodulation, it is recommended that a Gray's code be used to translate the groups of lb m bits in phase shifts (fig. 8.78).

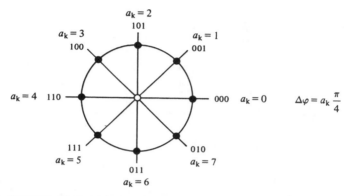

Fig. 8.78 DPSK modulation at $m = 8$ values (3 bits/symbol) according to Gray's code.

8.11.9 Minimum shift keying MSK

A particular case of FSK modulation which is theoretically interesting is that in which $\Delta f/\dot{D} = 1/\pi$ which leads to a spectral density which is virtually uniform between $f_p - \dot{D}/2$ and $f_p + \dot{D}/2$ (figs. 8.76 and 8.79). In practice, the case where $\Delta f = \dot{D}/4$ is even more interesting because its spectral density, although less uniform, is almost entirely concentrated (99% of the power) in a total bandwidth of 1.17 $\dot{D}$ (fig. 8.79). This case is termed *minimum shift keying MSK*.

$$u_2(t) = \pm U_p \cos(\omega_p t \pm \tfrac{1}{2}\pi \dot{D} t) =$$

$$\pm \underbrace{(U_p \cos \omega_p t \cdot \cos \tfrac{1}{2}\pi \dot{D} t)}_{v(t)} \mp \underbrace{(U_p \sin \omega_p t \cdot \sin \tfrac{1}{2}\pi \dot{D} t)}_{w(t)} \qquad (8.140)$$

We therefore see that the MSK signal decomposes into two DSBSC signals, $v(t)$ and $w(t)$, whose the carriers are in quadrature and modulated by two sinusoidal signals equally in quadrature, of frequency equal to one quarter of the bit rate. This statement permits the conception of an isochronous modulator and demodulator (figs. 8.80 and 8.81).

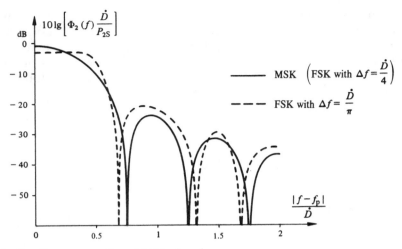

Fig. 8.79 Spectrum of an MSK signal.

The principal problem consists of isochronously extracting the carrier frequency f_p and the bit rate $\dot{D}$ from the received signal $u_2'(t)$.

8.11.10 Phasor diagrams

In the case of discrete modulations, the locus of the extremity of the secondary phasor $\underline{U}_2$ is reduced to m points at the characteristic moments of the symbols (fig. 8.82).

8.11.11 Effect of noise

As with any digital transmission, a regeneration (sect. 5.2) takes place at reception with the aid of thresholds. Although, in the case of a transmission with modulation, this regeneration is generally incorporated in the process of demodulation, we can analyze the operation according to the block diagram in figure 8.83. The noise added to the *modulated* signal during transmission in the channel is modified by the process of demodulaton and leads to regeneration errors, the probability ϵ of which must be evaluated.

The calculation is done with the following hypotheses:

- thermal white noise with a power spectral density $\Phi_N(f) = kT$ (unilateral)
- Gaussian noise
- binary transmission ($m = 2$)

ANALOG MODULATION

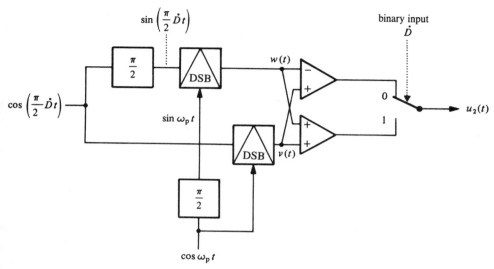

Fig. 8.80 Principle of an MSK modulator.

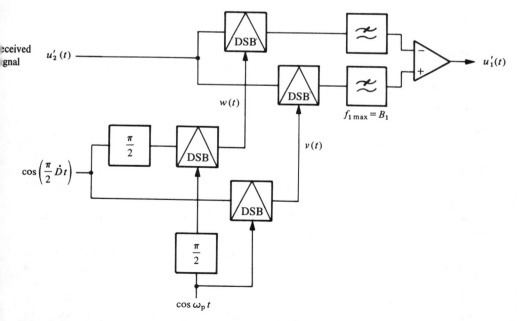

Fig. 8.81 Principle of an MSK demodulator.

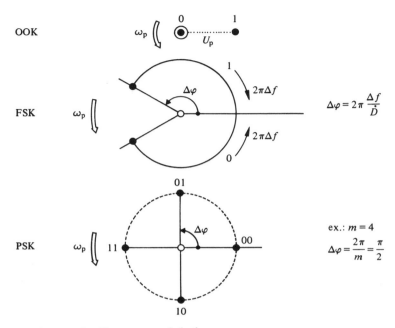

Fig. 8.82 Phasors in discrete modulations.

- equally probable binary values
- primary bandwidth evaluated according to (4.9) at $B_1 \cong 0.8\dot{M} = 0.8\,\dot{D}$.

The evaluation of the error probability of regeneration ϵ can be done in the same manner as in section 5.4.4. In particular, relation (5.25) becomes, with $m = 2$:

$$\epsilon = G_c(\tfrac{1}{2}U_{BR}/U'_{N\,eff}) \tag{8.141}$$

with the following conventions:

- U_{BR} represents the amplitude of the received, demodulated, and equalized elementary base signal (fig 8.79)
- $U'_{N\,eff}$ is the r.m.s. value of the noise collected by the demodulator, after equalization

By assuming that the demodulator does not introduce any amplification in relation to the case of figure 8.77, and that each DSBSC demodulator gives an rms noise output voltage proportional to $\sqrt{2B_1}$, we obtain the values summarized in table 8.84. In the case of FSK, the noise contributions of the two DSBSC demodulators are statistically independent and must be added in power.

ANALOG MODULATION

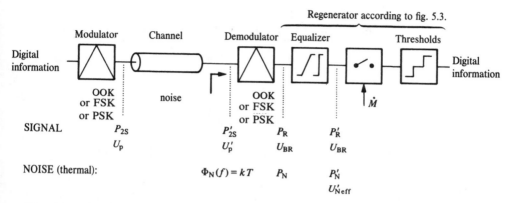

Fig. 8.83 Digital transmission with modulation.

By comparing the three discrete modulations for the *same error probability* ϵ and the *same bit rate* $\dot{D}$, it is found that OOK modulation requires an amplitude U'_p of the carrier

- $\sqrt{2}$ times greater than FSK modulation
- and 2 times greater than PSK modulation which is thus the most favorable of the three.

In certain cases, the power P'_{2S} of the received secondary signal is more significant than its amplitude U'_p. We can, for example, define a *reference signal-to-noise ratio* ξ_0 which contains the two characteristic parameters $\dot{D}$ and P'_{2S} of a binary modulated transmission:

$$\xi_0 = \frac{P'_{2S}}{kTB_1} \cong \frac{P'_{2S}}{0.8kT\dot{D}} \tag{8.142}$$

The transcription of the values of table 8.84 in relation (8.141) thus gives the following comparative values (table 8.85), to be considered in relation to

Table 8.84 Comparison of the three modulations OOK, FSK and PSK in the case of a binary signal, with identical amplitude U'_p at reception.

	U_{BR}	$U'_{N\,eff}$	$U_{BR}/U'_{N\,eff} = 2G_c^{-1}(\epsilon)$	P'_{2S}	B_2 (approx.)
OOK	U'_p	$\sqrt{R}\,\sqrt{1.6\,kT\dot{D}}$	$U'_p/\sqrt{R}\,\sqrt{1.6\,kT\dot{D}} = x$	$U'^2_p/4R$	$1.6\dot{D}$
FSK	$2U'_p$	$\sqrt{2}\,\sqrt{R}\,\sqrt{1.6\,kT\dot{D}}$	$\sqrt{2}\,x$	$U'^2_p/2R$	$1.6\dot{D} + 2\Delta f$
PSK	$2U'_p$	$\sqrt{R}\,\sqrt{1.6\,kT\dot{D}}$	$2x$	$U'^2_p/2R$	$1.6\dot{D}/\mathrm{lb}\,m$

Table 8.85 Comparison of the three modulations with equal power ($m = 2$).

OOK	$\epsilon = G_c\left(\sqrt{\frac{1}{2}\xi_0}\right)$
FSK	$\epsilon = G_c\left(\sqrt{\frac{1}{2}\xi_0}\right)$
PSK	$\epsilon = G_c(\sqrt{\xi_0})$

expressions (5.40) and (5.41) as well as figure 5.20, valid for a binary baseband transmission.

The comparison with equal power is also in favor of PSK, but makes OOK and FSK equal, as a result of the intermittent character of OOK emission, and the continuous character of FSK and PSK.

8.11.12 Applications of discrete analog modulations

When the properties of the available channel preclude digital baseband transmission, the discrete analog modulations allow the translation of frequency characteristics of the signal to adapt it to those of the channel. This is notably the case for

- *data transmission* in an analog channel of bandpass type, for example a telephone channel of a carrier system (section 11.4)
- *radioelectric transmission* of digital information (data or results of a digital modulation, for example PCM) by microwave link or satellite (section 12.3)
- *optical transmission* on fibers (OOK modulated light emission, section 14.3).

The favorable properties of the PSK modulation must be balanced with a greater complexity of the modulation equipment. This is why it is reserved for critical cases (data transmission at high rates, digital radio links).

Due to the economy of the secondary band which it offers, MSK modulation is of particular interest for radio links. Its performance in the presence of noise is similar to that of PSK modulation.

Chapter 9

Digital Systems

9.1 PRINCIPLES AND STRUCTURE

9.1.1 Characteristics of a digital system

Digital system is understood to mean the ensemble of means which permit us to convey information, *in digital form*, from one point to another. This information can be digital in nature from the source (data) or, conversely, it can come from an analog source (speech, music, images, *et cetera*) after having undergone an analog-digital conversion by means of digital modulation (chapter 7). Digital systems therefore make use of the principles of digital transmission (chapter 5).

Most often, we are concerned with systems that allow the simultaneous transmission of several pieces of information, carried by several channels grouped in a *multiplex* mode of operation. Due to the nature of digital transmission (sequence of symbols), the multiplex is a *time-division multiplex* (sect. 4.4.3).

The principal characteristics of a digital system are, in order of importance:
- the *total bit rate* $\dot{D}$;
- the number of channels z (it will be assumed we are discussing channels which are identical with respect to their bit rate $\dot{D}_i$);
- the sequential organization of the symbols corresponding to the different channels and of the auxiliary symbols necessary for signaling and framing, i.e., the *frame structure*;
- the digital *modulation parameters* used (sampling frequency, quantizing law, code), in particular the number q of quantizing steps and the number of bits b which represent them;
- the digital *transmission parameters* used (transmission medium, mode, symbol rate, error probability, *et cetera*).

9.1.2 Advantages of digital systems

The principal advantage of digital systems is tied to the *possibility of regeneration* (section 5.2) of the information which they transmit. Even in the presence of significant noise, the transmission quality, expressed as the probability of regeneration error ϵ, can remain excellent.

Digital systems are also of economic interest due to the better use of transmission media that *multiplexing* allows. This advantage is further emphasized by the technological evolution of microelectronics toward more complex and cheaper integrated circuits.

Furthermore, the digital form lends itself well to the coexistence of services as different as telephony and data transmission in the same system, thus opening the way for a future *integrated services digital network* (section 15.8).

However, these advantages must be paid for by the requirement of a considerable secondary bandwidth to implement digital transmission. In effect, while an *analog* telephone channel conventionally occupies a bandwidth of aproximately 3.1 kHz (sect. 2.4.9), the same channel in *digital* form requires a rate of 64 kbit/s (sect. 7.5.12), i.e., with binary transmission, a bandwidth of approximately 51 kHz (according to the expanded Nyquist's criterion, sect. 5.3.4). This is approximately 16 times greater than that for analog transmission.

In summary, it can be said that digital systems can use a transmission channel of mediocre quality (high attenuation, noise, crosstalk), provided that it offers the necessary bandwidth. They thus make good use of media and circumstances which would be very critical in an equivalent analog transmission.

9.1.3 Structure of a digital system

A digital system installed on a line is composed of the following principal elements (fig. 9.1):

- *terminal equipment* at the two ends of the line, whose essential function is to constitute the time division multiplex with z channels and to convert the analog information into digital messages (coding) and the reverse (decoding). All the interface functions (signaling, adaptation of levels and impedance, monitoring, synchronization, *et cetera*) are also carried out by the terminal equipment;
- *line equipment* (regenerators), distributed along the length of the line at regular intervals. They operate a regeneration of the signals carrying digital information.

In the case of digital radio systems, the structure is similar, except that there is only a single regenerator per hop and the transmission equipment also includes the modulators and demodulators necessary to adapt the signals to the available frequency range (section 12.3).

Digital systems are always, by nature, "four-wire" systems (sect. 4.6.4).

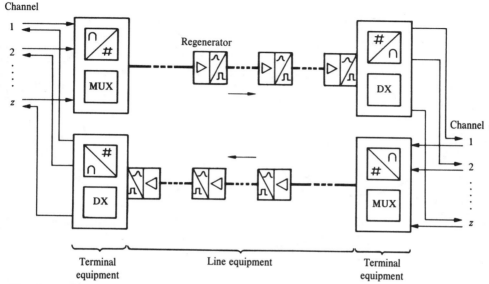

Fig. 9.1 Structure of a digital system.

9.2 STRUCTURE OF THE DIGITAL TIME-DIVISION MULTIPLEX

9.2.1 Organization of the frame

For each digital channel there is assigned a group of b bits, called a *"word"* which corresponds to the quantized and coded value of a sample, or to a digital message element (data). The groups belonging to the same channel succeed each other at a frequency f_e, equal to the sampling frequency. The bit rate of each channel i is thus

$$\dot{D}_i = f_e b \qquad (9.1)$$

In the case of digital PCM telephony (sect. 7.5.12), these words are composed of 8 bits and called *octets*. The bit rate of each channel is thus, from (7.45), equal to 64 kbit/s.

When z digital channels are assembled into a time division multiplex, the collection of the z words of b bits (and auxiliary bits which are added to them), within a period $T_e = 1/f_e$, constitutes a *frame*. The structure of the frame is strictly repetitive (but not its detailed content, because the channels contain variable digital information, of random character).

Two types of frame organization can be envisaged (fig. 9.2):

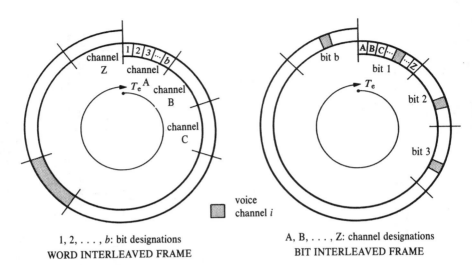

1, 2, ..., b: bit designations
WORD INTERLEAVED FRAME

A, B, ..., Z: channel designations
BIT INTERLEAVED FRAME

Fig. 9.2 Frame structures.

- *word interleaved*: the frame is subdivided into $y \geq z$ *time slots* each containing b *grouped* bits corresponding to the same digital channel or auxiliary bits;
- *bit interleaved*: the frame is subdivided into b groups each containing z bits of the same order belonging to each of the channels.

The word interleaved structure corresponds well to the mode of functioning of a PCM modulator common to the z channels (fig. 9.7), while the bit interleaved structure has advantages for time division digital switching.

9.2.2 Frame alignment

The *frame alignment* or *framing* operation, typical of time division mutliplex, consists of synchronizing the receiving terminal equipment, in frequency and in phase, to the cycle of symbols it receives. This operation is evidently necessary each time the receiver is switched on (initially or after a break). Once aligned, the receiver nevertheless needs a further periodic time reference, so as to check its isochronism and detect eventual shifts.

The necessary time reference consists of a particular pattern of several bits, periodically carried by the frame according to one of the following organizations:

- *grouped* framing pattern, constituted of v consecutive bits at the beginning of the frame;

- *distributed* framing pattern, bit by bit, at regular intervals within the frame, or on several frames at one bit per frame.

When the receiver has *lost frame alignment*, it searches for this pattern so as to re-align itself with as short a delay as possible (a few frames). There is then a *danger of simulation* of the framing pattern by chance combination of other information-carrying bits. One can protect oneself from this by

- choosing a framing pattern with low autocorrelation, impossible to imitate by shifting and infringement on random neighboring bits, for example:
  $v = 3$:   110
  $v = 7$:   1110010
- blocking all the channels of the frame (replaced by the emission of a deterministic signal), when the framing is lost at reception. This necessitates an announcement warning in the opposite direction, which is supposed to be correctly aligned;
- confirming correct framing by a different criterion than the framing pattern (for example, absence of the pattern in one frame out of two).

However, *in an aligned situation* (monitoring), the reaction to an incorrect reception of the framing pattern must be delayed (hysteresis), in such a way as to avoid reacting to each transmission error. An example of the solution to the framing problem is given in the discussion of primary PCM system (sect. 9.3.3).

9.2.3 Signaling

Signaling has as its aim the transmission of auxiliary information, which is digital in character, (digits, commands, acknowledge signals, *et cetera*) from exchange to exchange, to control switching and management operation of the network (sect. 15.5.6).

Aside from analog signaling by sinusoidal carriers (simple or multiple), "in-band" or "out-band", usual in analog systems (sect. 10.2.5) and translatable by PCM modulation, digital systems lend themselves by nature much better to direct digital transmission of signaling information. Several solutions are possible:

- *in-octet signaling*, also called "bit stealing": the bit of lowest weight among the 8 bits representing a coded sample of speech is periodically assigned (for example, every 6 frames) to signaling. The result is an imperceptible degradation of the telephone transmission, but a very annoying restriction in the use of the digital channel for data transmission;
- *out-of-octet signaling, channel by channel*: each digital channel arranges, besides its octet, one or several signaling bits. These signaling bits of the

370   TELECOMMUNICATION SYSTEMS

z channels of the frame can be *distributed*, i.e., juxtaposed at each octet as a supplementary bit per time slot (possible signaling rate: 8 kbit/s per channel), or *grouped* in a time slot reserved for this effect and of which the bits are assigned in turn, cyclically, to the z channels of the frame (possible signaling rate: $64/z$ kbit/s per channel);
- *common channel signaling*: one time slot per frame is reserved for signaling and assigned *from case to case, according to need* to one channel, then to another. Signaling is then done with the aid of *labelled messages*, the label indicating which channel the message belongs to (instantaneous signaling rate available for one channel at a time: 64 kbit/s). This type of signaling, recently developed, is envisaged for direct transmission of auxiliary information between two processor-controlled exchanges.

9.3 PRIMARY MULTIPLEX SYSTEMS

9.3.1 European and American systems

The delay with which the European Conference of Posts and Telecommunications (CEPT) has undertaken the definition of a primary digital PCM system has allowed it to profit from the experience of American systems. The result is a different primary PCM system. On the international level, the two types of systems coexist although they are incompatible, and are the subject of two Recommendations of the CCITT. Table 9.3 compares the main characteristics of the two systems.

Only the European norm will be considered in the rest of this chapter.

9.3.2 Frame structure of the frame of the primary PCM system at 2,048 Mbit/s

The European primary PCM system is characterized by a total bit rate of 2,048 Mbit/s resulting from the subdivision of the frame of $T_e = 125$ µs in 32 time slots of 8 bits each (word interleaved frame). The structure of the frame is given in figure 9.4. The time slots nos. 1 to 15 and 17 to 31 are assigned to 30 channels, each corresponding to a rate $\dot{D}_i$ of 64 kbit/s, usable for telephone transmission with PCM modulation (according to sect. 7.5.12) or for any direct digital transmission (data, telecopying, *et cetera*).

The time slot no. 0 is assigned every other frame to the frame alignment pattern FRA. In alternate frames, it carries the alarm bit A announcing the loss of frame alignment in the other direction.

DIGITAL SYSTEMS

Table 9.3 Primary European and American systems.

	European system (Recommendation G 732)	American system (Recommendation G 733)
Sampling frequency		$f_e = 8\,\text{kHz}$
Number of quantizing steps		$q = 256$
Number of bits per sample		$b = \text{lb}\, q = 8$
Bit rate per channel		$\dot{D}_i = 64\,\text{kbit/s}$
Quantizing		nonuniform
Law	$A\,(= 87.6)$	$\mu\,(= 255)$
Compression characteristic	13 segments	15 segments
Number of time slots $y =$	32	24
Number of channels $z =$	30	24
Number of bits/frame	$32 \cdot 8 = 256$	$24 \cdot 8 + 1 = 193$
Total bit rate	$256 \cdot 8\,\text{kHz} = 2{,}048\,\text{Mbit/s}$	$193 \cdot 8\,\text{kHz} = 1{,}544\,\text{Mbit/s}$
Framing	grouped word of 7 bits in the 0 channel of odd frames	distributed sequence 101010... consisting of the 193rd bit of odd frames
Signaling	out-of-octet, grouped in channel 16, consisting of 4 bits per channel, distributed over 16 frames ($= 1$ multiframe) (fig. 9.5)	in-octet (bit stealing), 8th bit, one frame in 6

Signaling is of the out-of-octet type, channel by channel, consisting of 4 bits per channel, grouped in the two halves of the time interval no. 16. It therefore requires 15 frames to carry the signaling of the 30 channels. Completed by a 16th frame (frame no. 0) they constitute a *multiframe*. The time slot no. 16 of this 16th frame contains a multiframe alignment pattern which permits unambiguous numbering of the frames at the interior of the multiframe (fig. 9.5).

Each channel thus arranges the signaling of 4 bits every 16 frames, i.e., every 2 ms.

9.3.3 Framing method of the primary PCM system

Fig 9.6 gives the complete flow chart of the framing process, as it is specified in Recommendation G 732 of the CCITT. The following points should be noted:
- the hysteresis of the monitoring process: framing is only considered to be lost after three consecutive absences of the framing pattern FRA;
- the confirmation of framing by the presence of a different bit (B_2), at the

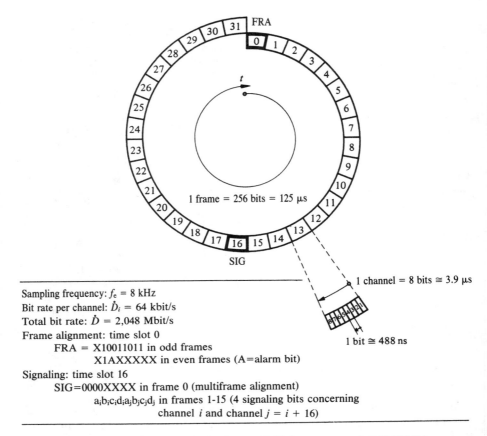

Fig. 9.4 Frame structure of the primary PCM system at 2,048 Mbit/s.

second place of the time interval no. 0 in the frame which follows that in which the FRA pattern was recognized;
- the announcement of framing loss by the alarm bit A, emitted in the opposite direction.

The reframing procedure takes between 250 μs and 375 μs depending on the case.

9.3.4 Primary PCM terminal equipment

The terminal equipment presents the inputs and outputs of the 30 analog telephone baseband channels (300 Hz - 3.4 kHz) on one side, and on the other

DIGITAL SYSTEMS

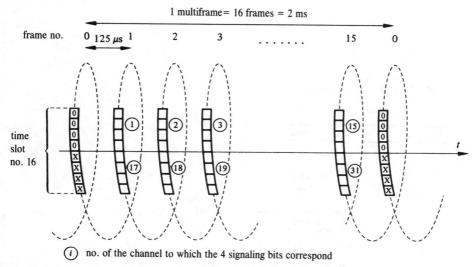

ⓘ no. of the channel to which the 4 signaling bits correspond

Fig. 9.5 Signaling multiframe.

the two multiplex lines (forward and return) of the primary digital system. It fulfills the following functions:

- PCM modulation and demodulation (according to section 7.5)
- time-division multiplexing and demultiplexing of the 30 channels
- composition of the frame according to figure 9.4
- generation of the emission clock (with a precision of $50 \cdot 10^{-6}$), restoration of the clock at reception, the two directions being possibly asynchronous
- frame alignment according to figure 9.6
- insertion and extraction of the signaling for each channel
- monitoring and maintenance.

The classical solution consists of multiplexing the analog samples (PAM) corresponding to the channels *before* converting them to PCM (fig. 9.7). The PCM coding and decoding device, also called a "*codec*," is unique and used in turn by all the channels. It must, consequently, be rapid (one sample to be coded or decoded every 3.9 µs). This structure, justified by the cost of the "codec", presents a danger of time crosstalk (sect. 8.10.7) during the assembly of the 30 channels in PAM (still analog!) on the same bus at the input of the encoder or at the output of the decoder.

Microelectronic technology presently allows the implementation of the "codec", including the nonuniform quantizing law, in the form of a single integrated circuit. It thus becomes possible to implement quantizing and coding

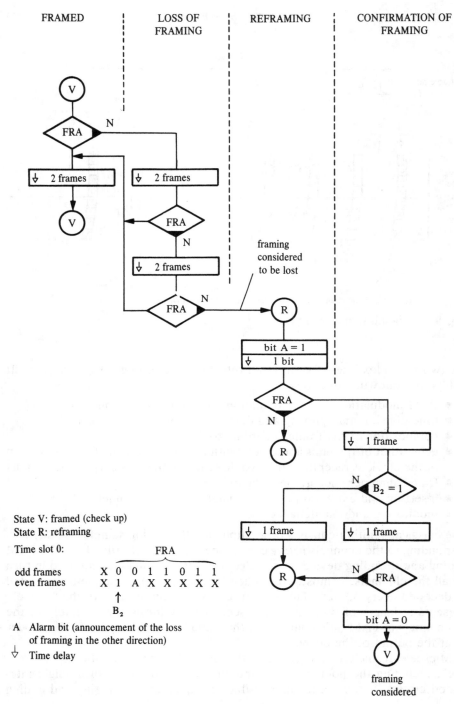

Fig. 9.6 Process of framing in a primary system.

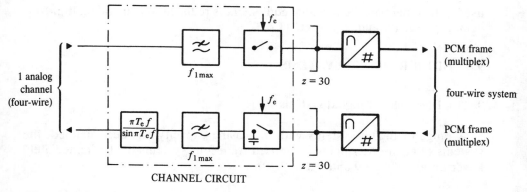

Fig. 9.7 Shared "codec" structure.

for each channel *individually* and *afterwards* to multiplex the samples in digital form. This solution (fig. 9.8) eliminates the problem of time crosstalk.

It is further noted in figures 9.7 and 9.8 that:

- filtering is indispensable *before* sampling (emitting side) in such a way as to satisfy the sampling theorem (sect. 4.3.5);
- filtering (on the receiving side) is necessary to the smoothing of the staircase signal (PAM demodulation);
- the attenuation distortion in $\sin\pi T_e f / \pi T_e f$ introduced by holding the samples (sect. 4.3.7 and 7.2.4) must be compensated for.

All or part of the channels can be used for data transmission instead of telephony. These channels are then inserted directly into the time-division multiplex, without going through a "codec." Ultimately, if all the channels are

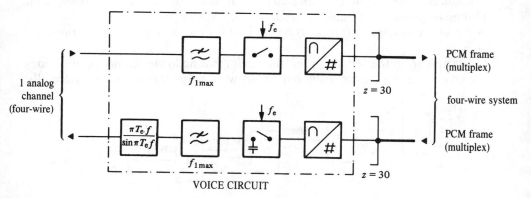

Fig. 9.8 Single channel "codec" structure.

used in this manner, the terminal equipment is reduced to a digital multiplexer-demultiplexer ("muldex") (sect. 11.4.7).

9.4 HIGHER ORDER SYSTEMS

9.4.1 Hierarchy of digital systems

A range of digital systems with increasing capacity has been defined, the systems of each order being composed from four systems or the immediately lower order (table 9.9 and figure 9.10).

Table 9.9 Characteristics of digital systems.

Order	Number of digital telephone channels (at 64 kbit/s)	Number of bits per sampling period ($T_e = 125$ μs)	Total bit rate $\dot{D}$ [Mbit/s]	Abbreviated designation [Mbit/s]
1 (Europe)	30	256	2,048	2
1 (USA)	24	193	1,544	1,5
2	120	1,056	8,448	8
3 (Europe)	480	4,296	34,368	34
4	1920	17,408	139,264	140
5	7680	70,624	564,992	565

With regard to the data mentioned in figure 1.10, we know that the primary system (2 Mbit/s) is capable of transmitting the signals corresponding to *one* visiophone channel, on the condition that it proceeds to a source coding with redundancy reduction, unless the secondary system at 8 Mbit/s is necessary. A 4th order system (140 Mbit/s) is required for the digital television transmission.

Starting from the second order, the equipment no longer contains an analog-digital converter ("codec") and only deals with digital frames. They essentially

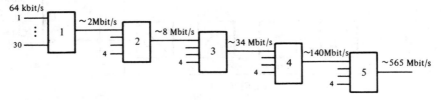

Fig. 9.10 Hierarchy of digital systems.

DIGITAL SYSTEMS

play the role of *multiplexers*, combining into a new frame the frames of four systems of the preceding order, called *tributaries*.

9.4.2 Multiplexing of anisochronous tributaries

During the construction of an *n*th order multiplex starting with frames of order $n - 1$, we are confronted with the problem of **anisochronism of the tributaries**. In effect, the frames to be grouped come from terminal equipments which are geographically distinct and often far away (fig. 9.11), and whose clocks have neighboring frequencies (*plesiochronous* tributaries) or, in the best case, equal (*synchronous* tributaries), but whose relative phases can certainly be anything and even vary, because the lines have different propagation times which furthermore depend on temperature.

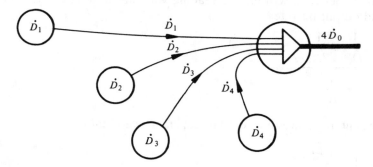

Fig. 9.11 Composition of a multiplex from four anisochronous tributaries.

The multiplexing of the four tributaries requires perfect isochronism between the bits. It is therefore necessary to bring them in beforehand to exactly the same rate $\dot{D}_0$. This is done with the aid of a **buffer store**, on each tributary, capable of storing an entire frame of duration T_e, i.e., of compensating for a time shift of

$$\Delta t \leq T_e \tag{9.2}$$

between the moment of arrival of the bit and its reading outside the buffer store.

The tributary k ($k = 1...4$) writes its bits with its own rate $\dot{D}_k$ in this memory from which they are read with the internal rate $\dot{D}_0$ (fig. 9.12).

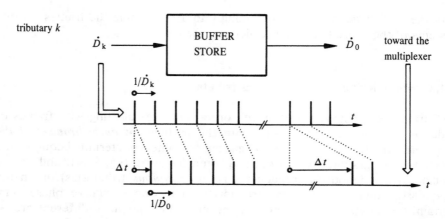

Fig. 9.12 Compensation of the time shift Δt by a buffer store.

The shift Δt between writing and reading varies according to the ratio of the input and output bit rates

$$\frac{d\Delta t}{dt} = \frac{\dfrac{1}{\dot{D}_0} - \dfrac{1}{\dot{D}_k}}{\dfrac{1}{\dot{D}_0}} = 1 - \frac{\dot{D}_0}{\dot{D}_k} \tag{9.3}$$

Assuming the rates are different but constant, we obtain

$$\Delta t(t) = \left(1 - \frac{\dot{D}_0}{\dot{D}_k}\right) t \tag{9.4}$$

9.4.3 Loss and repetition of information with asynchronous operation

If $\dot{D}_k < \dot{D}_0$, situations occur (fig. 9.13(a)) in which the reading should be made before writing ($\Delta t < 0$). Then the corresponding word of the *preceding frame* is read a second time (information repetition). Inversely, when $\dot{D}_k > \dot{D}_0$ (fig. 9.13(b)), Δt increases and if it exceeds T_e, a new word is written in the memory before the preceding word could be read. The latter is thus irreparably lost.

These *losses or repetitions* of information, called *slips*, occur with a period T_s which is derived from (9.4) with $\Delta t = T_e$

$$T_s = \frac{T_e}{\left|\dfrac{\dot{D}_0}{\dot{D}_k} - 1\right|} = T_e \left|\frac{\dot{D}_k}{\dot{D}_0 - \dot{D}_k}\right| = T_e \left|\frac{\dot{D}_k}{\Delta \dot{D}_k}\right| \tag{9.5}$$

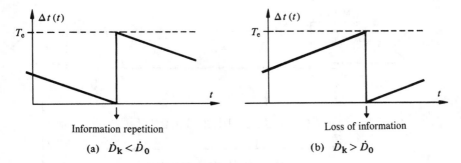

Fig. 9.13 Evolution of the time shift Δt between writing and reading of the buffer store.

where $\Delta \dot{D}_k$ expresses the difference between the bit rate of the tributary and the internal clock rate (table 9.14).

In telephony, these slips correspond to the loss or redoubling of a sample and produce parasitic pulses ("clicks") in the baseband, less perceptible when less frequent. However, in data transmission, the loss or repetition of an eight-bit word can have annoying consequences.

Table 9.14 Period of slips as a function of the relative difference of rates (for $T_e = 125$ μs).

$\dfrac{\Delta \dot{D}_k}{\dot{D}_k}$	T_s
10^{-3}	0.125 s
10^{-6}	2' 05"
10^{-9}	1.45 day
10^{-12}	3.96 yr

9.4.4 Synchronization by justification

To avoid these slips, the differences in rate can be compensated for by a process of *stuffing*, which consists of inserting a void (non-information carrying) bit from time to time in the fastest of the two rates $\dot{D}_k$ or $\dot{D}_0$, in such a way as to respect the continuity principle (fig. 9.15). The rate $\dot{D}_{jk}$ of such stuffing bits for the tributary k compensates for just the difference between the internal rate $\dot{D}_0$ and the rate $\dot{D}_k$ of the tributary

$$\dot{D}_{jk} = |\dot{D}_0 - \dot{D}_k| = \Delta \dot{D}_k \tag{9.6}$$

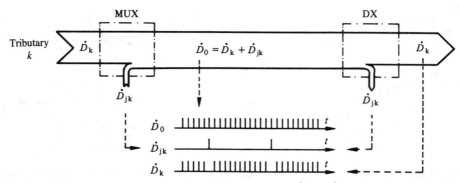

Fig. 9.15 Continuity of the rates (case in which $\dot{D}_0 > \dot{D}_k$).

The occasional insertion of a stuffing bit in the place of a normal information bit shifts all the rest of the frame by one bit. This must absolutely be signaled by a *stuffing control signal* which must be protected against transmission errors (for example, three identical bits 111 or 000, with majority decision). The demultiplexer can thus extract the void bit of the frame and restore an irregular frame with respect to tributary k, in which the bits follow each other with an *instantaneous rate* $\dot{D}_0$; however, holes appear in place of stuffing bits, thus leaving an *average rate* of $\dot{D}_0 - \dot{D}_{jk} = \dot{D}_k$.

The stuffing decision is taken with a view to the time shift Δt between writing and reading in the buffer store. On the first occasion, before Δt could be zero, a stuffing bit is inserted, which delays the reading of the previous bit by $1/\dot{D}_0$ and further increases Δt (fig. 9.16). A two-bit capacity (per tributary) is theoretically sufficient for the buffer-store.

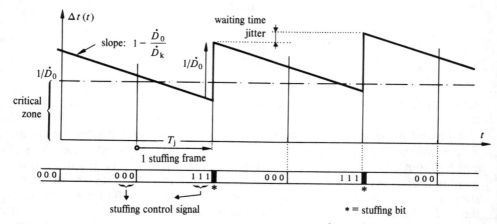

Fig. 9.16 Insertion of a stuffing bit (case in which $\dot{D}_0 > \dot{D}_k$).

DIGITAL SYSTEMS

The periodicity with which the stuffing bits can be inserted into predetermined places defines a *stuffing frame* of duration T_j (not necessarily equal to the frame with duration T_e related to the sampling). The maximum rate of stuffing bits for tributary k is thus

$$\dot{D}_{jk\max} = \frac{1}{T_j} \qquad (9.7)$$

From (9.3) we derive the maximum relative difference of rates which can be compensated for by this method, with a maximum of one bit per stuffing frame

$$\left(\frac{d\Delta t}{dt}\right)_{\max} = \left(1 - \frac{\dot{D}_0}{\dot{D}_k}\right)_{\max} = \left(\frac{\Delta \dot{D}_k}{\dot{D}_k}\right)_{\max} = \frac{\dot{D}_{jk\max}}{\dot{D}_k} = \frac{1}{T_j \dot{D}_k} \qquad (9.8)$$

The fact that the stuffing bit can only be inserted at a certain moment, predetermined for this effect in the frame, leads to a slight phase irregularity, called the *waiting time jitter*, which is of a random nature and difficult to eliminate because it contains very low frequencies.

In the preceding discussion, it has been implicity assumed that $\dot{D}_0 > \dot{D}_k$, which leads to the completion of the frame of tributary k by void bits (positive stuffing). The reverse can also be done, i.e., removing stuffing bits contained in the tributary k to adapt its rate to the condition $\dot{D}_k > \dot{D}_0$ (negative stuffing), or, finally, to combine the two procedures.

9.4.5 Frame structure of the second-order system at 8 Mbit/s

As an example of a multiplex with stuffing, the frame structure of the 8 Mbit/s second-order system (CCITT Recommendation G 742) is given in figure 9.17.
The characteristics of this system are:

- bit rate of the multiplex $\dot{D}_{\text{Mux}} = 8.448$ Mbit/s, tolerance: $\pm 30.10^{-6}$
- stuffing frame of 848 bits, thus $T_j \cong 100.4$ µs
- number of tributaries (at 2.048 Mbit/s):4
- frame structure: bit interleaving, of each tributary in turn.
- framing of the stuffing frame: a pattern of 10 bits grouped at the start of the frame
- number of significant bits belonging to each tributary per stuffing frame: 206 (205 in the case of emission of one stuffing bit)
- stuffing control signal: 3 bits per tributary, distributed in the stuffing frame, decision by majority.

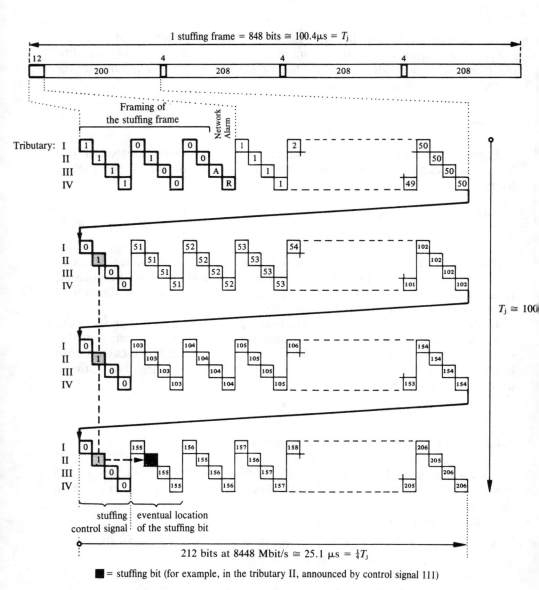

Fig. 9.17 Frame structure of the 8 Mbit/s system.

These characteristics bring up the following remarks:
- the stuffing frame ($T_j \cong 100.4$ μs) does not correspond to the sampling frame of the tributaries ($T_e = 125$μs). The secondary multiplex ignores the structure of the primary frames and considers only the bit flow;
- the maximum compensation capacity by stuffing is, for each tributary independently, of 1 bit every 212 bits, which allows the compensation of a maximum relative difference of 4.7⁰/oo between the rate $\dot{D}_k$ (tributary) and $\dot{D}_0 = \dot{D}_{Mux}/4$ (multiplex);
- for the nominal primary rate $\dot{D}_k = 2.048$ Mbit/s, it is necessary to emit an average of $\dot{D}_k T_j = \dot{D}_k(848/\dot{D}_{Mux}) = 205 \frac{19}{33}$ significant bits per stuffing frame for each tributary. This means that, on the average, 14 frames out of 33 contain one stuffing bit.

9.5 TRANSMULTIPLEXERS

9.5.1 Motivation

As digital systems are progressively introduced into the telecommunication network, the problem of their coexistence with present analog frequency-division multiplex FDM systems (carrier systems, section 10.1) occurs. It would be interesting to go *directly* from an analog frequency division multiplex (FDM) to a digital time division multiplex (TDM) and vice versa, using equipment called a *transmultiplexer*.

In particular, a transmultiplexer is designed to convert a supergroup (60 telephone channels) of a carrier system into two primary digital systems and the reverse, without using a baseband demodulation (fig. 9.18).

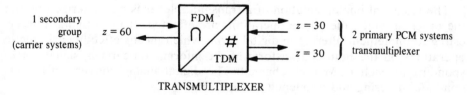

Fig. 9.18 Function of the transmultiplexer with 60 channels.

9.5.2 Principles and implementation

In the FDM → TDM direction, the following operations are necesary for each *i* channel:

- filtering $f_i...f_i + B_1$
- sampling
- analog-digital conversion (PCM modulation)
- time interleaving of samples corresponding to z channels

Conversely, in the TDM → FDM direction:
- time demultiplexing
- digital-analog conversion
- filtering $f_i...f_i + B_1$
- frequency juxtaposition of the z channels transposed in frequency.

The demultiplexed digital channel i presents a repetitive spectrum (sidebands of nf_e) from which the band corresponding to its position in the frequency multiplex can be extracted (by reversing it, if necessary). This is because the carriers of the FDM supergroup are, like the sampling frequency f_e (=8 kHz) of the digital system, multiples of 4 kHz (fig. 9.19).

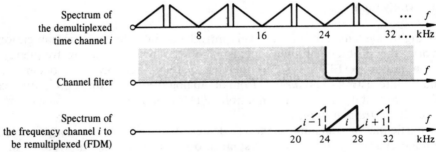

Fig. 9.19 Filtering of a sampled channel (conversion of TDM to FDM).

The technical implementation of a transmultiplexer is characterized by the the order in which the above operations are effected. It is particularly interesting to implement filtering in digital form, which means effecting arithmetic operations on the Fourier transform of the information-carryng signal corresponding to each PCM time channel. The digital-analog conversion is thus done *after* filtering and frequency multiplexing.

9.6 TRANSMISSION EQUIPMENT

9.6.1 Media

All transmission media are suitable for digital systems, but with different characteristics:

- ***balanced pairs:*** while they are difficult to use with analog systems outside the baseband, due to the degradation of the near- and far-end crosstalk (sect. 3.5.5) as the frequency increases, they are very suitable for *primary digital systems* (2 Mbit/s). By means of a primary PCM system on two balanced pairs ("four-wire" system), this solution allows the transmission capacity of one pair to be multiplied by 15 with respect to "two-wire" transmission in baseband. The limiting and determining factor for planning (section 9.7) here is the *crosstalk* between pairs of the same cable (in particular, near-end crosstalk, if the two directions are found in the same cable);
- ***coaxial pairs:*** a quasi-ideal medium for digital transmission, as for analog transmission, due to its very low phase distortions and excellent crosstalk qualities (sect. 3.6.5), the coaxial cable can serve as a support for 8 Mbit/s digital systems (mini-coaxial 0.7/2.9mm), 34 Mbit/s, 140 Mbit/s and 565 Mbit/s. The predominant source of disturbance is thus the *thermal noise* at the input of the regenerators (section 9.9);
- ***microwave links:*** translated with the aid of a discrete FSK or PSK multiphase modulation (section 8.11) in the suitable frequency band (1 ... 22 GHz), the signals of digital systems of order 1 to 4 (2 Mbit/s... 140 Mbit/s) also lend themselves to transmission over microwave links (section 12.3). In the case of satellite links, the digital systems are of interest for a frequency-division multiple access (FDMA) with a large number of individual PCM modulated carriers per channel (SPADE systems, sect. 13.5.4) or for a time-division multiple access (TDMA);
- ***optical fibers:*** The problem of linearity in optoelectronic emitters and receivers calls for on-off emission procedures. Digital, in particular binary, systems are therefore particularly suitable for optical transmission in the entire range of 2 Mbit/s to 565 Mbit/s.

9.6.2 Choice of a transmission mode

The requirements of digital line transmission are the following:
- avoidance of having to transmit and regenerate a *dc component* (so as to permit coupling by transformers and simplify regenerators);
- avoidance of the range of *low frequencies* in which linear distortions are particularly significant;
- limit the extension of the spectrum towards the *high frequencies* due to the increase of attenuation and crosstalk (balanced pairs);
- guarantee a *clock rate content* (frequency corresponding to the symbol rate $\dot{M}$) sufficient to assure the synchronization of regenerators and of the receiver, even when the channels are passive or very weakly modulated;

- allow *transparent transmission*, i.e., that for each binary combination at the input there is one and only one corresponding binary combination, depending on the code used, at the output.

Furthermore, the mode chosen must be reasonably simple to implement. Numerous propositions have been made to satisfy these requirements at best. All interpose a form of *redundancy*, i.e., an increase of the gross bit rate $\dot{D}'$, with the consequence of either an increase of the necessary bandwidth (higher symbol rate $\dot{M}$), or a decrease of noise immunity (symbols of order $m>2$).

9.6.3 Efficiency of a mode: definition

The *efficiency* of a mode is expressed by the ratio of the bit rate $\dot{D}$ actually used (bit rate of the system), to the gross maximum bit rate $\dot{D}_{max}$ possible with an m-ary symbol rate $\dot{M}$, given that, with (1.10)

$$\eta = \frac{\dot{D}}{\dot{D}'_{max}} = \frac{\dot{D}}{\dot{M} \text{ lb } m} \tag{9.9}$$

The modes with better efficiency are generally also more complex to generate and to interpret.

9.6.4 AMI mode for primary (2 Mbit/s) and secondary (8 Mbit/s) systems.

The elimination of the dc component is obtained by a *pseudo-ternary* mode or *Alternate Mark Inversion AMI* in which
- the binary state 0 is represented by a zero signal
- the binary state 1 is represented alternatively by a positive and negative signal, of duration equal to $1/\dot{D}$ (AMI-NRZ mode) or shorter, e.g., $1/2\,\dot{D}$ (AMI-RZ mode).

It is therefore a *ternary* ($m=3$) mode, in the sense that each symbol can take three values: $a_k = -1, 0, +1$. However, the sequential law of polarity alternation does not allow two successive symbols to have the same sign ($+1+1$ or $-1-1$). Of the nine possible combinations of two ternary symbols, only seven are used to represent the four groups of two binary elements (from which we get the qualifying designation pseudo-ternary).

This mode, which is very simple to implement, presents the following characteristics:
- the symbol rate $\dot{M}$ remains equal to the net decision rate $\dot{D}$ despite the fact that $m = 3$, because a pseudo-ternary symbol corresponds to each bit;

- its efficiency according to (9.9) becomes

$$\eta_{AMI} = \frac{1}{lb3} \cong 0.63 \tag{9.10}$$

- the power spectral density $\Phi(f)$ of a random binary sequence emitted according to this mode (fig. 9.20) is very weak at very low frequencies, zero at $\dot{D}$ and its multiples, and maximum for $f \cong 0.45\,\dot{D}$;

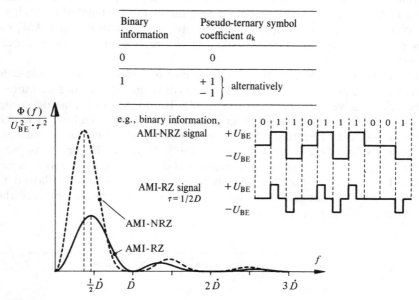

Fig. 9.20 Characteristics of the AMI mode.

- it contains no frequency corresponding to $\dot{D}$ (zero of the spectrum), but lines are present at multiples of $f_e = 8$ kHz (due to the repetitive structure of the frame, framing word, *et cetera*). In particular the line at $\dot{D}/2$ is important (near the maximum of the spectral density envelope) and can, after filtering and doubling of the frequency, serve in the synchronization of the regenerators;
- the rectification of the AMI-RZ signal at reception provides an RZ binary signal with a spectrum which cancels at multiples of $2\dot{D}$ and presents a line at $\dot{D}$, very favorable for the synchronization of regenerators;
- the clock rate content, in the case of a zero analog signal (passive channels), is notably increased by the choice of a code with inversion of even numbered bits (section 7.5.8 and fig. 7.19).

9.6.5 High density modes HDB

The inversions of even numbered bits in the process of coding is inefficient for increasing the clock rate content in the case of digital channels used for data transmission, because this information, which is digital by nature, is injected into the frame without passing through the encoder. Long series of zeros, which are impossible to prevent, may then deprive the regenerators of all synchronization information. To avoid this, a group of several consecutive zeros is replaced by a group containing a factitious 1, signaled as such to the receiver by a polarity in violation of the law of alternation of the AMI mode. This principle is systematically applied to all primary and second-order digital systems.

The European digital systems make use of a pseudo-ternary mode called a *high density bipolar code* HDB 3 which avoids the appearance of more than three consecutive zero symbols (fig. 9.21). It consists of replacing groups of four consecutive binary zeros by groups of four ternary symbols of which the last is non-zero and emitted with the *same* polarity as the last non-zero symbol, i.e., in *violation of the alternation law* of the AMI mode. This permits the easy identification of such a group at reception and its interpretation as four binary zeros. Furthermore, the first of the four ternary symbols is chosen to be positive, zero, or negative in such a way as to maintain or reset the dc component to a zero value.

	Binary	Pseudo-ternary symbols coefficient a_k	Mode	
	0	0	AMI	
	1	$+1/-1$ alternatively		HDB 3
Exception:	0000	B00V $B = -\Sigma a_k$ $V = \pm 1$ in violation of alternation		

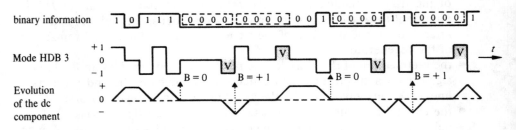

Fig. 9.21 HDB 3 mode.

The power spectral density of the HDB 3 mode differs very little from that of the AMI mode from which it is derived.

9.6.6 Modes for systems of an order higher than 2

The objective consists of searching for modes for these systems with better efficiency, while satisfying the requirements of section 9.6.3. This can be done by representing a group of *x binary elements* by a group of *y m-ary symbols*. We then choose among the m^y possible combinations of *y* *m*-ary moments, those which can represent the 2^x binary groups with a minimal fluctuation of the dc component. The symbol rate is thus

$$\dot{M} = \frac{y}{x} \dot{D} \tag{9.11}$$

and the efficiency of these modes according to (9.9) becomes

$$\eta_{xByM} = \frac{x}{y \text{ lb } m} \tag{9.12}$$

A typical example, that of 4B3T mode (x = 4, y = 3, m = 3), is presented in figure 9.22. Its efficiency is 0.84.

Transversal sum (indicating the dc component)	16 combinations of 3 ternary symbols (000 is not used)
± 3	+1 +1 +1 − 1 − 1 − 1
± 2	0 +1 +1 0 −1 −1 +1 0 +1 −1 0 −1 +1 +1 0 −1 −1 0
± 1	0 0 +1 0 0 −1 0 +1 0 0 −1 0 +1 0 0 −1 0 0
	+1 +1 −1 −1 −1 +1 +1 −1 +1 −1 +1 −1 −1 +1 +1 +1 −1 −1
0	0 +1 −1 0 −1 +1 +1 0 −1 −1 0 +1 +1 −1 0 −1 +1 0

Fig. 9.22 Example of 4B3T mode. The 16 combinations or pairs of combinations framed (with a zero average singly or in pairs) are assigned to the 16 groups of 4 bits.

Numerous other modes of this type have been proposed (6B4T, 5B6B, *et cetera*). They differ especially by the detection method of the groups of *y m*-ary symbols at reception (mode alignment).

For transmission at higher bit rates, notably with optical fibers, a mode called *coded mark inversion* CMI is envisaged. This mode is of type 1B2B, i.e., with **two binary symbols per bit** of information, according to the convention and with the characteristics given in figure 9.23.

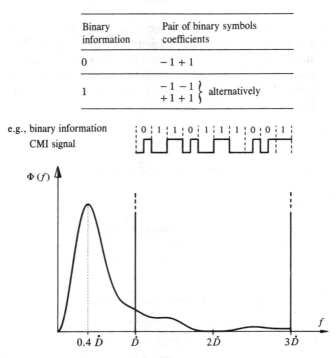

Fig. 9.23 Characteristics of the CMI mode.

The CMI mode has the advantage of being binary (only one decision threshold) and of canceling the components at very low frequencies, at the price, however, of a higher bandwidth than that of the AMI mode. Its clock rate content is excellent (line at $\dot{D}$). Its efficiency is, however, low:

$$\eta_{CMI} = 0.5 \tag{9.13}$$

9.6.7 Regenerators

While a digital system is installed on a line, regenerators distributed along the length of the line process the attenuated, distorted, and disturbed signals that they receive so as to extract, symbol by symbol, the digital information that they carry and to retransmit it with the regenerated signals. The principle and the effects of this regeneration have been discussed in detail in chapter 5. In the case of digital multiplex systems, it is, however, necessary to specify the following points:

- the regenerators are clocked (they extract the clock from the signal they receive);
- the regenerators ignore the frame structure. They work symbol by symbol and have no need for frame alignment;
- besides the functional blocks of figure 5.3, the regenerators must still be supported by circuits allowing them to be supplied with direct current from the ends of the line. On balanced pair lines, the *remote power supply* is done by the phantom circuit of the two pairs used for the two directions of transmission (fig. 9.24);

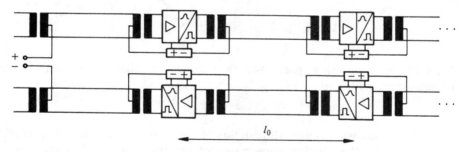

Fig. 9.24 Remote power supply of distributed regenerators.

- in the same way, because the regenerators are distributed along the length of the line, it is necessary to *monitor them from a distance* and to localize from a distance any defective regenerator;
- the regenerators of the two transmission directions are independent. They are clocked with very close, but generally different frequencies (plesiochronous);
- in the case of ternary modes, (AMI, HDB3, 4B3T, *et cetera*), the regeneration operates either on a *unipolar* signal obtained by rectification of the bipolar signal (a single decision threshold, then retransformation into a ternary mode), or directly on the *bipolar* signal with two decision thresholds and conservation of the ternary mode.

In the pseudo-ternary mode (AMI or HDB 3), the eye pattern after perfect equalization takes the form of figure 9.25. The rectangles of uncertainty sur-

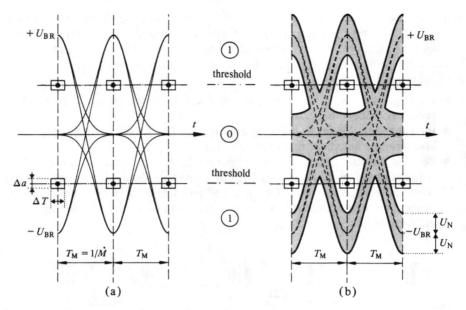

Fig. 9.25 Eye pattern in pseudo-ternary mode: (a) equalized, without noise; (b) equalized, with noise.

rounding the decision points (thresholds at characteristic instants iT_M) represent, vertically, the relative imprecision of the threshold, and, horizontally, the phase jitter (sect. 5.7.4).

9.7 PLANNING PROBLEMS

9.7.1 Given situation

The planning of a digital link between two points starts with the following data:

- hierarchical order of the digital system used (table 9.9), characterized by its bit rate $\dot{D}$
- distance l between these two points
- type of medium used
- envisaged transmission mode

9.7.2 Parameters

Planning interposes several interdependent parameters, to be considered as a function of the *transmission quality* required and *economic criteria* (sect. 2.1.3 and section 2.7). These parameters are:
- the cumulative regeneration error probability ϵ at the end of the link
- the near- and far-end crosstalk attenuations of the medium $A_{xp}(f)$ and $A_{xt}(f)$
- the attenuation coefficient $\alpha(f)$
- the number K of digital systems carried by the same cable
- the spacing l_0 of the regenerators from which we derive their number $N = 1/l_0$
- the interference with other services sharing the same medium.

Figure 9.26 illustrates the effect of these parameters on the cumulative error probability ϵ, which is the principal quality criterion of a digital transmission. The influence of a parameter can be compensated for by the antagonistic effect of another parameter to keep ϵ constant.

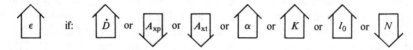

Fig. 9.26 Causes of error probability increase: ⇧ = increase ⇩ = decrease.

For example, we can space the regenerators on the condition of placing only a reduced number of systems on the same cable.

The causes of disturbances are, as always, crosstalk and noise (thermal noise and impulse noise, sect. 5.4.6). Depending on the type of medium used, one or the other of these sources is most important for planning.

9.7.3 Method

The calculation of the error probability ϵ_i per regeneration section is in principle done according to the method given in section 5.5 and that of the cumulative error probability ϵ after N sections, by relation (5.47). It is, in particular, convenient
- to express ϵ_i as a function of the signal-to-disturbance ratio ξ_R' at the output of the equalizer, which, according to (5.32), is a practical and realistic approximation of the signal-to-disturbance ratio ξ_R at the input of the regenerator;

- to roughly evaluate or, if necessary, to calculate more exactly the signal-to-disturbance ratio ξ_R as a function of the signals emitted, of their attenuation in the channel and of sources of disturbance (fig. 9.27).

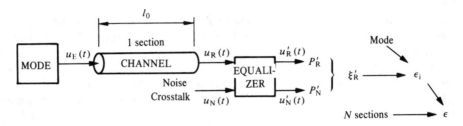

Fig. 9.27 Outline for the calculation of the error probability (notation conforms to chapter 5).

9.7.4 Objectives of quality

It has been established by subjective tests that an error probability of 10^{-5} does not noticeably affect the quality of digital telephone transmission in PCM with nonuniform 8 bit quantizing (section 7.7). However, data transmission on a digital system can require a higher quality. On this basis, the CCITT has defined a *hypothetical reference circuit HRC* characterized by the following objectives (Recommendation G 821):

- total link length: 25,000 km
- cumulative error probability: 10^{-5} in a channel with 64 kbit/s, to be distributed between the two national routes and the international route.

9.8 DIGITAL SYSTEMS USING BALANCED PAIRS

9.8.1 Relation between the signal-to-disturbance ratio and the error probability in pseudo-ternary mode

The pseudo-ternary modes AMI and HDB 3, usual for digital transmission on balanced pairs, do not satisfy the hypothesis of equal probability of the $m = 3$ states of each symbol, postulated in section 5.5.4 to estimate the relation between ξ_R' and ϵ_i. In fact, if we assume that the binary information to be transmitted presents equally probable values 1 and 0, the pseudo-ternary symbols which represent it in AMI mode according to the law of figure 9.20 have the following probabilities:

- Prob $(a_k = +1) = \frac{1}{2}$ Prob $(1) = \frac{1}{4}$
- Prob $(a_k = 0) =$ Prob $(0) = \frac{1}{2}$
- Prob $(a_k = -1) = \frac{1}{2}$ Prob $(1) = \frac{1}{4}$

These values are slightly different in HDB 3 mode.

The power P''_R of the received, equalized, and sampled-with-hold pseudo-ternary mode, is, in conformity with the general relation (5.33)

$$P''_R = \frac{1}{R}\left(\frac{1}{4} U^2_{BR} + 0 + \frac{1}{4} U^2_{BR}\right) = \frac{1}{R}\frac{1}{2} U^2_{BR} \qquad (9.14)$$

The signal-to-disturbance ratio after equalization ξ'_R is thus, from (5.32)

$$\xi'_R \cong \frac{P''_R}{P'_N} = \frac{1}{2}\frac{U^2_{BR}}{U^2_{N\text{eff}}} \qquad (9.15)$$

On the other hand, by introducing the probabilities of appearance of the three states in (5.19) and by keeping the hypothesis (5.21) of disturbance statistics with an even symmetry, the error probability ϵ_i after a single regenerator becomes

$$\epsilon_i = \tfrac{1}{4}\text{ Prob }(u'_N > \tfrac{1}{2} U_{BR}) + \tfrac{1}{2}\text{ Prob }(|u'_N| > \tfrac{1}{2} U_{BR}) +$$
$$+ \tfrac{1}{4}\text{ Prob }(u'_N < -\tfrac{1}{2} U_{BR}) = \tfrac{3}{2}\text{ Prob }(u'_N > \tfrac{1}{2} U_{BR}) \qquad (9.16)$$

If, further, the disturbance has a Gaussian distribution, we obtain, with (9.15), G_c being the complementary integral Gaussian function (section 16.2)

$$\epsilon_i = \frac{3}{2} G_c\left(\frac{1}{2} U_{BR}/U'_{N\text{ eff}}\right) = \frac{3}{2} G_c\left(\sqrt{\frac{1}{2}\xi'_R}\right) \qquad (9.17)$$

By comparison with (5.41), we can state that the pseudo-ternary mode AMI requires, for the same emission power, a ratio ξ'_R of approximately 3 dB higher than the binary antipolar mode. If compared at equal peak-to-peak voltages, the difference is approximately 6 dB.

9.8.2 Disturbance sources

As already stated in section 9.6.1, the principal cause of disturbances is *crosstalk* between balanced pairs, to such a degree that even the effects of thermal noise are negligible in comparison.

Often several digital systems are installed on different pairs of the same cable. The pulses of one of the systems are thus disturbed by those of all the others, in the same transmission direction (far-end crosstalk) and in the opposite direction (near-end crosstalk).

9.8.3 Evaluation of the signal-to-disturbance ratio

The problem is especially complicated due to
- the variation proportional to $\sqrt{f}$ of the composite attenuation A_{cp} of the line
- the phase distortions (the phase change b is also proportional to $\sqrt{f}$)
- the decrease (of approximately 15 dB per decade) of the near-end crosstalk attenuation A_{xp} between pairs, while the frequency increases (sect. 3.5.5)
- the great extension of the pulse spectrum (practically from 0 to $\dot{D}$, with a maximum toward 0.45 $\dot{D}$ in AMI mode, fig. 9.28)

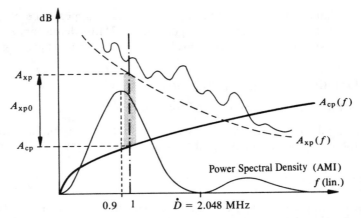

Fig. 9.28 Composite attenuation A_{cp}, near-end crosstalk attenuation A_{xp} and power spectral density of signals transmitted at 2.048 Mbit/s in AMI mode on balanced pairs.

A complete calculation must take into account all the factors needed to express the received and equalized form of the elementary base signals $u'_{BR}(t)$, as well as the form and relative importance of the parasitic pulses introduced with crosstalk.

For practical purposes, it is unrealistic to wish to calculate with precision the signal-to-disturbance ratio ξ'_R from which, according to (9.17), the error probability is derived. In effect, the values of near- or far-end crosstalk are submitted to a significant dispersion and depend very strongly on the relative position of the disturbing and disturbed pairs in the cable (same layer, neighboring layers, near the sheath, diametrically opposed pairs, table 3.21).

At first approximation, it is sufficient to *plan the link at a single frequency, equal to half the bit rate*, thus approximately 1 MHz for the primary PCM system, which approximately corresponds to the maximum power concentration in the spectrum of the AMI signal.

The level diagram of figure 9.29 illustrates the situation at frequency $f = \dot{D}/2$ when two identical transmissions in opposite directions disturb each other by near-end crosstalk. The signal-to-disturbance ratio at the input of the regenerator is thus equal to the near-end signal-to-crosstalk ratio A_{xp0}.

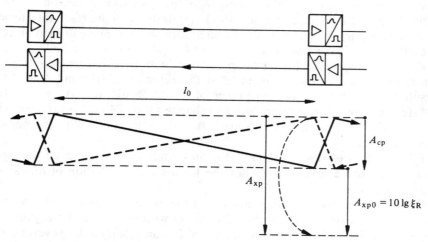

Fig. 9.29 Level diagram at $\dot{D}/2$.

If *K mutually asynchronous systems* disturb each other by near-end crosstalk, it may be assumed that the parasitic pulses are sufficiently incoherent to be added in power (in the case of synchronism, a voltage addition would be more realistic) The signal-to-disturbance ratio resulting from the K systems is thus (at $\dot{D}/2$):

$$10 \lg \xi_R = A_x - \alpha_{cp} l_0 - 10 \lg K \qquad (9.18)$$

where A_x represents the global crosstalk attenuation (near- and far-end crosstalk power added).

Because the disturbances have the same origin as the useful signals, they occupy the same spectral range and cross the equalizer of the regenerator in a similar manner. After equalization, therefore we still have approximately

$$\xi'_R \cong \xi_R \qquad (9.19)$$

9.8.4 Dimensioning of the link

Relation (9.17) allows the calculation of the ratio ξ'_R necessary so that the regenerator will not exceed an error probability ϵ_i, in AMI mode.

For example, (using figure 16.2), to guarantee $\epsilon_i \leq 10^{-8}$, it is necessary that $10 \lg \xi'_R \cong 10 \lg \xi_R \geq 18$ dB.

For a given cable, the linear attenuation α_{cp} at $\dot{D}/2$ is known; the crosstalk attenuation A_x at $\dot{D}/2$ is also known, but not very precisely.

In relation (9.18) there is still a degree of freedom for the choice of the regenerator spacing l_0 or of the number K of the systems installed on the cable. In the case of old loaded cables for which it is desirable to increase the capacity by installing primary systems, the regenerators take the place of loading coils. They thus inherit the historical spacing of $l_0 = 1830$ m (sect. 3.5.6).

By reviewing the realistic example above, with a cable (conductors of 0.6 mm in diameter), where α_{cp} is equal to 15 dB/km at 1 MHz and $A_x \cong 57$ dB (resulting from the addition in power of $A_{xp} \cong 58$ dB, and $A_{xt} \cong 65$ dB at 1 MHz), it is found that a signal-to-disturbance ratio of 18 dB is attained when $K = 14$ PCM systems at 2 Mbit/s are installed on the cable, with regenerators every 1830 m.

Effectively, near-end crosstalk, and in certain cases and in an analogous manner, far-end crosstalk, impedes the large-scale installation of digital systems on the same cable.

On the other hand, we must not forget the extreme sensitivity of the complementary integral Gaussian function $G_c(x)$ with respect to its argument x. For small values of ϵ_i, a small variation of ξ'_R can modify ϵ_i by several orders of magnitude. The dimensioning of a digital link must therefore be done with a generous reserve. Finally, the convenient hypothesis of Gaussian distribution is only a rough approximation of a very complex reality.

9.9 DIGITAL SYSTEMS USING COAXIAL PAIRS

9.9.1 Disturbance sources

The principal advantage of coaxial pairs is the almost complete disappearance of crosstalk coupling at high frequencies. The predominant source of disturbances is, due to this fact, the *background noise*, i.e., the thermal noise and the internal noise of the input stage of the regenerators.

The reception filter (sect. 5.4.5) allows the elimination of a large part of this noise while keeping only the contributions located in the same frequency range as the useful signals.

9.9.2 Calculation hypotheses

The calculation of the error probability ϵ_i per section according to the complete outline of figure 5.19 meets some difficulties which can be reduced by the following simplifying hypotheses:

- attenuation coefficient $\alpha(f)$ proportional to $\sqrt{f}$

$$\alpha(f) = \alpha_0 \sqrt{f/f_0} \tag{9.20}$$

(realistic as long as $\alpha_R \gg \alpha_G$: i.e., for $f \ll 1$ GHz)
- no phase distortions: phase change b proportional to f (very realistic)
- equalizer compensating exactly for the attenuation of the line and its distortions

$$E(f) = \frac{1}{H(f)} = 10^{\alpha_0 \sqrt{f/f_0} l_0/20} \tag{9.21}$$

(ideal, but not necessary, according to the remark of section 5.3.5)
- passing band of the equalizer limited to $f_{max} = \dot{M}$ (double of the theoretical minimum according to Nyquist)
- m-ary symbols with equally probable values, in antipolar mode
- effect of phase jitter neglected
- thermal noise with (realistic) Gaussian distribution and with uniform spectral density $\Phi_{Nth} = kT$ at the input of the equalizer.

The situation is schematized in figure 9.30.

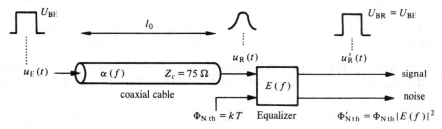

Fig. 9.30 Digital transmission on a coaxial cable with ideal equalization.

9.9.3 Dimensioning of the link

With these hypotheses, the error probability ϵ after a distance l (N regenerators spaced l_0 apart) is given by (5.26) and (5.47)

$$\epsilon \cong N \cdot \epsilon_i = \frac{l}{l_0} \frac{2(m-1)}{m} G_c \left(\frac{\frac{1}{2} U_{BR}}{U'_{Neff}} \right) \tag{9.22}$$

On the other hand, the thermal noise power at the output of the equalizer is, with (9.21) and (9.20),

$$P'_{\text{Nth}} = \int_0^{\dot{M}} \Phi_{\text{Nth}} \, [E(f)]^2 df = kT \int_0^{\dot{M}} 10^{0.1\alpha_0 \sqrt{f/f_0}} \, df \qquad (9.23)$$

As the equalizer is terminated on a resistance R, we have

$$U'_{\text{Neff}} = U'_{\text{Ntheff}} = \sqrt{P'_{\text{Nth}} R} \qquad (9.24)$$

Relations (9.22), (9.23), and (9.24) contain, in transcendental form, all the parameters permitting the dimensioning of the link, in particular the determination of the **regenerator spacing** l_0 as a function of the tolerated error probability ϵ, of the symbol rate $\dot{M}$ and of the characteristics of the cable (α_0 at f_0).

It is noted in particular that

- the quality can, in principle, be improved (lower ϵ) by increasing U_{BR}, however, within the limits of the available emission power P_E. This is equal, by analogy with (5.35), to

$$P_E = \frac{m^2 - 1}{12} \frac{U_{\text{BE}}^2}{Z_c} \qquad (9.25)$$

where U_{BE} is the amplitude of the elementary base signal at emission and Z_c the (real) characteristic impedance of the coaxial cable;
- the increase of the number m of values per symbol reduces $\dot{M} = \dot{D}/\text{lb}\, m$, thus $P'_{N\,\text{th}}$ and $U'_{N\,\text{eff}}$ for a given bit rate $\dot{D}$. Nevertheless, at a constant emission power P_E, U_{BE} must decrease, according to (9.25), which leads to a diminishing U_{BR} at reception. The antagonistic effects of the decrease of $U'_{N\,\text{eff}}$ and U_{BR} in (9.22) lead to a minimum of ϵ as a function of m. It is nevertheless not evident that the optimal value of m corresponds to the technological realization possibilities of the regenerators.

Concretely, table 9.31 gives an idea of the regenerator spacing l_0 envisioned for the digital systems of different orders, installed on normalized coaxial pairs

Table 9.31 Order of magnitude of the regenerator spacing for digital systems on coaxial pairs.

CCITT $\dot{D}$	normal 2.6/9.5	small ⌀ 1.2/4.4	mini-coaxial 0.7/2.9
8 Mbit/s	(12 km)	(6 km)	3.66 km
34 Mbit/s	(8 km)	4 km	1.83 km
140 Mbit/s	4.5 km	2 km	1 km
565 Mbit/s	1.5 km	–	–

(sect. 3.6.4). The exact value of l_0 is often determined by considerations of compatibility with the repeater spacing for the older analog systems (carrier systems, fig. 10.18) installed on other coaxial pairs of the same cable, or by the loading spacing of balanced pairs associated with this cable.

9.10 OPERATIONAL PROBLEMS

9.10.1 Measurement technique

Digital systems require a particular measurement technique, adapted to the digital and analog interfaces which characterize them (fig. 9.32).

Analog measurement from end-to-end (fig. 9.32(a)) allows us to monitor the quantizing noise (section 7.6), the background noise, the linearity of the input-output characteristic, the frequency response in the baseband, *et cetera*.

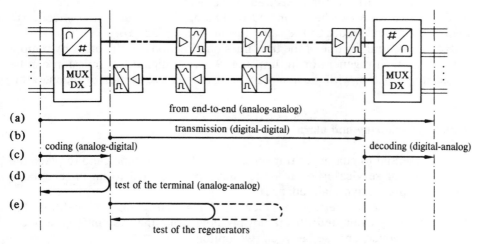

Fig. 9.32 Measurement possibilities in a PCM system.

The *digital transmission test* (fig. 9.32 (b)) has as its principal object the measurement of the *bit error rate BER*, representative of the error probability ϵ. To this end, a pseudo-random signal is emitted in place of a real signal and recognized by a measurement receiver. The periodicity of this signal must be long enough so as to not disturb the measurement (e.g., $2^{15} - 1 = 32767$ bits, giving 16 ms at 2 Mbit/s). The bit error rate thus measured is submitted to statistical fluctuations which depend on the duration of the tests (number of bits emitted n) and has a standard deviation σ which is that of a binomial distribution (n binary events with probabilities of ϵ and $1 - \epsilon$ respectively):

$$\sigma = \sqrt{\epsilon(1 - \epsilon)n} / n \cong \sqrt{\epsilon/n} \qquad (9.26)$$

Assuming a normal distribution of the results, the true value of ϵ is found with a probability of 95% (confidence level) in an interval of $\pm 2 \sigma$ (confidence interval) around the measured value. If we wished to know ϵ (e.g., estimated at 10^{-7}) with a relative precision of 10% and a confidence level of 95%, σ must be lower than $0.5 \cdot 10^{-8}$. From (9.26), it is therefore necessary to send $n \geq \epsilon/\sigma^2 = 4 \cdot 10^9$ bits, which requires more than half an hour at $\dot{D} = 2.048$ Mbit/s.

The *separated test of encoder and decoder* (fig. 9.32)(c)) allows us to guarantee the intrinsic performance of these two important pieces of equipment, independently of their remote partner, which can be variable (switched digital network). This test requires, on the one hand, a digital analyzer capable of examining the PCM coding result of a sinusoidal signal octet by octet, and, on the other hand, a digital generator simulating the coded samples of a test signal, which is afterwards measured at the analog output of the decoder.

These two tests can be combined by linking the output of the encoder to the input of the decoder of the same terminal (fig. 9.32 (d)).

By short-circuiting one of the transmission directions on the other by remote control, at each regenerator in turn, (fig. 9.32(e)), digital transmission can be tested step-by-step, with the aim of locating a defective section or regenerator.

9.10.2 Monitoring and alarms

It is essential to guarantee transmission quality at all times and to place the system out of service as soon as its quality is degraded to an unacceptable condition (preventive maintenance).

The error rate is an effective indicator of this quality, on the condition that it is continually evaluated; this is not possible while the system is in service. We can, however, choose between two options:

- observation at the framing pattern. The errors found on this pattern are considered to be representative of the overall error probability ϵ and their rate is consequently extrapolated;
- detection of irregularities in the sequential law of polarity alternation of the pseudo-ternary modes AMI or HDB 3. The violation rate of this law gives a good image of the error probabiltiy ϵ.

If the error rate evaluated in this way exceeds, e.g., 10^{-3}, access to the system must be forbidden (blocking of the channels).

The breakdown of a "codec" escapes this test. This is why routine tests, of the type of figure 9.32 (d), must be used, for example during the time intervals 0 or 16 in which the "codec" is unused.

The loss of framing is a serious accident which must be announced by an alarm in the opposite direction and leads to a blocking of the channels (sect. 9.3.3).

Remote power supply of the regenerators, on the other hand, offers a simple means of detecting large-scale defects (rupture, short circuit) on cables.

Chapter 10

Analog Systems

10.1 CARRIER SYSTEMS

10.1.1 Principle

Carrier systems, with a name sanctioned by historical use, are analog transmission systems with several channels, characterized by

- *frequency-division multiplexing*, FDM, of the z channels (sect. 4.4.2)
- the systematic use of *single sideband modulation*, SSB, for the composition of this multiplex (sect. 8.4)

Carrier systems are also used on balanced pairs or coaxial lines as well as for ground microwave links or satellite links.

Their properties are derived from the principles of SSB modulation, and from the general characteristics of analog transmission, which were presented in chapter 6.

10.1.2 History

The birth of carrier systems was made possible by the invention of electronic (vacuum) tubes (the diode in 1904, and triode in 1907) and their development followed these inventions very closely (the prototype was presented at the Brussels International Exposition in 1909). Toward 1920, they were placed into operation and have experienced spectacular development since then, being further favored by advances in semiconductor technology after 1948. Their capacity ranges from a few channels to a present and probably definitive maximum of 10,800 channels on the same physical support, while the bulk of corresponding equipment has decreased considerably (2 channels per rack in 1945; 480 channels per rack in 1975).

Currently, and in most countries, carrier systems still constitute the major part of the medium- and long-distance transmission systems, while waiting to be gradually supplanted by digital time-division multiplexed (TDM) systems. They have attained a remarkable degree of technical maturity and at the same time their economical use has been optimized.

10.1.3 Structure of a carrier system

A carrier system is composed of the following elements (fig. 10.1):

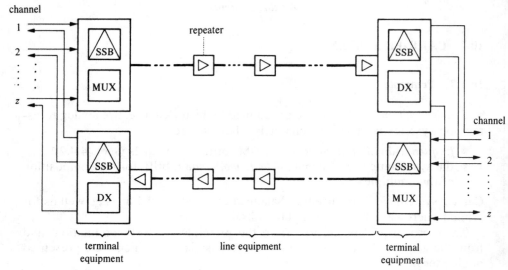

Fig. 10.1 Structure of a line carrier system.

- **terminal equipment** which carries out the frequency-multiplexing and de-multiplexing by SSB modulation and demodulation. Terminal equipment is independent of the medium used;
- **transmission equipment** adapted to the medium, i.e., **repeaters**, in the case of lines, to compensate for attenuation and for linear distortions at regular intervals (section 6.2). In the case of radio transmission, the **transmitters/ receivers** which transpose the signal into the appropriate frequency band, and the reverse, by means of a second modulation (generally FM or ΦM), are included, as well as **relay stations** (receivers-retransmitters) to overcome obstacles or long distances.

10.1.4 Carrier spacing

SSB modulation requires the elimination of one of the sidebands of DSBSC modulation. On the other hand, frequency-division multiplexing by juxtaposition of the channels using carriers at different frequencies requires a very strict delimitation of the band assigned to each channel. These two operations are carried out by *filters*, which are key elements of carrier systems (vol. XIX).

The spacing Δf_p of the carriers also determines the frequency spacing of the channels, and thus their number in a given total bandwidth. The choice of Δf_p results from an economic compromise:

- the spacing of the carriers evidently must be higher than the bandwidth of a channel

$$\Delta f_p > B_1 \quad (10.1)$$

- we seek to distribute the cost of the transmission medium between a maximum number of channels, which leads to spacing them as little as possible;
- the selectivity requirements of channel filters increase, as Δf_p approaches the limit (10.1). The result is a rapid increase in their cost.

Figure 10.2 illustrates the search for the economic optimum relating to Δf_p. It has been established that it is directly influenced by the type and cost of the transmission medium.

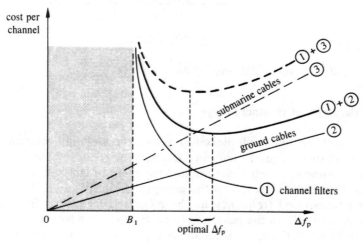

Fig. 10.2 Choice of the carrier spacing Δf_p.

Aside from a few special cases (systems on transoceanic cables with $\Delta f_p = 3$ kHz and B_1 reduced to 2.75 kHz), all the carrier systems for telephony use carrier frequencies which are multiples of 4 kHz, given

$$\Delta f_p = 4 \text{ kHz} \quad (10.2)$$

The consequences of this choice for the filter specifications are presented in figure 10.3.

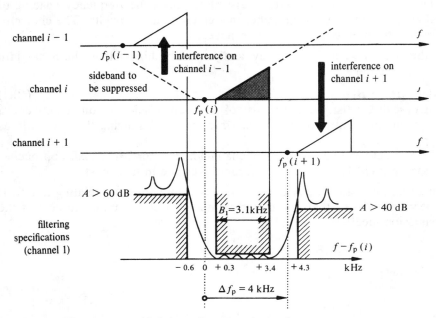

Fig. 10.3 Filtering specifications for one channel.

10.1.5 Hierarchy of channel groups

The channels are not multiplexed in a single modulation operation, but in several successive stages forming a hierarchy of groups with an increasing number of channels, which is determined by international convention. Figure 10.4 gives its structure. Note that the groups are alternately in direct position (increasing transposed frequency as the baseband frequency increases) and the reverse. This allows the passage of one group to the following by an SSB modulation in which the lower sideband is kept. The symbols used to represent these channels and groups are those which were introduced in figure 4.6 in section 4.2.2.

The advantages of such a hierarchy are the following:
- each group is set up by transposition and juxtaposition of several groups of the immediately lower order;
- a group represents an ensemble which can be treated as a whole—for example, extracted from a system, transposed, inserted into another

ANALOG SYSTEMS

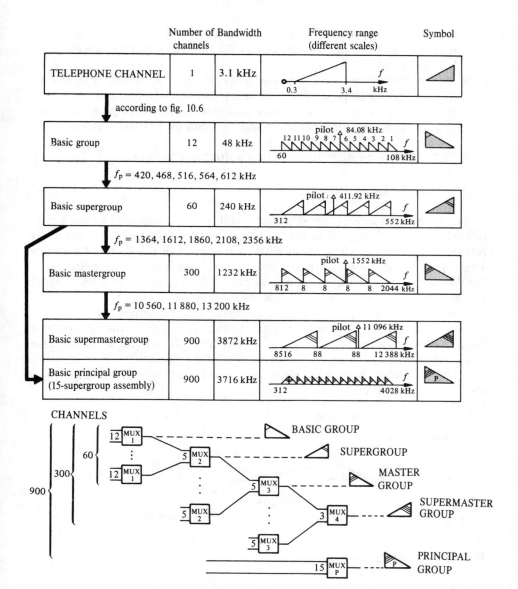

Fig. 10.4 Hierarchy of the channel groups.

system, demultiplexed, *et cetera*. This allows the "drop and insert" of whole groups during transmission, without having to demodulate to the baseband;
* the terminal equipment is modular and consists of an assembly of channel translators (basic groups) and of transpositions of groups to pass from one order to the next and finally compose a system of given capacity. The corresponding filters and modulators can thus be standardized.

10.2 TERMINAL EQUIPMENT FOR CARRIER SYSTEMS

10.2.1 Modulation by steps

The filtering of one of the sidebands with SSB modulation becomes more difficult as the primary band moves closer to zero. On the other hand, the relative slope of the edges of the filter increases with the carrier frequency. These reasons make the direct translation of a telephone channel from its baseband toward its position in a multiplex with a large number of channels extremely difficult and costly, or even impossible.

The principle of stepwise modulation consists of translating the channel by a first SSB modulation toward a frequency range in which filtering is favorable. This channel, which has first been shifted from the origin, is retranslated toward its desired position by one or several SSB modulations. The sidebands are thus separated from each other and discrimination by filtering is made much easier (fig. 10.5).

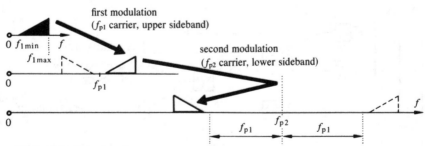

Fig. 10.5 Modulation in two steps.

10.2.2 Composition of the basic group

The assembly of 12 channels into a basic group, or *channel translation*, is the first indispensable step for all carrier systems. By its position at the base of the system and its repetitive character, it plays a very important role in the final cost. Several solutions have been proposed, with the principles presented in figure 10.6 and compared in figure 10.7.

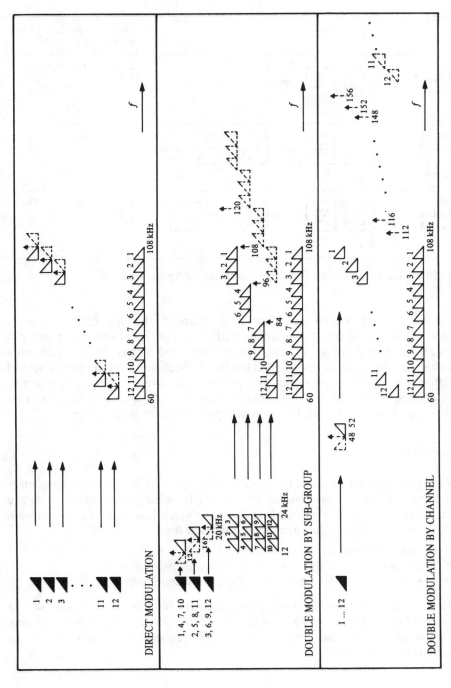

Fig. 10.6 Three methods of the basic group formation.

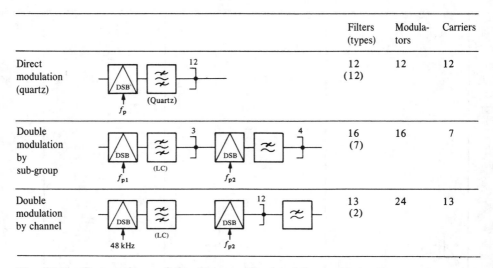

Fig. 10.7 Comparison of the three methods of figure 10.6.

In the case of double modulation per channel, the first filter (the most critical!) is identical for all the channels. It is worth the trouble to optimize it through the choice of a particular technology and a first carrier frequency adapted to this technology. In particular a mechanical filter with coupled metallic resonators is used for this purpose. It is excited by piezoelectric transducers, whose optimum functional range is located around 50 kHz. This technology lends itself well to automatic production in large series.

10.2.3 Pilots

Pilots are auxiliary sinusoidal signals of very precise and constant frequency and level. While the content of the channels is subjected to very strong fluctuations and can even cancel itself during slack hours, pilots constitute a permanent and stable reference. They are used for

- the regulation of the nominal signal level
- the surveillance of the system (detection of breakdowns)
- in some cases, the comparison of carrier frequencies

Each group contains a *group pilot*, near the middle of its band, in the interstice between two channels, with a level of -20 dBm0 (20 dB lower than the nominal level of the system). The exact frequencies of these pilots are indicated in figure 10.4.

Also, *line pilots* are inserted into the upper and lower ends of the packet of groups constituting the complete system to permit fine tuning of the repeater gain (section 10.3.3).

10.2.4 Carrier generation

Since telephone transmission does not require rigorous phase conformity, SSB demodulation need not necessarily be done isochronously with the modulation (sect. 8.4.3). An error of frequency Δf at the reconstruction of the carrier appears as a shift of Δf of the entire primary band. Especially considering the requirements of data transmission which can take place in the channels of a carrier system, a frequency shift of

$$|\Delta f| < 2 \text{ Hz} \tag{10.3}$$

is tolerated in a channel, after a path length of 2500 km and several translations.

The resultant precision for the carrier generation increases with the frequency and reaches for example 5×10^{-8} for $f_p = 12$ MHz, with three translations.

The carriers of the entire systems are generated in common, by division and multiplication of the frequency from a master oscillator of high precision and stability, at a very high degree of reliability.

10.2.5 Signaling

Each channel of the system must arrange its own signaling channel, so as to allow transmission of related auxiliary information for switching requirements (set up of connections, surveillance, taxation, freeing of temporary links related to this channel). As the carrier systems do not allow dc transmission, the signaling criteria must be carried by alternating signals.

Two procedures can be used:

- *"in-band" signaling*, carried by of audio-frequency signals (e.g., multifrequency coded signaling, MFC, by simultaneous emission of two frequencies among the six possible). The audibility of these signals precludes its the use of in-band signaling during conversation. This procedure has the advantage of processing (amplification, translation, switching) the signaling in the same way as with speech;
- *"out-of-band" signaling* in which the signaling pulses modulate a carrier at 3825 Hz by on-off keying (OOK). This is inaudible because it is cut by the channel filter. There is, therefore, no interference between speech and signaling. However, the signaling must be extracted and reintroduced each time the channel is transposed back to baseband.

10.2.6 Modulation plan of usual systems

By international convention, (CCITT, Recommendations G.322 to 346), a limited number of transmission systems of different capacities has been specified. They are set up by SSB modulation (lower sideband) from supergroups, mastergroups, supermastergroups, and 15-supergroup assemblies which are translated and juxtaposed. The *modulation plan* defines the exact position of these groups, which also determines the frequency of the carriers necessary to translate them (while reversing them), starting with the basic group. Figure 10.8 gives one of the possible modulation plans (without details) for the principal systems. Intervals are arranged among the groups involved in such a way as to facilitate filtering. Other modulation plans are used, notably from the 15-supergroup assemblies, with slightly different frequency boundaries.

The systems are designated either by their number of channels, or by their rounded-off maximum frequency.

10.3 LINE EQUIPMENT FOR CARRIER SYSTEMS

10.3.1 Systems installed on balanced pair

The principal obstacle to the installation of carrier systems on balanced pair cables is the rapid decrease of the signal-to-crosstalk ratio when the frequency increases. It is necessary either to take special precautions relating to the cables (polystyrene insulation, forward and return in separate cables), or to accept a very limited number of channels (e.g., 6) and to operate in "frequency-division pseudo-4-wire" (section 4.6.4), which eliminates the effects of near-end crosstalk.

As a result of the development of digital systems, this application of carrier systems loses much of its importance.

10.3.2 Systems installed on coaxial pair

The coaxial pairs specified by the CCITT (normal tube 2.6/9.5 mm, small diameter tube 1.2/4.4 mm, section 3.6.4) are the media of choice for medium- and long-distance ground systems, i.e., for systems from 300 to 10,800 channels, according to figure 10.8.

All these systems use two separate coaxial pairs, one for the forward direction, the other for the return (true "4-wire").

However, the systems installed on *transoceanic cables* are the subject of a specific planning from case to case. Due to its very high cost, the cable

ANALOG SYSTEMS

z	f_{max}	line	l_0	Modulation plan (different scales)	CCITT Rec.
120	552 kHz	sym.	8 km	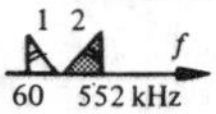 1 2 ... f; 60 552 kHz	G 322
300	1.3 MHz	coaxial 1.2/4.4	6 or 8 km	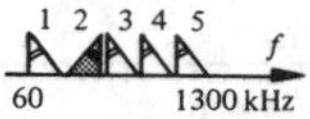 1 2 3 4 5 ... f; 60 1300 kHz	G 341
960	4 MHz	coaxial 1.2/4.4; coaxial 2.6/9.5	4 km; 9 km	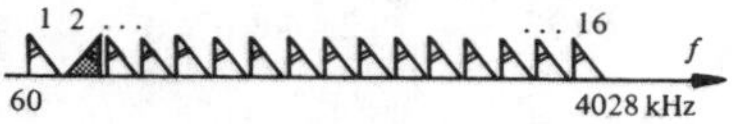 1 2 ... 16 f; 60 4028 kHz	G 343
1 260	6 MHz	coaxial 1.2/4.4; coaxial 2.6/9.5	3 km; 9 km	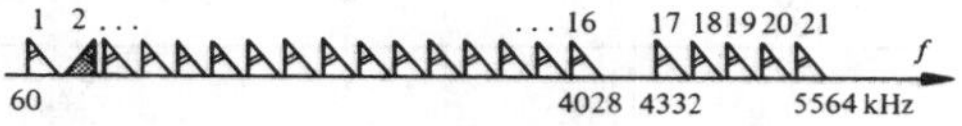 1 2 ... 16 17 18 19 20 21 f; 60 4028 4332 5564 kHz	G 344
2 700	12 MHz	coaxial 1.2/4.4; coaxial 2.6/9.5	2 km; ~4.5 km	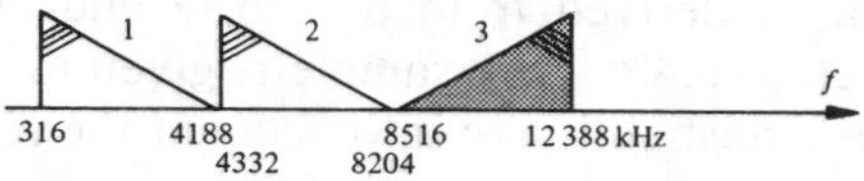 1 2 3 f; 316 4188 4332 8516 8204 12 388 kHz	G 345; G 332
3 600	18 MHz	coaxial 1.2/4.4; coaxial 2.6/9.5	2 km; ~4.5 km	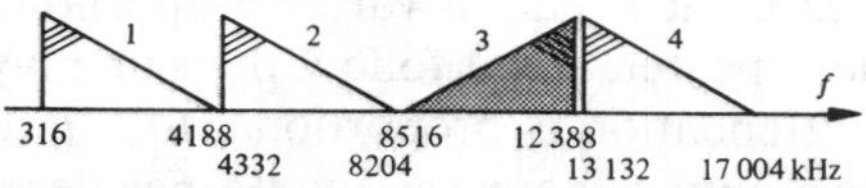 1 2 3 4 f; 316 4188 4332 8516 8204 12 388 13 132 17 004 kHz	G 346; G 334
10 800	60 MHz	coaxial 2.6/9.5	~1.5 km	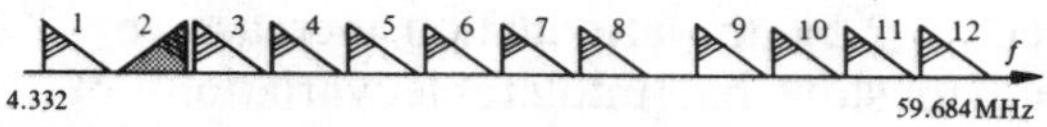 1 2 3 4 5 6 7 8 9 10 11 12 f; 4.332 59.684 MHz	G 333

Fig. 10.8 Principal carrier systems.

generally only contains a single coaxial pair and the system must work in "frequency-division pseudo-4-wire" (table 10.9).

The capacity of these systems on submarine cable can be doubled by a complex procedure of rapid switching using the pauses of one conversation to slip in other sections. This costly procedure, called *time assignment speech interpolation, TASI*, is only justified by the economy it allows due to the distribution of the enormous cost of the cable over a larger number of channels.

10.3.3 Repeaters

As has been shown in section 6.4.5, compensation of the line attenuation absolutely must be done section by section by *distributed repeaters* along the line and not simply in concentrated form at the end of the line. These repeaters are in principle and as far as possible identical and equidistant. The calculation

Table 10.9 Examples of systems on transatlantic cables.

Designation	Year of Installation	Length (km)	Dimensions of the coaxial pair d_{ext}/d_{int} (mm)	Number of circuits without TASI (Δf_p)	Number of repeaters	
TAT 1	1956 (-78)	3617	15.75/4.1	36 (4 kHz)	2 × 51	2 cables
CANTAT 1	1961	4452	25.4/8.6	80 (4 kHz)	90	
TAT 3	1963	6508	25.4/8.6	138 (3 kHz)	183	
⋮						1 cable
CANTAT 2	1974	5189	37.34/9.3	1840 (3 kHz)	489	
TAT 6	1976	6307	43.18/12.14	4200 (3 kHz)	693	

of their spacing is derived from the noise budget according to the method indicated in section 6.4.4. An example is given in section 10.5.5.

The repeaters must still take into account the following points:

- since the cable attenuation varies proportionally to $\sqrt{f}$ (sect. 3.6.3), the gain of the repeaters must follow the same law;
- the cable attenuation is proportional to $\sqrt{\rho}$ according to relation (3.73). As the resistivity ρ increases by 4% per degree of temperature increase (sect. 3.2.5), the attenuation increases by 2% per degree, which would lead to slow but intolerable variations of the output level and of the signal-to-noise ratio. The repeaters gain must thus be *regulated as a function of the temperature* of the cable. The best solution, but also the most costly, consists of taking as a reference criterion the level of a line pilot transmitted by the system and extracted in the repeater;
- the repeaters are *supplied remotely* with constant direct current (50 mA) by the internal conductors of the forward and returning coaxial pairs;
- surveillance and maintenance: the repeaters are observed at a distance, and in case of a breakdown the defective repeater must be able to be located without ambiguity.

The case of repeaters immersed along the length of submarine cables is particularly critical. They can be repaired, but the probability of such interventions is reduced by seeking as high an intrinsic reliablity of the repeaters (choice and testing of components, circuit redundancy) as possible.

10.3.4 Pre-emphasis

The variation in $\sqrt{f}$ of the cable attenuation has as a consequence that the higher channels of the multiplex are much more attenuated than the lower channels. The gain (also variable according to $\sqrt{f}$) of the repeaters compensates for these differences as far as level is concerned. Nevertheless, if the nominal level $L_{SE} = L_{SR}$ at the input and output of the system were the same for all the channels, the upper channels would have a distinctly worse signal-to-noise ratio despite everything (fig. 10.10(a)). This inconvenience is remedied by a *pre-emphasis*, which consists of emitting the channels with a nominal level which increases as a function of their frequency position in the multiplex. By a reciprocal operation of *de-emphasis*, the original levels, identical in all the channels, are re-established at reception (fig. 10.10(b)).

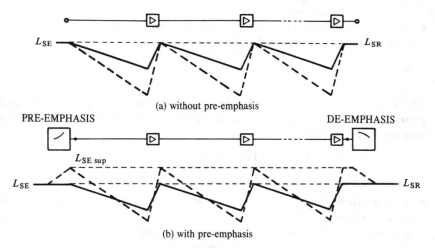

Fig. 10.10 Effect of a pre-emphasis: ——— lower channel; – – – upper channel

A compromise, however, must be found between the quality of the upper channels and the non-linearities of the repeaters (intermodulation noise) which impedes pre-emphasis as far as would be necessary to guarantee a uniform quality in all the channels.

10.4 MULTICHANNEL SYSTEM LEVEL

10.4.1 Characteristics of a multiplex signal

The signal transmitted by a carrier system is the sum
- of the (random and independent) signals of the channels which compose it, translated and juxtaposed in frequency;

- of some auxiliary signals such as pilots, signaling, carrier residues, *et cetera*.

This signal depends on the traffic; it becomes very low during the slack hours and attains a maximum during the busy hour. On the other hand, it is the sum of several random signals with a zero average and strong variance. In the case of telephone conversations, the statistical distribution of signals in each channel can be appproximately schematized by figure 1.9. As a consequence of the central limit theorem of the statistics, we can however say that

- the resulting signal of z channels approaches Gaussian statistics more closely as z increases
- the average power (variance) P_z of the signal is equal to the sum of the average powers P_i in each channel

$$P_z = \sum_{i=1}^{z} P_i \tag{10.4}$$

In telephony, the calculation of P_z, and thus the real level $L_{z\,tph}$ corresponding to z busy channels, is based on convention (2.55) relative to the *real level of a communication referred to zero*:

$$L_{ztph0} = L_{tph0} + 10 \lg z = -15 + 10 \lg z \qquad \text{dBm0} \tag{10.5}$$

It must be emphasized that this value

- is an *average* which takes into account the pauses (channel activity rate of approximately 25%)
- also includes the power of the auxiliary signals
- assumes that there is no pre-emphasis

From (10.5) we find, for example, that the real level in a system with 2700 channels is located at +19.3 dBm0, i.e., at any point of the system at 19.3 dB above the nominal level which is, let us recall, the level of a sinusoidal test signal (section 2.4.14). Knowing this real level and the fact that the resulting signal of the z multiplexed channels has a Gaussian distribution, we can evaluate the probability of exceeding a certain value (e.g., the clipping threshold of the repeaters).

10.4.2 Simulation of the multiplex signal

For certain calculations and especially for practical tests of the system, it is necessary to use a model of the signal resulting from the occupation of the z channels by telephone conversations. From the preceding paragraph, this signal can be simulated by *noise*

- with a *Gaussian* statistical distribution
- with a *uniform* power spectral density in the range occupied by the system

ANALOG SYSTEMS

The level (rms value) of this Gaussian white noise is given in relation (10.5), provided that z is higher than 240 (limit fixed by convention).

For $z < 240$, it is found that the central limit theorem is not quite satisfied and that, for the same level, the Gaussian statistics give rise to clipping probabilities which are too low with respect to the real cases observed. For practical reasons, the Gaussian model is retained anyway, but with a level which is empirically increased according to the relation

$$L_{ztph0} = -1 + 4 \lg z \quad \text{(for } 12 < z < 240\text{)} \qquad \text{dBm0} \qquad (10.6)$$

10.4.3 Determination of the nominal level as a function of the clipping probability.

The parameters of the problem are the following:
- a carrier system with z channels
- repeaters with a maximum output voltage located, for technological reasons, at $\pm U_{S\,max}$, i.e., at a level $L_{S\,max}$ (clipping limit)

At what value should the nominal level of the system $L_{S\,nom}$ be placed (at the output of the repeaters) with respect to this maximum to guarantee a probability of clipping lower than $ê$?

The probability of clipping is

$$ê = \text{Prob}\,(|u_S| > U_{Smax}) = 2\,\text{Prob}\,(u_S > U_{Smax}) \qquad (10.7)$$

In the case of a Gaussian distribution, the clipping probability is given by the complementary integral Gaussian function $G_c(x)$ (appendix, sect. 16.2).

$$ê = 2\,G_c\,(U_{Smax}/U_{zeff}) \qquad (10.8)$$

U_{xeff} being the rms value of the resulting signal from z channels. It corresponds to the real level $L_{z\,tph\,real}$ at the output.

$$L_{ztphreal} = L_{Snom} + L_{ztph0} \qquad \text{dBm} \qquad (10.9)$$

Knowing the tolerated value of $ê$, one can deduce, by the function G_c (fig. 16.2), the ratio U_{Smax}/U_{zeff}, and thus the difference between the clipping level $L_{S\,max}$ and the real level $L_{x\,tph\,real}$:

$$L_{Smax} - L_{ztphreal} = 20 \lg (U_{Smax}/U_{zeff}) \qquad (10.10)$$
$$= 20 \lg[G_c^{-1}(\tfrac{1}{2}ê)]$$

From (10.9) and (10.5) we obtain the nominal level sought

$$L_{Snom} = L_{ztphreal} - 10 \lg z + 15 \qquad \text{dBm} \qquad (10.11)$$

so that with (10.10)

$$L_{Snom} = L_{Smax} - 20 \lg[G_c^{-1}(\tfrac{1}{2}\hat{e})] - 10 \lg z + 15 \qquad \text{dBm} \qquad (10.12)$$

This relation is expressed graphically in figure 10.11. Thus, for example, if the clipping level is found at +20 dBm at the output of the repeaters for a system with 1260 channels, and if we tolerate a clipping probability of 10^{-6}, the real level must be located at $20 \lg 4.9 = 13.8$ dB lower than the clipping level, thus at +6.2 dBm. This leads to fixing the nominal level of the system (at the output of the repeaters) at -9.8 dBm.

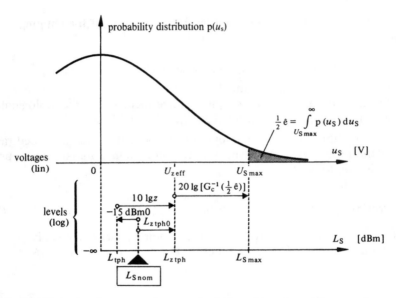

Fig. 10.11 Relation between the nominal signal level and the maximum signal level.

It must be emphasized that
- the calculations above do not take into account any possible pre-emphasis. The nominal level is assumed to be the same in all the channels;
- the clipping limit is rather in reality a zone in which the nonlinearities increase progressively before ending with the sharp clipping (saturation). The choice of the nominal level finally requires a compromise between the background noise and intermodulation products (section 6.2.9).

10.5 PLANNING PROBLEMS

10.5.1 Parameters

The principal planning criterion is the *signal-to-noise ratio* ξ_{Ri} in the channel *i* at the output of the system. As the SSB demodulation does not modify this ratio (section 8.9.3), it is also the signal-to-noise ratio ξ'_{1i} in the channel *after* demodulation (return to the baseband). As was shown in section 6.4.4, in particular in relation (6.30), this ratio essentially depends on

- the nominal level $L_{SR\,nom}$ at the output (reception level)
- the transmission distance *l*
- the attenuation coefficient α of the cable (at the frequency in which the channel *i* is located in the multiplex)
- the number of repeaters *N*

The objective of planning is primarily to determine the *repeater spacing* l/N, for a given system and on a given cable, with a view toward guaranteeing a certain signal-to-noise ratio.

10.5.2 Hypothetical reference circuit

The distance parameter intervening in the calculation of ξ_{Ri} (either directly, or indirectly by the intermediate value of the number *N* of repeaters) must be carefully examined. In effect, in a shared network each transmission system of given length can become a link in a chain of much greater length. A minimum quality level must be guaranteed even for the longest links.

With this aim, the CCITT has defined a *hypothetical reference circuit, HRC*, to serve as a basis for the planning of analog systems. There are in fact several HRCs, each adapted to a type of system and medium. We must specify the number and type of frequency translations that a channel undergoes in the course of transmission. Without going into details, it must be remembered that the HRCs have in common

- a reference length of 2500 km
- an overall signal-to-noise ratio at the end of the link *in a telephone channel* of

$$10 \lg \xi'_1 \geq 50 \text{ dB} \qquad (10.13)$$

which corresponds to a noise power referred to zero $P'_{1\,N0}$ of

$$P'_{1N0} = \frac{1 \text{ mW}}{\xi'_1} \equiv 10{,}000 \text{ pW0} \qquad (10.14)$$

This value takes into account the psophometric weighting (sect. 2.4.13).

10.5.3 Distribution of the tolerated noise among the different sources

The noise tolerated at the end of the link includes *all* the sources of noise in the system, i.e.,

- terminal equipment noise
- transmission equipment noise, with accumulating contributions along the link

In an analog system, each of these noise sources is the result of three principal causes, already mentioned in section 2.4.11, i.e.,

- *background noise* (thermal noise, increased by the noise factor of the amplifiers, according to section 6.2.5)
- *intermodulation noise* (self-interference from section 6.2.8) due to nonlinearities
- *crosstalk* between channels of the same system or between neighboring circuits.

The distribution of the 10,000 pW0 of total noise power referred to zero after 2500 km between these categories of noise is a delicate enterprise which can be full of economic consequences for system planning. As a first approximation for rough planning, the distribution of table 10.12 can be assumed, when dealing with a telephone channel. From case to case, these values can be subject to discussion provided that their sum remains constant.

Table 10.12 Typical distribution of total admissible noise in the hypothetical reference circuit.

		Systems on balanced-pair lines	Systems on coaxial lines	Systems on microwave links
Noise of the terminal equipment		2500 pW0	2500 pW0	2500 pW0
Transmission noise	Background noise	2500 pW0	5000 pW0	3750 pW0
	Intermodulation	2500 pW0	2500 pW0	3750 pW0
	Crosstalk	2500 pW0	–	–
	Total	10000 pW0	10000 pW0	10000 pW0

10.5.4 Methodology

Whatever the real length of the link to be planned, it is *extrapolated to 2500 km* in order to compare it with the hypothetical reference circuit.

ANALOG SYSTEMS

The calculation of the repeater spacing rests on the *background noise* budget alone. The condition related to *intermodulation noise* allows us to fix the nominal and maximum levels, relative to another, according to a calculation similar to that given in section 10.4.3.

If the repeaters are spaced at regular intervals of $l_0 = l/N$, it requires a number of repeaters $N_{ref} = 2500 \text{ km}/l_0$ to cover the reference distance. Relation (6.30), giving the signal-to-background noise ratio, thus becomes

$$10 \lg \xi_{Rref} = L_{SEnom} - \alpha l_0 - (L_{Nth} + F) - 10 \lg N_{ref} \qquad (10.15)$$

In a multichannel carrier system, the calculation of ξ_{Rref} concerns *any channel* (channel i). The parameters intervening in relation (10.15) are thus

- the nominal level of emission L_{SEnom} in the channel i, which can be different from one channel to the next in the case of pre-emphasis (section 10.3.4);
- the attenuation coefficient α of the cable at the frequency corresponding to the position of the channel i in the multiplex;
- the level of thermal noise L_{Nth}, increased by the noise factor F to give L_N and calculated in the bandwidth of one channel. In telephony, the psophometric weighting $L_{Nthtph} = -141.5$ dBmp at 290 K is taken into account, from (2.45);
- relation (10.15) thus gives the signal-to-noise ratio $\xi'_{li\ ref}$ in the channel i after SSB demodulation.

With the hypothesis of an admissible total noise distribution from table 10.12, the signal-to-background noise ratio alone, after the reference distance of 2500 km, needs be lower than the following values in each channel (particularly in the upper channel, generally the most critical, since it is the most attenuated):

$$\text{balanced pairs: } 10 \lg \xi'_{liref} \geq 10 \lg \frac{1 \text{ mW}}{2500 \text{pW0}} = 56 \text{ dB} \qquad (10.16)$$

$$\text{coaxial lines: } 10 \lg \xi'_{liref} \geq 10 \lg \frac{1 \text{ mW}}{5000 \text{pW0}} = 53 \text{ dB} \qquad (10.17)$$

10.5.5 Example of the calculation of the repeater spacing

Figure 10.13 illustrates the graphic calculation of repeater spacing l_0 in the case of a 1.3 MHz carrier system (300 channels, fig. 10.8) installed on a small diameter coaxial cable, where the attenuation coefficient is equal to 6 dB/km at 1.3 MHz. The calculation is made for the upper channel in which the nominal level is assumed to be -13 dBm. A noise factor of 7 dB is assumed for the repeaters.

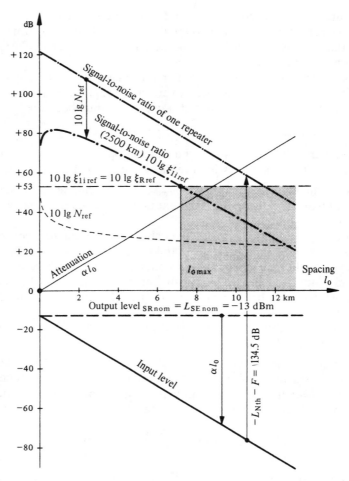

Fig. 10.13 Calculation of the repeater spacing in the 1.3 MHz system (example).

The longer the distance l_0 between the repeaters, the lower their input level becomes and approaches the level of thermal noise. The result is a diminution of the signal-to-noise ratio of each repeater. The cumulative effect of the N_{ref} repeaters necessary to cover the 2500 km appears as a lowering of $10 \log N_{\text{ref}}$ on the signal-to-noise ratio $\xi'_{1i\,\text{ref}}$ which must, according to (10.17), remain higher than 53 dB. Thus, we obtain a maximum spacing of approximately 7 km.

ANALOG SYSTEMS

It is noted that there is a spacing (approximately 0.7 km) which optimizes the signal-to-noise ratio (>80 dB), conforming to relation (6.33). For economic reasons, this optimum signal-to-noise ratio is nevertheless without practical interest.

10.6 OPERATIONAL PROBLEMS

10.6.1 Measurement of background noise

The measurement of background noise in a channel must be made when all the channels of the system are idle. A psophometer (section 2.4.13) gives the rms value of the weighted noise U_{Neff}, or its level L_N. The signal-to-background-noise ratio is derived with reference to the nominal signal level in this channel

$$10 \lg \xi = L_{Snom} - L_N \qquad (10.18)$$

1.6.2 Measurement of intermodulation noise

Because the appearance of this noise is linked to the presence of a signal in the system, its measurement is more delicate. On the other hand, it cannot be measured alone, but is always superimposed on the background noise.

The measurement procedure generally used is the following, *noise-in-slot method*, presented in figure 10.14: a white noise generator simulates the real signal of the ensemble of z channels of the multiplex, with a level conventionally referred to zero according to (10.5) or (10.6), depending on the number of channels. A band-stop filter eliminates the band corresponding to the channel i to be measured. After traversing a channel studded with nonlinearities, the

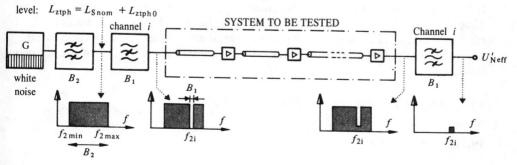

Fig. 10.14 Intermodulation noise.

intermodulation noise diffuses to the interior of this band. At reception, a filter extracts the noise from the band in which the channel i is located.

The effective value of the noise U'_{Neff}, measured in the channel i includes the contributions of the background noise U_{Neff} and the intermodulation noise U_{IMeff}

$$U'_{\text{Neff}} = \sqrt{U^2_{\text{Neff}} + U^2_{\text{IMeff}}} \qquad (10.19)$$

If the nominal level L_{Snom} is varied, we observe that part of U_{IMeff} increases noticeably when L_{Snom} approaches the saturation level (fig. 10.15, to be compared with figure 6.6).

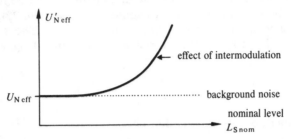

Fig. 10.15 Results of the measurement of noise when the nominal signal level varies.

10.6.3 In-service system monitoring

The presence and the level of the group and line pilots (section 10.2.3) allow permanent monitoring of the proper functioning of the system. However, a variation of the noise level would escape monitoring. This is why a noise measurement similar to that described in the preceding section is also made, either in an unused channel, or in two narrow frequency bands pre-assigned for this purpose at the upper and lower ends of the multiplex band. In this case, however, it is the whole of the system occupied by telephone communications which plays the role of noise generator. Its level fluctuates according to the intensity of traffic and allows only a qualitative evaluation of the noise level.

10.7 ANALOG BROADCASTING SYSTEMS

10.7.1 Objective

Here, analog broadcasting systems is understood to mean the transmission procedures used in *audio broadcasting* (musical programs) and *television*. Only

ANALOG SYSTEMS

certain aspects of these systems will be discussed here, as illustrations of applications of analog modulation. These systems are principally characterized by

- a *frequency-division multiplexing* of a large number of transmitters within a limited frequency range. The assignment of frequencies to each transmitter and the re-use of the same frequency by distant transmitters are subject to very strict international conventions, which take into account propagation conditions on the ground and in the ionosphere, during the day and at night;
- a much larger number of *universal receivers*, i.e., capable of receiving the emission of one of several transmitters at will.

10.7.2 Principle of superheterodyne reception

The congestion of the frequency ranges reserved for radio and television broadcasting requires very high receiver selectivity. On the other hand, the universality of tunable receivers requires that this selectivity guarantees the reception of each transmitter. The *heterodyne receiver* resolves this double difficulty by using *frequency translation* by means of DSBSC modulation before demodulation. The carrier used for this purpose is provided by a local oscillator (in the receiver) with a frequency f_{loc} which is tuned to the receiver in such a way that it presents a constant difference f_{IF}, called the *intermediate frequency, IF*, with respect to the frequency f_E of the desired transmitter. It is preferable to choose a local frequency **higher** than that of the transmitter (*super*heterodyne receiver):

$$f_{loc} = f_E + f_{IF} \quad \text{with } f_{IF} = \text{const.} \tag{10.20}$$

To cover a given range of wavelengths, this choice leads to a range of relative variation of f_{loc} lower than, and thus easier to implement, than if f_{loc} had been chosen lower than f_E. All the commercial radio and TV receivers make use of this principle (fig. 10.16).

With this method, the problem of selectivity is carried onto a filter (incorporated into an amplifier) centered on the *fixed* frequency f_{IF}.

Another transmitter of frequency f'_E such that

$$f'_E = f_{loc} + f_{IF} \tag{10.21}$$

would give rise, due to the lower sideband of the DSBSC modulation, which is folded around the origin, to a signal that would also cross the filter centered on f_{IF} and thus would unacceptably disturb the reception of the transmitter at f_E. This frequency f'_E is called the *image frequency*. It is determined by (10.20) and (10.21):

$$f'_E = f_E + 2f_{IF} \tag{10.22}$$

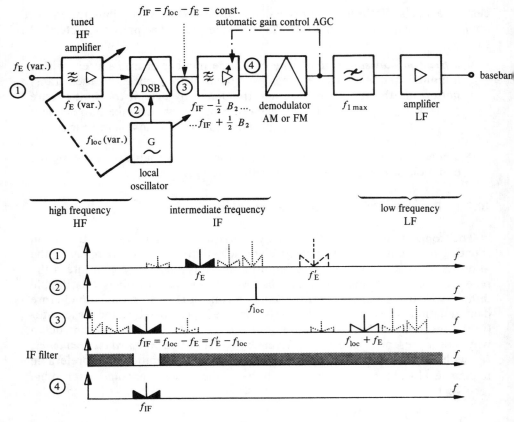

Fig. 10.16 Superheterodyne reception

To avoid this disturbance, it is absolutely necessary to introduce frequency selectivity in the HF amplifier, in the form of a circuit tuned to f_E and blocking f'_E. The adjustment is made simultaneously and in parallel with the frequency variation of the local oscillator, in a way in accord with (10.20).

The value of the intermediate frequency is not subject to precise norms. It is located in the following ranges, according to the type of modulation and signal:

- AM radio broadcasting: $f_{IF} = 440\text{-}490$ kHz
- FM radio broadcasting: $f_{IF} = 10.7$ MHz
- television (VSB): $f_{IF} = 32\text{-}40$ MHz.

10.7.3 Characteristics of radio broadcasting systems

With long, medium, and short waves, the transmitters are *amplitude modulated* with the following parameters:
- primary bandwidth $B_1 \cong 4.5$ kHz
- carrier spacing of neighboring transmitters $\Delta f_p = 9$ kHz

The larger bandwidths available with ultra-shortwave permit *frequency modulation* with the following characteristics:
- primary bandwidth $B_1 \cong 15$ kHz
- maximum frequency deviation $\Delta f_{max} = 75$ kHz, given a modulation index $\delta = 5$
- secondary bandwidth $B_2 \cong 180$ kHz
- carrier spacing of the neighboring transmitters (channels) $\Delta f_p = 300$ kHz
- pre-emphasis of 17 dB at 15 kHz to compensate for the variation of the modulation index (and thus for the signal-to-noise ratio) inside the primary band. This effect is achieved by a high-pass RC circuit with a time constant of 75 µs (fig. 10.17). It is compensated for by a reciprocal characteristic (de-emphasis) in the receivers

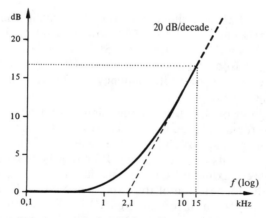

Fig. 10.17 Pre-emphasis curve for FM radio transmission.

10.7.4 Stereophonic FM broadcasting

Stereophonic radio broadcasting in principle requires the transmission of two signals (left channel L and right channel R) of the same quality. Nevertheless, two important conditions of *compatibility* must be satisfied:

- the stereo program must be able to be received in the monoaural mode by an ordinary receiver;
- the secondary bandwidth required in the stereophonic mode must remain equal to that necessary in monoaural.

The solution adopted consists of not transmitting the original signals of the two channels L and R, but rather their sum and their difference. The sum is offered in the baseband to the FM modulator, while the difference is preliminarily translated by DSBSC modulation with a carrier at 38 kHz (fig. 10.18).

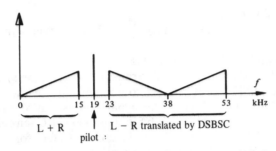

Fig. 10.18 Spectral composition of the stereophonic signal.

At reception, the FM demodulator restores this signal from which the sum of the two channels can be extracted by low-pass filtering (sufficient for a monaural receiver) as well as, by inverse DSBSC modulation, their difference. For this purpose, an isochronous carrier at 38 kHz is necessary (sect. 8.3.3). It is obtained by doubling the pilot frequency at 19 kHz, which is transmitted with the signal (fig. 10.19).

While transmission of the stereophonic signal from figure 10.18 with AM modulation would requires a secondary bandwidth of at least 3.5 times higher than monaural transmission, such is not the case with FM modulation. While it is noted that the signals L + R and L − R cannot be maximum at the same time, if we keep the maximum frequency deviation at $\Delta f_{max} = 75$ kHz, usual in monoaural mode, for the sum of the two channels L + R, this maximum is not increased by the addition of the signal L − R and the real secondary bandwidth in FM stereo does not exceed that of a monaural transmission $B_2 \cong 180$ kHz).

In any case, since the effect of noise with FM is more noticeable the further the baseband frequency is from zero (fig. 8.49), the L − R signal is noticeably more disturbed than the sum L + R. The result is that, despite the application of a pre-emphasis according to figure 10.17 to the two signals, the stereophonic transmission is of lower quality than the monoaural transmission under the same conditions.

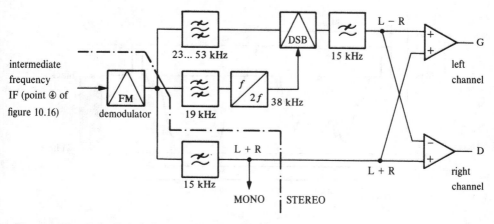

Fig. 10.19 Principle of the stereophonic receiver.

10.7.5 Video signals

Video signals represent the information contained in an image, analyzed line by line. Video signals also contain synchronization information necessary to determine the succession of images and the scanning of each line.

The *U.S. standards*, specify the following characteristics:

- gross number of lines per image: 525 (including the "lines" devoted to the synchronization of the image and to the return of the beam);
- 30 complete images per second, by means of 50 half-images with interleaved lines (even lines, odd lines) so as to avoid a scintillation effect;
- resulting line frequency f = 15,750 lines/s (63.5 µs per line);
- bandwidth of the video signal B_1 = 4.2 MHz.

For reasons mentioned in section 8.5.1, television transmission uses VSB modulation. To this end, the video signal applied at the input of the modulator (primary signal) generally represents the luminance of the image by a voltage in such a way that the lower the voltage the clearer the image (fig. 10.20). This convention, designated by the term *"negative modulation"* presents the following advantages:

- the superposition of parasitic pulses produces black traces, less annoying than bright points;
- nonlinear distortions first affect the synchronization pulses, which are not very sensitive, and not the luminance information.

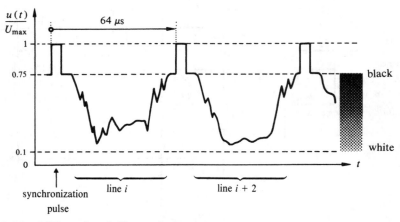

Fig. 10.20 Video signal ("negative modulation").

10.7.6 Characteristics of television systems

After VSB modulation, the translated video signal occupies a bandwidth of approximately 6 MHz (fig. 10.21). The sound is transmitted simultaneously by FM modulation by means of an auxiliary carrier located 4.5 MHz higher than the principal carrier and with a maximum frequency deviation of 50 kHz.

Consequently, the spacing of the channels assigned to the TV transmitters is around 7 MHz in bands I and III (VHF, fig. 3.41) and 8 MHz in bands IV and V (UHF).

At reception, the signal which is preliminarily translated to a fixed intermediate frequency (e.g., 38.9 MHz) is demodulated so as to extract the video

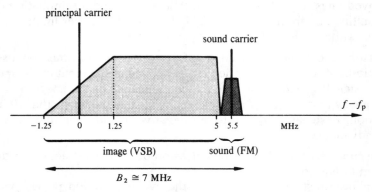

Fig. 10.21 Spectrum of the television signal (black & white) in transmission.

signal in the baseband and the sound carrier (which is still FM modulated) from it. The sound carrier is then applied to an FM demodulator which extracts the sound in the baseband (fig. 10.22).

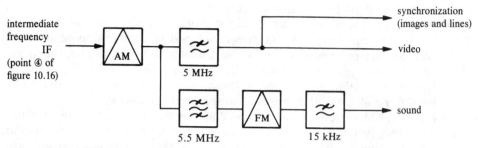

Fig. 10.22 Principle of a TV receiver.

10.7.7 Color television

Three conditions of *compatibility* between black-and-white and color television must be satisfied:
- a color emission must be capable of suitable reception by a receiver only equipped for black-and-white;
- the color television receiver must give a satisfactory black-and-white image in the case of a monochromatic emission;
- the bandwidth occupied by the signal with or without color must be the same.

The first condition is fulfilled by the separate transmission of the luminance and chrominance parameters (section 1.7.8), which are obtained, in the color video camera, by weighted combination of the three monochromatic components (red, green and blue) according to relations (1.14), (1.15), and (1.16). These relations permit the reconstruction of an artificial white upon reception of a black-and-white emission by a receiver equipped for color.

The luminance parameter Y modulates the carrier with VSB in the same way as in the case of black-and-white television (fig 10.21). For physiological reasons mentioned in section 1.7.8, the two chrominance parameters I and Q can be transmitted with noticeably smaller bandwidths than that of the luminance parameters.

The insertion of signals carrying chrominance into the frequency band devoted to luminance (in such a way as to not increase the total bandwidth required) has given way to three very elaborate and technically perfected solutions, of which only the fundamental principles are mentioned here:

- National Television Systems Committee (NTSC), USA: the two chrominance parameters I and Q modulate in amplitude and phase an auxiliary carrier located in the upper part of the secondary band at a frequency such that it is interposed between the spectral lines corresponding to the multiples of the line scanning frequency;
- the Phase Alternation Line (PAL) system: very close to the American NTSC principle, the PAL system developed in Germany avoids the inconvenience of a sensitivity to phase distortions. With this aim, the phase of one of the chrominance signals is inverted on every other line. For the European system at 625 lines/s, the auxiliary chrominance carrier has been fixed at 4.43 MHz higher than the principal carrier;
- the SECAM system: used in France and in the Eastern European countries, this system is based on the redundancy of chrominance information from one line of the image to the other. The two chrominance signals are transmitted sequentially, alternately to each line, then delayed by a line at reception before being recombined. In contrast to the two other procedures, a frequency modulation is used here for the chrominance, with two different auxiliary carriers (4.4 MHz and 4.25 MHz).

FM modulation provides favorable color quality properties in SECAM, as long as the signal-to-noise ratio at reception is sufficient to avoid the threshold effect (section 8.8.11). Beyond this point, the other procedures are better. In any case, an objective technical and economic comparison is difficult in this area in which criteria of politics or national prestige often predominate.

Chapter 11

Telex and Data Communication

11.1 INTRODUCTION

11.1.1 Types of services

This chapter will discuss *services* (as defined in section 1.2.1) in which the information to be transmitted and routed is of a ***digital nature*** from the source: alphanumeric text, fixed black-and-white pictures, commands, quantized results of measurements, alarms, *et cetera*. These types of information are often included under the general term of ***data*** (sect. 1.7.9).

This principally includes the following services:

- *telegraphy* in the general sense of electrical transmission of any type of text and graphical information, and in the historical sense restricted to the transmission of alphanumeric *telegrams* by electrical means.
- *telex*, a modern and automatized form of telegraphic service, including transmission and switching (routing) of alphanumeric text, composed on the keyboard of a ***teleprinter*** to be sent to another teleprinter. The code (generally CCITT no. 2, sect. 11.2.3) and the symbol rate (50-300 baud) are fixed;
- *teletex*, an extension of telex service in the sense of its speed, its similarity to the typewriter (a more extensive alphabet), and its adaptability to remote word processing;
- *telecopying*, or facsimile transmission of fixed pictures, generally in black and white, or possibly with grey shades, or even in color;
- *telecontrol*, a general term covering *telemetry*, *remote control*, and *telemonitoring*.

All these services are often considered to be examples of *data communication*, in the broadest sense. In this context, we can also speak of ***telecomputing*** or ***telematics*** to designate services combining telecommunicaions and computing [50-52]. Here, in effect, the sources and receivers of the information are machines, generally automatic and, more and more frequently, computerized. Nevertheless, human beings can still intervene as a primary source and impose a random timing in the production of information (for example, manual input on a keyboard).

11.1.2 Methodology

Data communication is an application of the principles of digital transmission (chap. 5), either in the baseband, or in a form translated by means of a discrete analog modulation (sect. 8.11).

Very frequently, the public telecommunication network, originally conceived for telephony, serves as a support for data services, which are considered to be guests in the network and which must consequently cope with its nonoptimal characteristics.

The introduction of digital transmission and switching techniques (PCM) into the telephone network favors sharing the infrastructure with other uses and leads to the possible idea of an integrated services digital network, ISDN (sect. 15.8.3).

11.2 CODES

11.2.1 Alphanumeric codes

For the transmission of text and numbers, the source makes use of a finite number of *characters*: 26 letters (Roman alphabet), 10 digits, and auxiliary signs (spaces, punctuation, line feed, *et cetera*). These characters must be translated into logical expresssions (symbols), which are generally binary, with a correspondence table called a *code*.

The different alphanumeric codes are distinguished by

- their *richness*: availablity of upper and lower case letters, accenting of the lower case letters, special signs;
- their *conciseness* in the sense of the manageable number of elementary symbols (bits or moments) required to represent a character of the alphabet, which can be expressed by the ratio between the character rate and the symbol rate;
- their *immunity*, i.e., their ability to detect, or even correct, isolated errors in the bits representing a character.

The principle of error detecting or correcting codes consists of deliberately introducing a certain *redundancy* into the code, designed in such a way that the reception of an erroneous bit does not necessarily lead to a false character, but to the refusal of the error-containing character or to the automatic correction of the bit identified as false. Certain errors nevertheless, remain undetected. Their probability of appearance, or *residual probability error* ϵ_r per character, is expressed by the ratio of the mean number of false characters to the total number of transmitted characters. The *immunity* of the code in the presence of bit errors is greater the lower ϵ_r is.

Immunity and conciseness are two contradictory requirements between which a compromise must be found.

11.2.2 Morse code

While the code proposed by Samuel Morse in 1837 (fig. 11.1) has lost much of its current interest for commercial telegraphy, it remains the universal code of amateur and professional radiotelegraphers (seagoing links), primarily by manual operation. It has the particular quality of representing the alphanumeric characters by a sequence of *binary* states (black, white) and by the *relative and quantized duration* of these states (dashes have triple the duration of a dot, differentiated intervals). Further, as an intuitive precursor of modern coding theories, Morse had the idea of encoding letters by shorter characters when the letter is more frequent (in English!). This reduces the average duration of the transmission of a coherent text (concise code).

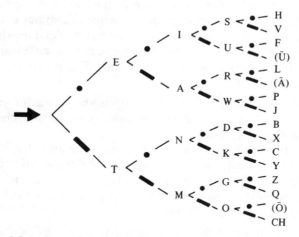

Fig. 11.1 Decision tree of the alphabetic part of the Morse code.

11.2.3 Five-bit telex code

Morse code has the drawback of being difficult to apply to automatic decoding, which is necessary in telex to restore an intelligible alphabetic text to the receiver (teleprinter). For this reason, the codes used in telex have a fixed format (and thus are less concise).

438   TELECOMMUNICATION SYSTEMS

Telex uses 51 characters, consisting 26 lower-case letters, 10 digits, 12 signs and 3 commands (spacing, carriage return, line feed). A binary code of 5 bits (32 characters) is sufficient to represent them as a result of the following principle, called an *escape code*: two meanings are assigned to each character ("letter" or "figures") and two particular characters (numbers 29 and 30) are used to indicate a change of meaning for the next characters. This procedure is only effective if the letters and figures appear in groups, which is the case in normal text. Otherwise, it tends to double the transmission time. The 5-bit code of figure 11.2 is the subject of an international convention (CCITT code no. 2) for telex transmission and the recording of text on punched tape. The two states of the binary code are conventionally designated A and Z.

11.2.4 Seven-bit error detecting code

The 5-bit code has no redundancy since all the symbols (with the exception of 00000, number 32, which is unused) correspond to characters. A transmission error in one of the bits simulates the presence of another character of the code, and thus gives rise to an erroneous character. For long-distance radio communication, subjected to strong interference, it is necessary to protect against these errors, which are then frequent. The principle of *automatic request, ARQ* consists of

- using a code in which an error transforms the character emitted into a combination of bits which does not correspond to any character used; the error is thus detectable;
- requesting (via a channel in the opposite direction) that the erroneous character be repeated. Taking into account the propagation time between transmitter and receiver, the transmitter must keep the last symbols emitted in memory so as to be able to respond at any time to a possible request for a character to be repeated.

The *7-bit CCITT code no. 3* (fig. 11.2), also called the van Duuren code, uses only the $\binom{7}{3} = 35$ characters composed of four bits in A state and three bits in Z state. It thus allows easy control at reception. All the simple errors (a single false bit per character) are detected. However, the double errors which exchange an A state and a Z state pass unnoticed. Their probability gives a good approximation of the residual error probability ϵ_r per character, as a function of the error probability per bit ϵ

$$\epsilon_r \cong 4\epsilon(1-\epsilon)^3 \cdot 3\epsilon(1-\epsilon)^2 \cong 12\epsilon^2 \qquad (11.1)$$

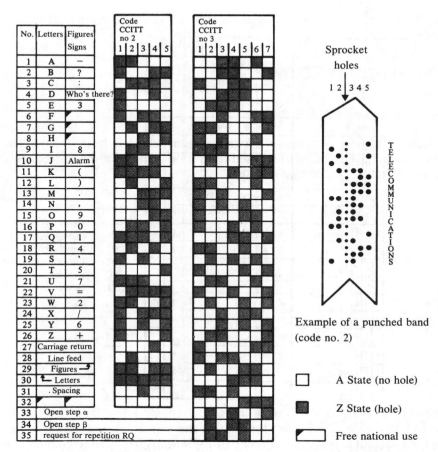

Fig. 11.2 Five-bit CCITT code no. 2 (telex) and seven-bit CCITT code no. 3.

11.2.5 Seven-bit ISO code

The extension of the code to an ensemble of 128 characters allows the representation (without key shift) of upper-case and lower-case letters, digits, and numerous special signs and commands (fig. 11.3).

The different versions of this code are called the *International Standard Organization (ISO) code, CCITT code no. 5,* and *ASCII (American Standard Code for Information Interchange) code,* and differ only in a few details concerning the auxiliary characters.

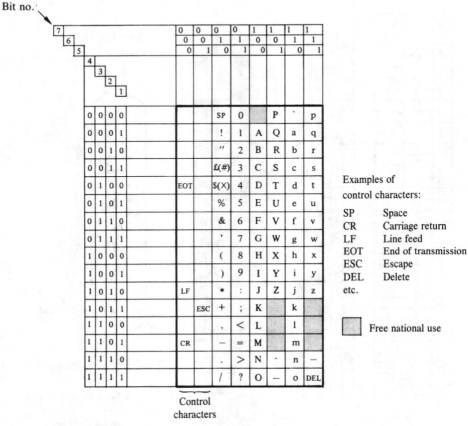

Fig. 11.3 Seven-bit ISO code (CCITT no. 5, ASCII).

This code is often used in data communication and also forms the basis of teletex. The 7 bits strictly necessary for the representation of the 128 characters are frequently completed by an eighth bit, which is redundant, and allows the detection of simple errors by parity control.

11.3 TELEX

11.3.1 Start-stop operation

With manual input at the keyboard of the teleprinter, the characters are produced in a more or less irregular manner. At rest, the Z state is emitted. It is, therefore, necessary that the receiver be able to recognize the arrival of a character at any instant, even if it starts with the Z state. To this end, we can precede the character, strickly speaking (5 bits in code no. 2), by a *start bit*, always in A state. This intermittent procedure is called *start-stop* operation.

11.3.2 Continuous operation

If the characters are emitted from a prerecorded text (e.g., in an electronic memory or on a punched tape), they appear at regular intervals. The receiver distinguishes them by

- the presence of a Z state or *stop bit* of minimum duration between any two characters, allowing the detection of the transition to the start bit (A state) of the following character;
- or by a particular code detail which makes it *self-synchronizing*, for example, the search for symbols comprising four A's and three Z's in the case of CCITT code no. 3.

11.3.3 Format of the telex signal

Telex transmission is internally *clocked* (sect. 5.1.3) for each character, in the sense that the significant bits which comprise each character always have the same duration $1/\dot{M}$, fixed by a clock at emission and reception. The start bit has the same duration as the significant bits, while the stop condition must last at least $1.5/\dot{M}$ (fig. 11.4).

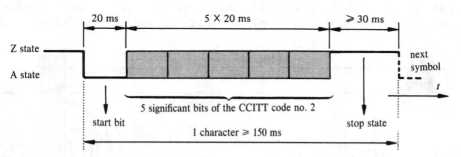

Fig. 11.4 Format of the telex signal at 50 baud.

The usual symbol rate in telex is $\dot{M} = 50$ baud. At this rhythm, each character lasts at least 150 ms, which brings the maximum character rate of $6\frac{2}{3}$ characters/s.

With modern electronic teleprinters, it is also possible to have continuous (automatic) operation, at rates of 100 baud, 200 baud, or 300 baud.

11.3.4 Baseband transmission

Two binary states A and Z are represented by signals in unipolar mode (transmission is said to be "single current," A state: $i = 0$, Z state: $i = 40$ mA) or antipolar ("double current" transmission, $i = \pm 20$ mA). The second solution allows the line to be terminated on a defined impedance, which is identical in the two states and lends itself better to medium and long distance transmission on balanced-pair lines.

The bandwidth necessary for $\dot{M} = 50$ baud is evaluated according to (4.9) at $B_1 \cong 40$ Hz, which allows it to be placed, if necessary, below the normal telephone band from which it can be separated by means of adequate filters (infra-acoustic telegraphy).

11.3.5 Telex transmission in an analog transmission channel

Aside from subscriber lines and some lines specifically dedicated to it, telex frequently borrows shared transmission means from the long-distance telephone network (sect. 15.8.2). To this end, certain carrier system channels are devoted to telex and are thus lost for telephony. In the available band from 300 to 3400 Hz, it is possible to multiplex several telex channels translated by modulation. The procedure called *voice frequency telegraphy* uses a discrete analog modulation (OOK or FSK with $\Delta f = 0.6 \dot{M}$) to form a multiplex of 24 telex channels at 50 baud (or 12 channels at 100 baud) within a telephone channel (fig. 11.5). The total bit rate thus obtained (1200 bit/s,) however,

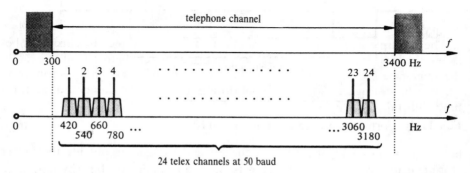

Fig. 11.5 Voice-frequency telegraphy: 24 telex channels at 50 baud in a telephone channel.

remains very modest with respect to the theoretical capacity of the conventional telephone channel, evaluated at approximately 50 kSh/s according to (4.12)!

11.4 DATA TRANSMISSION

11.4.1 Dedicated circuits and switched circuits

According to the needs of the user, three types of circuits are proposed for data transmission:
- *dedicated circuit*, i.e., a permanent link **between two fixed terminal points** which is reserved exclusively for this private use. It has the advantage of being available at any time and of presenting known and stable transmission characteristics. However, its high cost of operation is only justified in the case of intensive use. It can use telegraphic circuits (50–300 bit/s) or telephone circuits (2400–9600 bit/s);
- *switched circuit* of the type used in telephony, set up from case to case for the duration of each communication, across the shared telephone network. This is an economical and flexible solution, which can reach different terminals, at the cost, however, of a preliminary selection operation. Its principal inconvenience is the variable and unpredictable quality of the circuits thus assigned;
- *special switched data circuit*, established on demand by circuit or packet switching across a digital network especially designed for this purpose (sect. 15.6).

11.4.2 Requirements

Data transmission is distinguished from telex by
- much higher bit rates
- richer codes (e.g., CCITT no. 5) or more specialized codes
- more stringent quality requirements

Table 11.6 gives an idea of the residual error probabilities ϵ_r (after possible detection of bit errors) required per character, according to the type of service, compared with the error probabilties ϵ per bit offered by the different circuits (table 11.7).

The comparison of these figures demonstrates the necessity of using error detecting, or even correcting codes in critical cases.

Table 11.6 Probability of residual error required per character.

	ϵ_r
Redundant messages (conversational mode)	$10^{-3} \ldots 10^{-4}$
Critical messages, non reproducible	10^{-7}
Messages of vital importance (e.g., air navigation services)	10^{-9}

Table 11.7 Probability of error per bit observed (orders of magnitude).

	$\dot{D}$ bit/s	ϵ
Switched telex circuit	50	$\sim 10^{-5}$
Dedicated telex circuit	50 ... 200	$\sim 10^{-6}$
Switched telephone circuit	200 ... 1200	$\sim 10^{-4}$
Dedicated telephone circuit	2400 ... 9600	$\sim 10^{-5}$

11.4.3 Baseband transmission

Balanced pair lines constitute a medium in which the attenuation increases regularly with frequency, without presenting a definite upper limit. They lend themselves to direct data transmission, said to be in baseband, i.e., without preliminary modulation, up to high bit rates. The case of digital systems at 2 Mbit/s, or even at 8 Mbit/s (chap. 9) installed on such lines illustrates their possibilities. The criteria for choosing the transmission method are those already mentioned in section 9.6.2.

Among the large number of methods proposed, the most frequently used are the *pseudo-ternary AMI mode* and especially its variant, the *HDB 3 mode* (sect. 9.6.5), as well as the diphase code.

The *diphase code* (Manchester code) is a binary antipolar mode with a zero-mean elementary signal (fig. 11.8). Its power spectral density strictly cancels at $f = 0$, whatever the statistical distribution of the binary states. By virtue of its binary character, it requires only a single decision threshold for regeneration, independently of the received signal amplitude. The presence of at least one transition per bit guarantees a high clock rate content (frequency $\dot{D}$). However, this mode requires a bandwidth twice that of the AMI mode (compare figures 11.8 and 9.20).

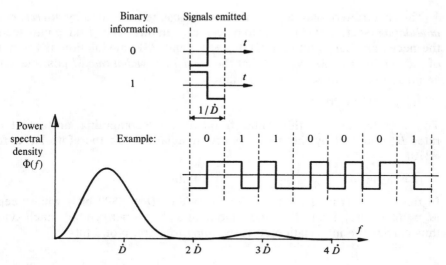

Fig. 11.8 Diphase (Manchester) code.

11.4.4 Transmission in a telephone channel

The telephone network has the triple advantage of being in place, being worldwide in scope, and of being very reliable. It is, therefore, tempting to use it also to transmit other information, especially data.

In any case, the telephone channel, whatever the nature of the multiplex (frequency-division and analog, or time-division and digital) in which it is included, presents a band-pass filter characteristic, with *attenuation distortion* between 300 and 3400 Hz (fig. 2.9). *Phase distortion*, not critical in telephony, is largely tolerated, and thus presents serious problems for data transmission. Further, the telephone channel is the location of *disturbances* of all kinds: thermal noise, quantizing noise (PCM), external influences (crosstalk, heavy current, *et cetera*), impulse noise (crosstalk coupling of signaling pulses on neighboring pairs or regeneration errors in PCM).

11.4.5 Case of the analog telephone channel

Taking into account the significant phase distortions in an analog telephone channel (due to the filters of the carrier systems), the practical usable band for data transmission is limited to between approximately 500 and 2800 Hz, given a bandwidth of $B_2 \cong 2300$ Hz.

The spectrum of data to be transmitted must be translated by *discrete analog modulation* (sect. 8.11) in order to be centered in this band. Taking into account the necessary bandwidths in OOK, FSK, and PSK modulation as a function of the symbol rate $\dot{M}$, we find that the *maximum symbol rate* $\dot{M}$ possible across an analog telephone channel is roughly

$$\dot{M}_{max} \cong 1500 \text{ Bd} \tag{11.2}$$

The rates higher than this value, frequent in telecomputing, are actually bit rates $\dot{D}$, obtained by the use of m-ary modulation. We therefore have, from (1.10)

$$\dot{D} = \dot{M} \text{ lb } m \qquad \text{bit/s} \tag{11.3}$$

In this way it is possible to achieve a rate of $\dot{D} \cong 4800$ bit/s while keeping $\dot{M} = 1600$ baud, due to PSK modulation at 8 phases per symbol. Each symbol thus carries the information corresponding to a group of 3 bits.

11.4.6 Standardized modems

It is common to distribute data transmission functions between two distinct pieces of equipment (fig. 11.9):

- the data terminal equipment, DTE, more commonly called the *terminal*;
- the data circuit-terminating equipment, DCE, more often called a *modem* because it contains the *mo*dulator and the *dem*odulator necessary for transmission in an analog channel.

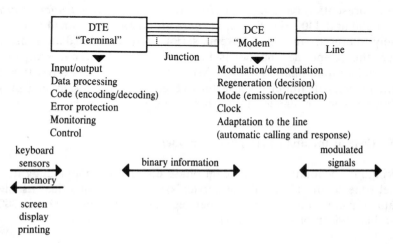

Fig. 11.9 Distribution of functions between the modem and the terminal.

TELEX AND DATA COMMUNICATION

The modem belongs to the field of telecommunications, while the terminal is considered to be computing equipment. The compatibility between these two pieces of equipment, generally of different origin, requires a very careful definition of the *interface* separating them (e.g., CCITT Recommendation V.24).

The CCITT has defined a series of modems which can be used in analog telephone channels. The standardized rates correspond to different applications and modes of operation (simplex, duplex, sect. 4.6.1). The return channel with a low rate (75 bit/s) of half-duplex modems can fulfill the character repetition requirement for error detection in the forward channel. Table 11.10 summarizes and compares the principal characteristics of these modems.

Table 11.10 Modems standardized for an analog telephone channel (CCITT).

$\dot{D}$ bit/s	m	$\dot{M}$ Bd	Modulation	Spectrum	Mode of operation	CCITT Rec.
200 (×2)	2	200 (×2)	FSK $\Delta f = \frac{1}{2}\dot{D} = \pm 100$ Hz	1080 f_{p1}, 1750 f_{p2}	full duplex mode (frequency-division pseudo-4-wire)	V 21
600 or 1200	2	600 1200	FSK $\Delta f = \frac{1}{3}\dot{D} = \pm 200$ Hz $\Delta f = \frac{1}{3}\dot{D} = \pm 400$ Hz	420, 1500; 420, 1700	half-duplex mode + ← 75 bit/s	V 23
1200	4	600	DPSK $\Delta\varphi = i\frac{\pi}{2}$ with $i = 0, 1, 2, 3$	1200 f_{p1}, 1800, 2400 f_{p2}	full-duplex mode (frequency-division pseudo-4-wire)	V 22
2400	4	1200	DPSK $\Delta\varphi = i\frac{\pi}{2}$ with $i = 0, 1, 2, 3$ or $i = \frac{1}{2}, \frac{3}{2}, \frac{5}{2}, \frac{7}{2}$	420, 1800	half-duplex mode + ← 75 bit/s	V 26
4800	8	1600	DPSK $\Delta\varphi = i\frac{\pi}{4}$ with $i = 0, 1 \ldots 7$	420, 1800	half-duplex mode + ← 75 bit/s	V 27

→ forward channel
← return channel

0, 300 .... 3400 Hz

The 9600 bit/s modem (CCITT Recommendation V.29) for dedicated circuits uses a modulation with $m = 16$ values resulting from the combination of discrete amplitude modulation ASK and 8 phase DPSK modulation (fig. 11.11). This modulation can also be considered to be an ASK modulation with two simultaneous carriers in quadrature (quadrature amplitude modulation, QAM). The symbol rate is thus brought to 2400 baud, a rate which is only possible in a telephone channel if the phase distortions are carefully corrected by an appropriate adaptive equalizer.

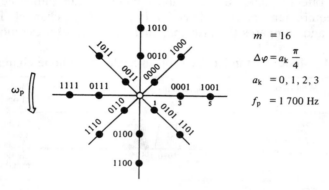

Fig. 11.11 Phasor diagram for transmission at 9600 bit/s and 2400 baud (QAM modulation).

11.4.7 Data transmission in a digital telephone channel

The digital nature of PCM multiplex systems (chap. 9) lends itself particularly well to data transmission, which is also digital, and all the more so since a high bit rate is required for a good quality telephone transmission ($\dot{D} = 64$ kbit/s).

It is certainly possible to connect a standardized modem to the analog end of a PCM channel (before the encoder or after the decoder), which is then considered to be like an ordinary telephone channel (fig. 11.12). This solution, however, is only of temporary interest in a partially digital network. It is, in fact, absurd in principle, because it entails the following inconveniences:
- contradictory modulation sequence (FSK or PSK, then PCM and the reverse)
- addition of quantizing noise
- under-utilization of the available bit rate (maximum data transmission rate of 2400 bit/s in a 64 kbit/s channel)

On the other hand, the direct access to a digital channel, without analog-digital conversion, allows us to make use of the following characteristics:

TELEX AND DATA COMMUNICATION

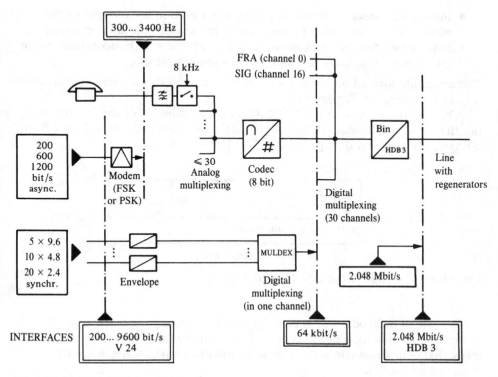

Fig. 11.12 Possibilities for data transmission in a 2 Mbit/s digital PCM system.

- rate of 64 kbit/s can be directly used
- low error probability ($\epsilon \cong 10^{-7}$)
- clocked operation, the channel being connected to a high precision clock

Clocked operation requires that the terminals produce their data at clock frequency imposed by the digital network.

For present needs, a rate of 64 kbit/s is generally exaggerated. It is often distributed by sub-multiplexing between several users at 2400, 4800, or 9600 bit/s. At the opposite extreme, it is also possible to place the entire bit rate of a primary digital system (2.048 Mbit/s) available for high speed data transmission by giving up the frame structure and PCM terminal equipment.

11.4.8 Format adaptation

Before being inserted into the octets (groups of 8 bits) of a digital channel, the data must be completed by

- *signaling information* destined for the other end, to be transmitted "out of band", i.e., thus avoiding disturbance of the data and code restriction;
- *sub-channel framing indications*, which allow them to be identified inside the channel that has been divided by sub-multiplexing.

The ensemble formed by a group of data bits, from a signaling bit or status bit S, and a framing bit F, constitutes an *envelope*.

A possibility, according to CCITT Recommendation X.50, consists in adopting the envelope format to that of the octet by placing the bits F and S on bothsides of a six-bit data group (fig. 11.13).

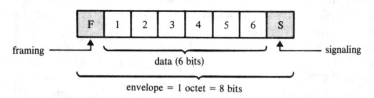

Fig. 11.13 Format of the envelope (6 + 2).

The rate of 8000 octets/s, imposed by the 8 kHz sampling frequency of PCM systems, allows a total net rate of 48 kbit/s per digital telephone channel; for example, 5 sub-channels at 9.6 kbit/s, 10 sub-channels at 4.8 kbit/s, or 20 sub-channels at 2.4 kbit/s.

11.4.9 Connection of asynchronous terminals to a digital channel

Low speed terminals generally set their clock at frequency $\dot{D}$ themselves and do not accept external synchronization. The result is a conflict at the interface between the terminal and the digital channel (or sub-channel), each having its own clock, which is asynchronous with respect to the other. The data emitted by the terminal at a rate $\dot{D}_T$ are thus sampled by the channel clock with a frequency $\dot{D}_C$ (fig. 11.14). The result is a time quantizing of the data bits, i.e., an approximation obtained by rounding off of their original duration which appears as a *time distortion* Δ or relative error in the duration of the bits

$$\Delta \leq \frac{\pm \dfrac{1}{\dot{D}_C}}{\dfrac{1}{\dot{D}_T}} = \pm \frac{\dot{D}_T}{\dot{D}_C}$$

(11.4)

This distortion appears at reception as a *phase jitter* which degrades the regeneration performance and increases the error probability (sect. 5.7.4 and 5.7.6). It is, therefore, necessary to limit its significance by the choice of a channel rate $\dot{D}_C$ which is distinctly higher than $\dot{D}_T$. This leads, however, to wasted capacity. It is necessary, for example, to devote a digital sub-channel with $\dot{D}_C = 4800$ bit/s to an asynchronous transmission of $\dot{D}_T = 200$ bit/s to guarantee a time distortion lower than 5%.

It is possible to improve the efficiency of a digital channel without deteriorating the time distortion by more elaborate procedures, such as, for example:

- *transition coding* (or sliding index coding): each binary transition of the data signal is announced synchronously with the channel clock (at $\dot{D}_C$), but specifying the fraction of the preceding clock period in which it took place. The time distortion is thus reduced with respect to (11.14) to a factor equal to the number of partial intervals defined in the clock period;

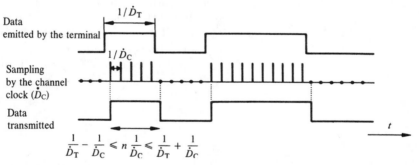

Fig. 11.14 Sampling of asynchronous data with respect to the digital channel.

- storage and *justification*: by a similar procedure to that described in section 9.4.4, the stuffing bits, which carry no information, are inserted into the binary flux of the channel at $\dot{D}_C$ so as to compensate for the difference $\dot{D}_C - \dot{D}_T \ (>0)$.

Chapter 12

Microwave Links

12.1 PRINCIPLES AND STRUCTURE OF MICROWAVE LINKS

12.1.1 Principle of the microwave link

A *microwave link* is a transmission system between two fixed points by radioelectric waves which are very strongly concentrated by means of directive antennas.

Depending on the case, it transmits:

- telephone conversations grouped into a frequency-or time-division multiplex;
- television programs;
- data.

Depending on the form (digital or analog) in which this information is presented, different types of modulation are used, firstly, to form the multiplex, and secondly, to translate the spectrum of signals into the frequency range appropriate for emission:

- **digital microwave links**: time-division multiplexing of digital telephone channels (PCM modulation), or data, then translation into microwaves by discrete analog modulation of a sinusoidal carrier with OOK, PSK, MSK, *et cetera*
- **analog microwave links**: frequency-division multiplexing of analog telephone channels (SSB modulation, according to the principle of carrier systems, section 10.1) or baseband video signal, then translation into microwaves by modulation of a sinusoidal carrier in FM/ΦM. In exceptional cases: timedivision multiplexing by PPM modulation, followed by an OOK modulation

12.1.2 System structure

The microwave link is a "frequency-division pseudo-four-wire system" (sect. 4.6.4) because the two directions of transmission are carried by different frequencies. The antennas are generally common to the two directions.

The general structure of a microwave link (analog or digital) is given in figure 12.1, in simplified form.

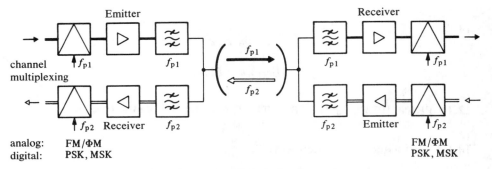

Fig. 12.1 Structure of a microwave link (a hop).

The microwave path between an emitter and a receiver constitutes a *hop*. The propagation conditions (distance, visibility) often require the division of a link into several hops separated by *relay-stations* which receive the microwave signal, amplify it, and retransmit it, generally with another carrier, in the directon of the next station. In exceptional cases, passive relays (reflective plan) can allow them to go around an obstacle.

12.1.3 Choice of carrier frequencies

As mentioned in section 3.8.4, the range of microwave links extends from 250 MHz to approximately 22 GHz. In this range, only certain well-defined bands of frequencies have been assigned to terrestrial microwave links. The lower part of the range offers only relatively narrow bands and is suitable only for low-capacity systems. The majority of microwave links are located above 1.6 GHz. However, from 12 GHz, the absorption due to rain leads to an increasing attenuation. This range is only suitable in practice for digital links. By international convention (CCIR), the assigned bands have been divided into adjacent radioelectric channels (e.g., 8 pairs of channels, 29.65 MHz apart, in the 6 GHz band, i.e., from 5.9 to 6.4 GHz).

The choice of channels for the two directions of transmission of each hop of a microwave link in a dense network (frequency plan) is a delicate operation which must take into account

- the parasitic couplings possible between antennas located on the same support;
- the interferences between neighboring links due to the imperfect directivity of the antennas;
- the frequency selectivity of the receivers;

- the possibility of alternating the polarizations (horizontal or vertical) in adjacent channels;
- a concern for optimal utilization of the available frequency range.

12.2 PROPAGATION CONDITIONS

12.2.1 Atmospheric refraction and hypothetical radius of the earth

The decrease of the refractive index with altitude has as a consequence the curving of the path towards the ground, which adapts the transmission better to the curvature of the earth and increases the range of the link (vol. XIII, section 8.2).

It is more convenient, for planning, to assume rectilinear propagation and to consequently correct the curvature of the earth by artificially increasing the earth's radius R (fig. 12.2). For the temperate zone, the virtual earth's radius R' has been fixed by convention, on the basis of numerous observations, at the value of

$$R' = \frac{4}{3} R = 8500 \text{ km} \qquad (12.1)$$

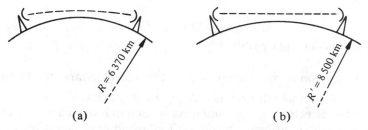

(a)   (b)

Fig. 12.2 Correction of the earth's radius: (a) real conditions; (b) rectilinear link and virtual earth's radius.

12.2.2 Diffraction on obstacles. Fresnel ellipsoids

An obstacle X (land, building, vegetation, *et cetera*) located on the path of the link between A and B (fig. 12.3), becomes the source of a secondary emission which reaches the receiver by a roundabout route AXB (diffraction). If the direct signal (path AB) and the signal due to diffraction (path AXB) are

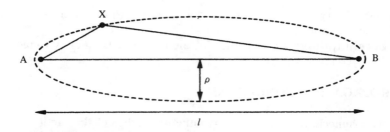

Fig. 12.3 Effect of diffraction on an obstacle in X.

opposed in phase, the reception level can be considerably lowered. This is the case if

$$\text{AXB} - \text{AB} = n\frac{\lambda}{2} \qquad (12.2)$$

with odd $n > 0$.

The geometrical locus of the points X satisfying (12.2) is a family of ellipsoids which are confocal with foci A and B. They are called *Fresnel ellipsoids*.

With $\text{AB} = l \gg \lambda$, the radius ρ (half the small axis) of the first of these ellipsoids ($n = 1$) is

$$\rho = \frac{1}{2}\sqrt{l\lambda} \qquad (12.3)$$

12.2.3 Free line-of-sight path

The transmission is considered to be made under optimal conditions:
- if the first Fresnel ellipsoid is free of any obstacles;
- and if the directivity of the antennas is such that any causes of reflection (notably, water surfaces) outside this ellipsoid have negligible effects.

To ensure these conditions, a profile of the land between the two stations is established, taking into account the corrected curvature of the earth ($R' = 8500$ km), and the ellipse corresponding to the frequency used is placed over it (fig. 12.4). The height of the antenna supports is determined in such a way as to avoid the ellipse touching the ground or other obstacles (trees, buildings, *et cetera*).

Let us note that the arrow y due to the curvature of the earth is proportional to the square of the distance l:

$$y \cong \frac{l^2}{8R'} \qquad (12.4)$$

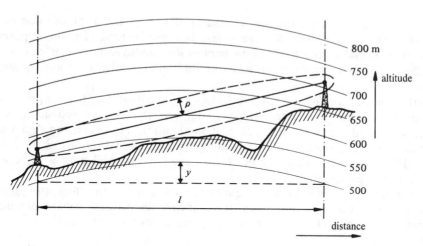

Fig. 12.4 Profile of the land and first Fresnel ellipse (the scale of altitudes is highly exaggerated with respect to that of the distances).

The minimum height h of the antennas above a flat ground on flat ground is the sum of the arrow y and the radius ρ of the ellipse. Thus, for example, for $l = 50$ km, and $f = 6$ GHz ($\lambda = 5$ cm), the calculation by (12.3) and (12.4) gives $y = 37$ m and $\rho = 25$ m, and therefore towers of $h = 62$ m. Hence the interest in placing the microwave stations on high detached sites.

12.2.4 Fading and multiplicative noise

Despite the precautions taken so that the path is clear, it often happens that passing inhomogeneities of the atmosphere (layers with different temperatures) are the cause of partial reflections and of multiple path propagation. The signal reaching the receiver is thus the vector sum of several components of different and variable amplitudes and phases. The resulting reception level can be submitted to sharp and unpredictable decreases, called *fading*.

This phenomenon depends very strongly on the frequency. One can protect from this by *diversity* techniques (sect. 3.8.5).

Other causes can also lead to significant momentary increases in the attenuation of a hop, notably

- a variation of the refractive index of the atmosphere and of its gradient, modifying the curvature of the path, and, consequently, the optimal tracking of the antennae;
- absorption losses in the path, due to strong rain or snowfall (especially noticeable at very high frequencies).

Due to the unpredictable random character of fading, it can be considered a kind of noise, called *multiplicative noise*, because instead of being superimposed on a signal as the usual disturbances are (additive noise), it works on the intensity of the signals received by multiplying them by a variable random factor.

In the case of **analog** microwave links (FM modulation), the output level after demodulation is not affected by fading, because the received signal is in any case clipped (sect. 8.6.10), before demodulation. However, the signal-to-noise ratio ξ'_1, according to (8.106), undergoes the same variations as the level of the FM signal. It would therefore be necessary to design in a sufficient reserve at the planning stage.

The situation is different with **digital** microwave links. In effect, the highly selective character of fading leads to significant distortions, causes of intersymbol interference. The error probability increases very noticeably.

12.2.5 Antennas

Ground microwave links are studded with **parabolic reflector antennas**. Their surface is a portion of the paraboloid whose focus is occupied by the radiating source, generally a horn reflector antenna (vol. III sect. .7.3.3) at the end of a waveguide. The spherical wave issued from the horn is thus transformed into a plane wave.

Depending on the relative position of the horn and the reflector, and on the solution adopted for the connection of the waveguide, several types of antennas have been proposed (fig. 12.5). They are distinguished by their space requirements and certain radiation properties (importance of secondary lobes).

The gain g of such an antenna is calculated by (3.99) and (3.101) from its effective area A_e, which is itself proportional to the surface A of the geometrical aperture of the antenna, measured perpendicularly to the main radiation axis

$$g = r \frac{4\pi}{\lambda^2} A \tag{12.5}$$

The yield factor r (sect. 3.9.4) is of 0.5 order. It depends on construction details. It is noted that the gain g is proportional to the square of the frequency. The choice of a type of antenna and of their dimensions depends on

- the **gain** necessary to bring the attenuation of the link to an acceptable level (planning as a function of distance and frequency);
- the **directivity** (related to the gain) indispensable to avoid interference between neighboring links and undesirable reflections;
- the carrier *frequencies* envisaged;
- the *space* available and the cost.

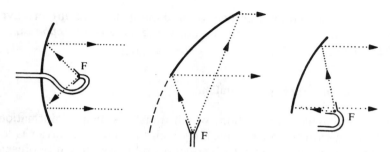

Fig. 12.5 Examples of antennas with parabolic reflectors: F = focus of the paraboloid.

The formation of frost on the antenna noticeably disturbs its characteristics. This is why, when climatic conditions necessitate it, the aperture of the antenna is protected by a polyester membrane or the entire antenna is placed in a hole covered by a polyurethane wall.

12.3 DIGITAL MICROWAVE LINKS

12.3.1 Planning objectives

A digital microwave link is characterized essentially by the *bit rate* $\dot{D}$ that it conveys, independently of the fact that this rate results from the time-division multiplexing of z telephone channels converted into digital by a PCM modulation or that it corresponds to a fast transmission of data.

Due to the congestion of the radioelectric frequency range and to the danger of interference between neighboring links, it is essential that only a *minimum bandwidth* be used to transmit this output. The band necessary for a digital system is not necessarily narrow. It is therefore necessary to choose a favorable modulation from this point of view.

The principal criterion of quality is the *error probability per symbol* ϵ, which depends on

- the intersymbol interference (signal distortion)
- additive noise and multiplicative noise (fading)
- the type of modulation

Contrary to line transmission, the radioelectric channel does not itself present noticeable distortions. Nevertheless the strict limitation of the pass band in the terminal equipment is responsible for significant linear distortions, leading to intersymbol interference. On the other hand, selective fading is also the cause of distortion and interference, of an unpredictable nature. To avoid the accumulation of these distortions and noise effects, a *regeneration* takes place *after each hop*.

12.3.2 Choice of a type of modulation

The discrete analog modulation, which translates digital information from its baseband toward the microwaves in which the emission carrier is located, must satisfy the criteria of spectral economy and noise immunity presented in section 12.3.1.

Phase shift keying PSK (sect. 8.11.5) is particularly favourable due to:
- the low error probability that it produces in the presence of a given signal-to-noise ratio, as relation of table 8.85 demonstrates;
- the possibility of implementing an *m*-ary transmission by defining *m* phases per symbol, which reduces the symbol rate $\dot{M} = \dot{D}/\text{lb } m$;
- the relatively modest bandwidth required.

For reasons mentioned in section 8.11.8, its differential form DPSK is most often used for bit rates up to 140 Mbit/s, with $m=2$, 4, or 8 phases.

Numerous variants of the PSK and FSK modulation have been proposed for digital microwave links, always with the goal of simultaneously optimizing the bandwidth and the error probability. MSK modulation (sect. 8.11.9) is an example.

For fourth-order digital systems ($\dot{D} = 140$ Mbit/s), a *quadrature modulation of amplitude and phase* QAM at $m = 16$ values is also recommended, according to the same principle as the modems at 9.6 kbit/s (fig. 11.11).

12.3.3 Necessary bandwidth

There is no direct relation between the bit rate $\dot{D}$ to be transmitted and the secondary bandwidth B_2 necessary for this transmission by microwave link with PSK modulation, because the ratio $B_2/\dot{D}$ can be modified by the choice of the number of values *m* per symbol. The determining element for the evaluation of B_2 is thus the symbol rate $\dot{M} = \dot{D}/\text{lb } m$.

Independently of the number of phases *m* per symbol, we have, in *m*-ary PSK, conforming to the expanded Nyquist criterion (sect. 5.3.4) applied to the baseband:

$$B_{2\text{PSK}} = 2B_1 \cong 1.6\dot{M} \tag{12.6}$$

MSK modulation ($m = 2$, $\dot{M} = \dot{D}$) permits the reduction of B_2 to approximately 1.2 $\dot{D}$.

12.3.4 Error probability

The error probability ϵ per symbol and per hop is a function of the signal-to-noise ratio at reception ξ'_R (after equalization). Furthermore, this function depends on:
- the type of discrete analog modulation;
- the number m of values per symbol;
- the principle of demodulation (coherent or not).

The details of these dependences have been given in sections 5.4.3 and 8.11.11.

The increase of m allows a reduction of the secondary band, but it is accompanied by a noticeable increase in the error probability, as the spacing U_{BR} between the regeneration thresholds diminishes (at a constant received power).

12.3.5 Planning of a digital microwave link

The degree of freedom in the choice of the principal interdependent parameters, namely,
- power of the signal at emission P_E;
- secondary bandwidth B_2;
- number m of values per symbol;
- error probability ϵ;

is bounded by the fact that B_2 is limited by the presence of other microwave links at neighboring frequencies, that P_E is linked to technological limits and that the error probability must not exceed the value tolerated in the international hypothetical reference circuit (sect. 9.7.4).

The reception level L_R (corresponding to the received power P_R), on which the probability of error finally depends, is derived from the attenuation A of the hop. It is thus very important to insert a sufficient margin to *take into account the fading* which gives rise to decreases of L_R. In effect, due to the sensitivity of the complementary integral Gaussian function at its argument, a variation of a few decibels of the reception level (and thus of the ratio ξ'_R) can lead to a variation of several orders of magnitude of the error probability.

However, the probability of simultaneous fading on several consecutive hops is minimal. It is therefore sufficient to take the margin for substantial fading into account only for *one of the hops* whose error probabilities add approximately to give the overall error probability for the complete link, according to (5.47).

12.4 ANALOG MICROWAVE LINKS

12.4.1 Level diagram of a hop

The evolution of the level along the length of the link is given in figure 12.6. This level diagram brings up the following remarks:

- the attenuation A_{iso} is that which would occur between two isotropic antennas in free space (first Fresnel ellipsoid free), in a lossless medium without fading. It is expressed by relation (3.106);
- A_{iso} does not increase linearly but logarithmically with distance l;
- the virtual power available at the output of the emission antenna, including its gain, is sometimes termed the *equivalent isotropic radiated power EIRP*. It is equal to the product of the power P_E provided to the antenna by its gain g;
- the ideal attenuation of the link is $A = A_{iso} - G_E - G_R$, conforming to relation (3.110);
- to take the fading into account, it is necessary to raise the attenuation A by a *safety margin* of approximately 6 dB;

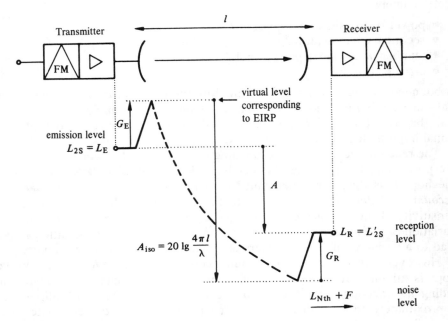

Fig. 12.6 Level diagram of a hop.

- the noise factor F is that of the receiver; it partly expresses the noise of the receiving amplifier itself (e.g., shot noise), and partly the cosmic and atmospheric noise captured by the antenna;
- in figure 12.6 and the rest of this chapter, the attenuation of waveguides between radioelectric equipment and antennas (approximately 2 dB) will be ignored.

12.4.2 Choice of modulation type

The choice of an *angular modulation* (FM/ΦM) is justified for three reasons:
- large attenuation variations due to fading should not lead to fluctuations of the demodulated signal level; the angular modulations are insensitive to these variations (sect. 8.6.11);
- angular modulation allows us to obtain a satisfactory transmission quality (signal-to-noise ratio after demodulation) despite the low level of reception and superimposed noise;
- the bandwidth required by these modulations is available in the range of microwaves.

12.4.3 Signal-to-noise ratio after demodulation

After reception, the signal is demodulated (FM demodulator), then, if required, demultiplexed (SSB demodulator), to restore the primary signal whose quality and signal-to-noise ratio ξ'_1 are important for the user. The nominal value of this ratio depends on the nominal modulation index δ_{nom} according to relations (8.108) or (8.110). Translated into decibels, with the notations $L_R = L'_{2S} = 10 \lg(P'_{2S}/1mW) = L_E - A$ and $L_{Nth1} = 10 \lg(kTB_1/1mW)$ and by raising the thermal noise level by the noise factor F, these relations become, in the case of a microwave hop:

$$10 \lg \xi'_1 = L_E - A - (L_{Nth1} + F + 20 \lg \delta_{nom} + 10 \lg (3/2) \quad \text{if } f_{1max} = B_1$$
$$\text{dB} \qquad (12.7)$$

$$10 \lg \xi'_1 = L_E - A - (L_{Nth1} + F) + 20 \lg \delta_{nom} + 10 \lg (1/2) \quad \text{if } f_{1max} \gg B_1$$
$$\text{dB} \qquad (12.8)$$

where $f_{1\,max}$ is the maximum frequency of the signal at the input of the FM modulator. While relation (12.7) is valid, e.g., in the case of a point-to-point television transmission, relation (12.8) corresponds *to each telephone channel* of a multichannel microwave link, with each having the same primary bandwidth $B_1 = 3.1$ kHz and nominal indices of modulation which are different from one channel to the next.

If the link includes several hops, without intermediate demodulation, the contributions of the noise of each receiver, each corresponding to a level $L_{Nth1} + F$, must be referred to zero then power added to determine the power (and thus the level) of the noise at the input of the final demodulator.

12.4.4 Pre-emphasis

As uniform a quality (ξ'_i) as possible is desired for all the channels when the microwave link is occupied by a frequency-division multiplex of z telephone channels, as is most often the case. In relation (12.8), the only parameter which can vary from one channel to another is the modulation index δ. The ideal situation would be to keep δ identical for all the channels. Because they are located at different frequencies in the multiplex, the frequency deviation Δf_i would have to vary from one channel i to the other, proportionally to its frequency f_i in the multiplex. We would thus obtain a *pure phase modulation* ΦM for the entire multiplex (fig. 12.7).

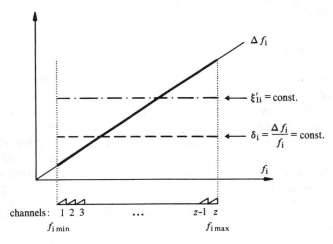

Fig. 12.7 Ideal pre-emphasis (pure ΦM modulation).

This solution would lead to very reduced deviations Δf_i, and thus to very low levels after demodulation, for the lower channels, which would expose them to significant intermodulation products from the upper channels, at a high level (in the case of nonlinearities, notably of phase, in the modulators and demodulators). This is why it is necessary to be content with a compromise (as in the case of carrier systems on lines, sect. 10.3.4, but with a different law). The CCIR has defined a moderate pre-emphasis law which does not completely make ξ'_{1i} uniform (fig. 12.8). This law approaches a phase modulation for the upper part of the multiplex and a frequency modulation, but with a deviation reduced by 4 dB for the lower part.

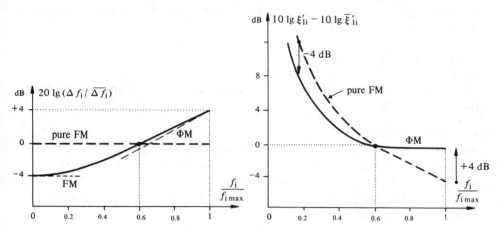

Fig. 12.8 Pre-emphasis compromise according to the CCIR (mixed FM/ΦM modulation): ——— with pre-emphasis; - - - - - without pre-emphasis.

This pre-emphasis is carried out by an element whose gain increases with the frequency before the FM modulator and an exactly reciprocal element after the FM demodulator (fig. 12.9).

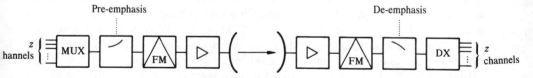

Fig. 12.9 Implementation of pre-emphasis.

According to this law, the signal-to-noise ratio ξ'_{1i} is increased by 4 dB in the upper channel with respect to the case without pre-emphasis (pure FM). However, it is decreased by 4 dB in the lower channels, which nevertheless remain better than the others.

12.4.5 Nominal bandwidth

The *overall* nominal frequency deviation $\Delta f_{z\,\text{nom}}$, resulting from the simultaneous occupation (entirely unrealistic!) of the z channels, each by a *sinusoidal* signal, incoherent with respect to the others and of a level equal to the nominal level, is given, from the addition law for independent signals (2.33), by the quadratic sum of the invidivual nominal deviations of the z channels (different due to the pre-emphasis)

$$\Delta f_{z\text{nom}} = \sqrt{\sum_{i=1}^{z} \Delta f_{i\text{nom}}^2} \tag{12.9}$$

The *average* nominal deviation $\overline{\Delta f}_{i\,\text{nom}}$, which leads to the same overall deviation, can be defined by

$$\overline{\Delta f}_{i\text{nom}} = \Delta f_{z\text{nom}} / \sqrt{z} \tag{12.10}$$

According to the pre-emphasis law of figure 12.8, it is the deviation obtained when the channel at $f_i = 0.608 f_{i\,\text{max}}$ is only occupied, with the nominal level. This channel is not affected by pre-emphasis.

The *nominal bandwidth* $B_{2\text{nom}}$ occupied by the microwave link can then be estimated by Carson's rule (8.66):

$$B_{2\text{nom}} \cong 2(f_{i\text{max}} + \Delta f_{z\text{nom}}) \tag{12.11}$$

i.e., with (12.10), we have

$$B_{2\text{nom}} \cong 2(f_{i\text{max}} + \sqrt{z}\,\overline{\Delta f}_{i\text{nom}}) \tag{12.12}$$

12.4.6 Real bandwidth

In operation, the channels are never occupied simultaneously by the nominal (sinusoidal!) signal, but by a random telephone signal whose real average level is located by convention (sect. 2.4.19) at 15 dB lower than the nominal level. The addition (in power) of these z random signals gives a signal with noticeably Gaussian statistics, of a level referred to zero equal to $-15 + 10 \lg z$ dBm0 (sect. 10.4.1). In this case, the *real* overall instantaneous frequency deviation due to z simultaneous telephone conversations is a random variable

- with a zero average
- normally distributed (Gaussian law)
- with a standard deviation (rms value) of 15 dB lower than the rms value of the overall nominal deviation $\Delta f_{x\text{nomeff}}$

$$\Delta f_{z\text{realeff}} = 10^{-15/20}\, \Delta f_{z\text{nomeff}} \tag{12.13}$$

Because the nominal signal is sinusoidal by definition, we have

$$\Delta f_{znomeff} = \Delta f_{znom}/\sqrt{2} \qquad (12.14)$$

The *real bandwidth* is thus also a random variable. Its rms value is, from Carson's rule:

$$B_{2realeff} \cong 2(f_{imax} + \Delta f_{zrealeff}) \qquad (12.15)$$

and its distribution is approximately unilaterally Gaussian (fig. 12.10), which gives the probability of exceeding a value B_{2x}:

$$\text{Prob}(B_{2real} > B_{2x}) = 2G_c \left(\frac{B_{2x} - 2f_{imax}}{2\Delta f_{zrealeff}} \right) \qquad (12.16)$$

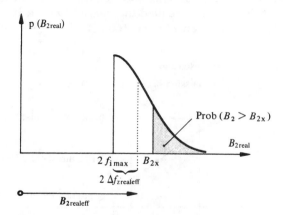

Fig. 12.10 Probability density of the real bandwidth of a multichannel microwave link.

In particular, the rms value from (12.15) is exceeded with a probability of 32%. Taking (12.10), (12.13), and (12.14) into account, it is found that

$$\Delta f_{znom} = \sqrt{2} \, 10^{15/20} \, \Delta f_{zrealeff} = 7.95 \, \Delta f_{zrealeff} \qquad (12.17)$$

The result is that the nominal bandwidth according to (12.11) is only exceeded with a probability of $2G_c(7.95) \cong 10^{-15}$, which makes the interest of this nominal value relative.

12.4.7 Background noise

The background noise has two principal sources, as with any analog transmission (sect. 10.5.3):

- background noise of the terminal equipment (modulators, demodulators, amplifiers,) is generally very weak
- transmission background noise, principally due to the thermal noise at the input of the receiver. Background noise has a relative importance which is greater when the signal is more attenuated at this point.

These two causes combine their effects by power addition, which means, from section 6.3.2, that the overall noise-to-signal ratio is the sum of the noise-to-signal ratios corresponding to the two sources separately.

For a certain type of transmitter (emission level L_E given) and for a given modulation (by its modulation index δ), relations (12.7) and (12.8) show that the sum of the transmission signal-to-noise ratio 10 lg ξ'_i and of the link attenuation A is constant (fig. 12.11). This sum, characteristic of a given type of terminal equipment, is sometimes called the *figure of merit*. Its value, depending on the systems, lies between 140 and 160 dB.

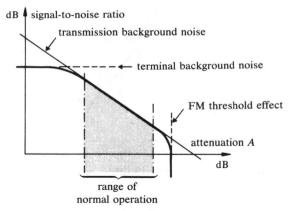

Fig. 12.11 Variation of the signal-to-noise ratio after demodulation, as a function of the attenuation.

In reality, for low values of A, the background noise due to terminal equipment (independent of A) predominates. Further, the maximum tolerable attenuation is limited by the FM *threshold effect* (sect. 8.8.11) which leads to a sharp decrease of ξ'_i. The optimal functioning range is indicated in figure 12.11; it must include a reserve for fading.

The outline for the dimensioning of the link on the basis of given terminal equipment is thus the following:

- calculation of A by (12.1) or (12.8) from ξ'_i required (including the fading margin);

- calculation of A_{iso} by (3.106), knowing the distance l and the wavelength λ;
- determination of the necessary antennas, by calculation of their gain $G = \frac{1}{2}(A_{iso} - A)$.

12.4.8 Planning of a multichannel analog microwave link for telephony

Planning is done by the overall noise budget in a channel; at the end of a *hypothetical reference circuit* of 2500 km (sect. 10.5.2). According to the distribution suggested in section 10.5.3 (table 10.12), the power referred to zero of the background noise due only to tranmission equipment must not exceed 3,750 pW0, i.e., on the average 1.5 pW0/km.

This limiting value, multiplied by the length l of the hop and carried over the power (1 mW) of the zero level signal, gives the signal-to-background noise ratio $\xi'_{li\ min}$ tolerable in channel i after transmission and demodulation. From (12.8), we have

$$10 \lg \xi'_{limin} \leq 10 \lg \xi'_{li} = L_E - A - (L_{Nth1} + F) + 20 \lg \delta_{inom} - 3 \quad \text{dB} \tag{12.18}$$

with

$$\delta_{inom} = \frac{\Delta f_{inom}}{f_i} \tag{12.19}$$

where Δf_{inom} is the amplitude (peak value) of the frequency deviation when only the channel i is occupied by the nominal signal, and the other channels are unoccupied.

The calculation can be made first without pre-emphasis, with Δf_{inom} identical for all channels. The effect of *pre-emphasis* from figure 12.8 then makes it gain 4 dB on the upper channels and lose 4 dB on the lower channels.

It is usual in the calculation to take into account the *psophometric weighting* (sect. 2.3.13) by reducing the noise level L_{Nth1} by 2.5 dB with respect to its value calculated from kTB_1 in the telephone band.

12.5 COMPARISON BETWEEN DIGITAL AND ANALOG MICROWAVE LINKS

12.5.1 Typical systems

Tables 12.12 and 12.13 give an outline of some present or projected systems, both analog and digital.

Table 12.12 Digital links (examples).

Bit rate	Number of telephone channels	f_p GHz	Modulation	m	$B_2 \cong 1.6 \dot{M}$ MHz	B_2/z kHz/channel
2 × 8	2 × 120	~ 15	PSK	4	~ 13	54
34	480	~ 1,9/~ 13	PSK	4	~ 27	56
140	1920	~ 11/~ 19	ASK + PSK	16	~ 56	29

Table 12.13 Analog links (examples).

Number of telephone channels	$f_{i\,min} \ldots f_{i\,max}$ kHz	f_p GHz	$\overline{\Delta f_i}$ nom eff [1] kHz	$B_{2\,real}\ 1^{0/oo}$ [2] MHz	$B_{2\,real}/z$ kHz/channel
300	60 ... 1300	~ 2.2/~ 7.5	200	~ 5.6	18.5
960	60 ... 4028	~ 4	200	~ 13	13.5
1800	312 ... 8204	~ 6	140	~ 23	12.7
2700	312 ... 12 388	~ 6,7	140	~ 33	12.2

[1] for a nominal level of 0 dBm
[2] value exceeded with a probability of $1^{0/oo}$

The emission powers are located, depending on the technology adopted and the range of frequencies, between 0.2 W and approximately 15 W.

12.5.2 Necessary bandwidths

Although the approximation $B_2 \cong 1.6 \dot{M}$ for digital systems is pessimistic (we can, with adequate filtering, approach $1.2\ \dot{M}$), it is clear from the comparison of tables 12.12 and 12.13 that the digital links require a per channel bandwidth which is 2 to 3 times greater than analog links.

Due to the necessity of using FM modulation in analog links, this ratio is nevertheless definitely less unfavorable than that of approximately 16 which is found (sect. 9.1.2) by comparing the analog and digital systems on lines.

12.5.3 Noise sensitivity

The important criteria are
- the signal-to-noise ratio ξ'_i after demodulation with analog links. It varies linearly with the attenuation A according to relations (12.7) or (12.8), illustrated by figure 12.11;

- the regeneration error probability ϵ with digital links. It depends on the power of the received signal (and thus on the attenuation A) by the intermediary of a complementary integral Gaussian function.

The behavior of these two types of systems as a function of the reception level L_R and its fluctuations (fading) is qualitatively compared in figure 12.14.

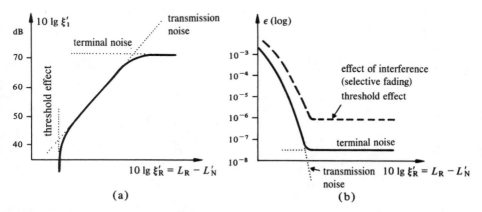

Fig. 12.14 Noise sensitivity. (a) analog microwave link; (b) digital microwave link.

The comparison shows the consistency of the quality of digital links in the presence of fading within a rather large range, beyond which the error probability increases very rapidly, conforming with the strong dependence between the function $G_c(x)$ and its argument.

In digital transmission, it is nevertheless necessary to also consider frequency selective fading which can provoke a momentary increase of the error probability of several orders of magnitude. This effect cannot be compensated for by a larger margin for the reception level. It is therefore necesary to use *space or frequency diversity reception* (sect. 3.8.5) or *adaptive equalization*, i.e., automatically variable equalization as a function of the momentary channel characteristics.

Chapter 13

Satellite Links

13.1 PRINCIPLES AND CONDITIONS OF SATELLITE LINKS

13.1.1 Development of satellite communications

From the beginnings of telecommunications, the crossing of oceans has held a technical fascination and represented an economic challenge. Several solutions have marked this adventure, the latest being the use of an artificial satellite as a microwave relay.

1858 First transatlantic telegraphic cable
1866 Second transatlantic telegraphic cable
1901 Marconi transmits wireless telegraphy from England to Newfoundland
1927 Transatlantic telephone link by short wave
1957 First transatlantic telephone coaxial cable (TAT1)
1961 Second transatlantic telephone cable
1960 Passive satellite "Echo 1" (metallized balloon of 30 m in diameter, at 1600 km high, used as a microwave reflector)
1962 First active satellite "Telstar 1," first transatlantic transmisson of television (variable altitude between 950 and 5650 km)
1965 First geostationary satellite "Early Bird," or Intelsat I, (240 telephone channels, 38 kg)

Since 1965, the generations of geostationary satellites (Intelsat II, III, IV, IVA, V, Symphony, OTS, *et cetera*) have succeeded each other with increasing performance (from 240 to 15,000 channels).

In addition to telecommunications satellites proper (telephony, telematics, television transmission), launched or planned satellites for radio broadcasting, meterological observation, remote detection of natural resources, maritime navigation *et cetera*, must also be mentioned.

13.1.2 Choice of orbit

The orbit of an artifical satellite is an ellipse, the plane of which passes through the center of the earth and which obeys the laws of celestial mechanics, in particular, the three laws of Kepler. The third determines the period of revolution T of the satellite as a function of half the long axis a of its elliptical orbit or its altitude $h = a - R$ (fig. 13.1)

$$T = 2\pi \sqrt{\frac{a^3}{GM}} = 2\pi \sqrt{\frac{(h+R)^3}{GM}} \qquad \text{s} \qquad (13.1)$$

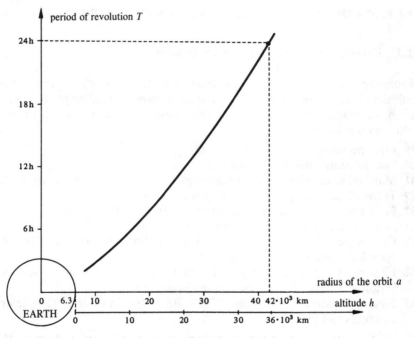

Fig. 13.1 Period of revolution as a function of altitude.

where $M(\cong 6.10^{24}$ kg) is the mass of the earth, assumed to be homogeneous, $G(= 6.67 \times 10^{-11}$ Nm^2/kg) the universal gravitational constant and $R(= 6378$ km) the average radius of the earth.

Satellites at low altitude are *moving satellites*, which shift rapidly with respect to a point on the earth's surface.

Contrarily, however, a satellite is said to be geostationary if it appears to be immobile when seen from the earth. For this the following orbital conditions must be met:

- synchronicity with the earth's rotation ($T = 1$ sidereal day $\cong$ 23 hours 56′4″) which imposes, from (13.1) an altitude of

  $$h = 35,786 \text{ km} \qquad (13.2)$$

SATELLITE LINKS

- circular orbit (otherwise the angular velocity is not constant);
- equatorial orbit (otherwise the apparent movement of the satellite describes an "eight" around the average position on the equator);
- placement into orbit in an easterly direction.

In the exact calculation of the orbit further secondary effects intervene, such as inhomogeneity and non-sphericity of the earth, or the influence of the moon and sun, which disturb the position of the satellite.

13.1.3 Evaluation of geostationary satellites

With respect to moving satellites, geostationary satellites present the following advantages:

- coverage of approximately a third of the earth's surface in real time and without interruption;
- simplification of satellite tracking by terrestrial antennas;
- less frequent and shorter eclipses by the shadow cone of the earth (thermal shocks and absence of solar power);
- avoidance of the van Allen belt (altitude: 10,000 to 14,000 km) where the radiation density is high;
- no Doppler effect, and no noticeable variation of the propagation time.

However, their disadvantages are related to their high altitude:

- very significant attenuation in free space;
- propagation time of 240 to 275 ms for the earth-satellite-earth path, at the limit of psychologically tolerable values in telephony; echo suppressors (sect. 15.3.2) are necessary;
- necessity for a precise and permanent correction of orbital position to conserve the geostationary orbit (remote control of rockets on the satellite);
- extra launching difficulties;
- blackout of earth stations when the satellite passes in front of the sun at equinoxes.

Despite these inconveniences, the majority of telecommunications satellites are geostationary, except for those destined to serve polar regions or when worldwide coverage at different times is desired (e.g., data collection).

13.1.4 Frequency range

The choice of an appropriate frequency range for satellite links is determined by the following criteria:

- the ionosphere must not represent an obstacle
- the absorption in the troposphere must be minimal
- the range with strong atmospheric or cosmic disturbances must be avoided

The majority of present-day satellites use the band at 6/4 GHz, assigned by the CCIR (but shared with ground microwave links), namely

- f_{pm} = 5.924–6.425 GHz for the ascending path (ground-satellite)
- f_{pd} = 3.700–4.200 GHz for the descending path (satellite-ground)

given an available band of 500 MHz in each direction.

The assignment of two distinctly separated frequency bands for the two paths facilitates separation between the emitted and received signals on the same ground antenna ("frequency-division pseudo-four-wire", section 4.6.4) at extremely different levels. This solution requires a frequency translation of -2.225 GHz on the satellite, but, on the other hand, permits a unique type of earth station.

The traffic growth necessitates finding other frequency bands for satellites:

- 14/11 GHz with a bandwidth of 500 MHz
- 30/20 GHz with an available band of 3500 MHz, but extra technological problems and atmospheric attenuation appear

13.1.5 Structure of a link

The structure of a satellite link is similar to that of a terrestrial microwave link with two hops, the satellite playing the role of relay-station (fig. 13.2).

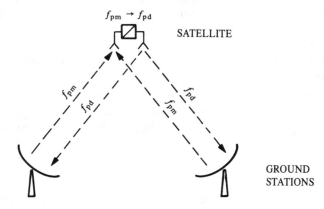

Fig. 13.2 Structure principle of a satellite link.

SATELLITE LINKS   477

The information transmitted can be analog or digital, which leads to different types of modulation:
- analog transmission: FM modulation with pre-emphasis, as in ground microwave links (section 12.4.4);
- digital transmission: DPSK and MSK modulation and numerous variants, with very elaborate coding procedures.

13.1.6 Noise temperature of the receiver

At reception, three principal sources of noise disturb the signal:
- the *thermal noise*, essentially that of the input resistance of the receiving amplifier, with a power which depends directly on the *ambient temperature* T_a at this point and on the bandwidth, according to (2.4.1);
- the *internal noise* of the receiver, expressed by its noise factor F (sect. 6.2.5), or by its *noise temperature* ΔT_{NR} according to (6.9);
- the external noise *captured by the antenna* (cosmic noise, radiation of the earth itself, *et cetera*) whose power contribution can also be expressed by a virtual increase of temperature ΔT_{NA}, called the *noise temperature of the antenna*. It depends on the frequency and the elevation angle of the antenna above the horizon (vol. XIII sect. 7.6.4).

These three sources combine their effects by power addition. The total noise power P_{NRx} collected at the input of the receiver can be calculated with an equivalent receiver noise temperature T_N, obtained by increasing the ambient temperature by the receiver noise temperature and by that of the antenna

$$P_{NRx} = kT_N B \tag{13.3}$$

with

$$T_N = T_a + \Delta T_{NR} + T_{NA} \tag{13.4}$$

13.2 PLANNING OF A LINK

13.2.1 Comparison with a terrestrial microwave link

With respect to a terrestrial microwave link, conceived according to the same principles, the satellite link is confronted with much more unfavorable conditions:
- the length of the hop is 36,000 km (zenithal) to 41,000 km (minimal elevation of 5 °) instead of approximately 50 km;
- the emission power from the satellite is limited;

- the satellite antenna has a modest gain (approximately 20 dB) if it must cover all the visible parts of the earth.

However, the satellite link makes good use of the following favorable circumstances:
- the fading margin can be reduced, because the troposphere is crossed obliquely and not horizontally;
- the reference distance of 2500 km (sect. 10.5.2) requires about 50 hops with microwave links, whereas the satellite link covers a much longer distance with only two hops.

13.2.2 Noise budget

The value of 10,000 pW0 (10 lg ξ = 50 dB), tolerated in a telephone channel at the end of a hypothetical reference circuit for terrestrial links (sect. 10.5.2), is also accepted for the planning of satellite links at any distance.

Part of this noise power referred to zero, for example 7500 pW0, is attributed to the background noise and distributed unequally between the ascending path (e.g., 1500 pW0) and the descending path (e.g., 6000 pW0), the latter end being by far the most critical due to the low emission power of the satellite.

13.2.3 Dimensioning of a multichannel analog link

Telephone channels, grouped into a frequency-division multiplex by SSB modulation, modulate the transmitter in frequency [41]. The signal-to-noise ratio after demodulation in the channel i is given, as in the case of terrestrial microwave links, by relation (12.18)

$$10 \lg \xi'_{1i} = L_E - A - L_{N1} + 20 \lg \delta_{inom} - 3 \text{ dB} \tag{13.5}$$

where L_{N1} is the noise level *in the primary bandwidth of a channel*, calculated at the noise temperature T_N.

The calculation must be made separately for the two paths, then the two noise-to-signal ratios $1/\xi'_{1im}$ (ascending path) and $1/\xi'_{1id}$ (descending path), i.e., the corresponding noise powers referred to zero are added to give the noise-to-signal ratio ($\geq$ 7500 pW0) at the end of the link.

By assuming the indicative noise distribution given in the preceding section, we would have:

$$10 \lg \xi'_{1im} \geq 10 \lg \frac{1mW}{1500pW} = 58.2 \text{ dB for the ascending path} \tag{13.6}$$

$$10 \lg \xi'_{1id} \geq 10 \lg \frac{1mW}{6000pW} = 52 \text{ dB for the descending path} \tag{13.7}$$

SATELLITE LINKS 479

The attenuation A of each path includes, from (3.110) the free space attenuation between two isotropic antennas A_{iso} and the gain of the antennas (satellite and ground station). For the maximum path ($l = 41{,}200$ km) at 6 and 4 GHz, the attenuation A_i is equal to (from 3.106)

$A_{isom} = 200.3$ dB for the ascending path (6 GHz) (13.8)

$A_{isod} = 196.8$ dB for the descending path (4 GHz) (13.9)

The emission powers are about a few hundred watts (+50 to +65 dBm) in ground stations, several watts (+30 to +40 dBm) in the satellite.

Requirements (13.6) and (13.7), introduced in (13.5) with (13.8) and (13.9), cannot be satisfied except by optimizing all the other parameters, i.e.,

- the gain of ground antennas: by parabolic reflectors of very large dimensions, gains of the order of 60 dB can be achieved;
- the nominal modulation index $\delta_{i\,nom}$ for a channel is chosen around 0.2, i.e., at a much greater value than with terrestrial links, at the cost, of course, of a corresponding enlargement of the occupied frequency band;
- the noise level L_{N1} is reduced by using input stages with very low noise ($\Delta T_{NR} \cong 50$ K) in the terrestrial receiver and even, in certain cases, less and less frequently, by cooling these stages at the boiling point of helium ($T_a = 18$ K).

13.2.4 Level diagram of a link

Figure 13.3 gives, as an example, the level diagram of an Intelsat IV A link in the band of 6/4 GHz, for one of the 20 repeaters of the satellite (bandwidth of 36 MHz), used for a single group of $z = 972$ telephone channels.

13.2.5 Planning of a digital link

The determining criterion here is the error probability per symbol ϵ. On the other hand, the necessary bandwidth for a bit rate $\dot{D}$ and the complexity of the coding and decoding procedures are also elements to be appraised [42,43].

An analogous calculation to that made in section 5.5.3 allows us to evaluate the signal-to-noise ratio ξ'_R at reception after equalization. The parameters which intervene in this calculation are

- the power of the received signal
- the power spectral density of the noise, including the noise temperature of the receiver

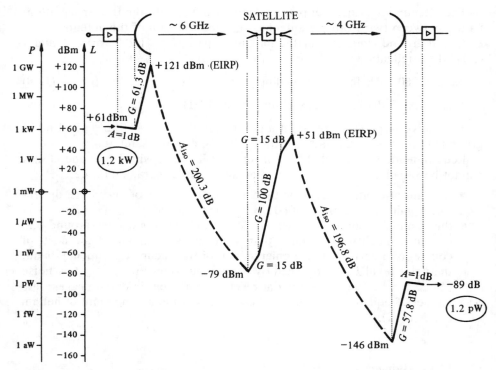

Fig. 13.3 Level diagram of a link.

The relationship betwen ξ'_R and the error probability ϵ depends on the type of discrete modulation chosen. Section 8.11 provides indications on this subject, as far as the basic modulations such as FSK or PSK and some of their variations are concerned. The bandwidth economy and the performance improvement motivate a lot of research in the direction of special modulations and coding techniques suitable for preventing errors.

13.3 EQUIPMENT ON THE SATELLITE

13.3.1 Antennas

The apparent diameter of the earth as seen from a geostationary satellite is around 17°24′ (fig. 13.4). Taking into account the minimal elevation of the ground antennas of 5°, the satellite antenna must radiate in a cone of 17°18′ wide.

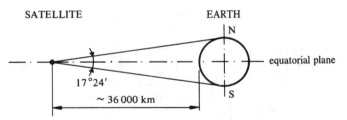

Fig. 13.4 Apparent diameter of the earth as seen from a satellite.

Several types of antennas can be considered:
- approximately isotropic antenna: it requires no particular orientation of the satellite, but wastes emission power ($G = 0$ dB);
- dipole (table 3.46): its radiation diagram presents a maximum in the plane perpendicular to its axis. This plane must be oriented toward the earth while stabilizing the satellite by spinning around an axis parallel to the dipole;
- global beam horn antenna, with a radiation diagram which coincides with the cone of 17°. The gain is around 20 dB, but the antenna must remain pointed toward the earth. In recent satellites, it is made part of the satellite which is stabilized along three axes by a gyroscope;
- antennas with a parabolic reflector giving a spot beam. The zone covered is thus limited to a continent, a region, or even a country, and the gain increases inversely as the opening of the beam decreases, according to (3.102) and (3.103).

13.3.2 Repeaters

Several identical repeaters (e.g., 20 in Intelsat IV A) share the available bandwidth (500 MHz) with 36 MHz per repeater. Certain repeaters thus operate on the same frequency, but with different (spot beam) antennas. Each repeater amplifies the appropriate signal ($G = 100$ dB) and transposes it by SSB modulation (transponder) from the carrier frequency of the ascending path toward that of the descending path (fig. 13.5).

If the same repeater is used by several carriers simultaneously (frequency division multiple access section 13.5.2), the repeater must then satisfy very severe linear requirements to avoid intermodulation effects.

13.3.3 Remote control, telemetry

Different functions of the satellite are remotely controlled from the earth (e.g., repeater gain, pointing of directed-beam antennas, orbit corrections, *et*

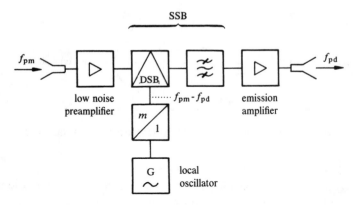

Fig. 13.5 Block diagram of a satellite repeater.

cetera). The parameters of the satellite must be observed by telemetry from the earth. These transmissions are effected by discrete modulation (PSK) of an auxiliary carrier with time-division multiplexing of data.

On the other hand, the satellite continually emits **beacons signals** which allow pointing and exact tracking by ground stations.

13.3.4 Power supply

Solar energy is the preferred source for powering satellites. Photo-voltaic cells cover the cylindrical surface of the satellite, or are grouped on enfolded panels that can be oriented toward the sun. Solar radiation furnishes 1400 W/m^2, but the yield of the cells is only approximately 10%. The necessary electrical power is around several hundred watts and tends to increase with the new generations of satellite.

The eclipses of the satellite, when it passes in the shadow of the earth, necessitates a reserve of energy. They take place in the six weeks around the equinoxes and last for a maximum of 70 minutes each.

13.3.5 Reliability of satellites

The satellite occupies a particular position in the reliability of the link, for the following reasons:
- it constitutes, at least at present, a *non-repairable unit*;
- its duration of proper functioning is not determined only by the random occurence of a breakdown, but also by the exhaustion of possibilities of orbit correction (available propulsive mass);

SATELLITE LINKS

- it is placed in a very harsh environment: thermal shocks, solar and cosmic radiation, meteorite impact.

For these reasons, the reliablity of the satellite must be the object of extreme precautions: choice, manufacturing and testing of its components, internal equipment redundancy, *et cetera*

For calculation of economic profitability, the mean duration of good functioning of a satellite is estimated at 5–7 years. It is further necessary to take into account the risk of failure at launching. The presence in orbit of a reserve satellite significantly increases reliability, but also cost.

13.3.6 Satellites Intelsat IV and V

The International Telecommunications Satellite Organization (Intelsat), with more than 100 member countries, has as its aim the conception, implementation and management of geostationary satellites for international telecommunication. It possesses a dozen satellites distributed in three regions: the Atlantic Ocean, Indian Ocean, and Pacific Ocean (fig. 13.6).

As examples, table 13.7 compares two types of present Intelsat satellites.

13.4 EARTH STATIONS

13.4.1 Choice of site

The placement of an earth station deserves careful consideration, dictated by the following imperatives:
- good visibility of the celestial equator down to the minimum elevation of 5° above the horizon at east and west;
- danger of interference by microwave links of the same frequencies and pointed in the direction of the site, even hundreds of kilometers away;
- distance from heavy-current sources of disturbance (industrial centers, high voltage lines, *et cetera*);
- risk of interception of the beam by regular airplanes;
- price and availability (reserve for expansion) of the land;
- good accessibility;
- compatibility with site protection.

13.4.2 Antennas

The large dimensions of the ground antenna (transmitting and receiving) are necessitated by

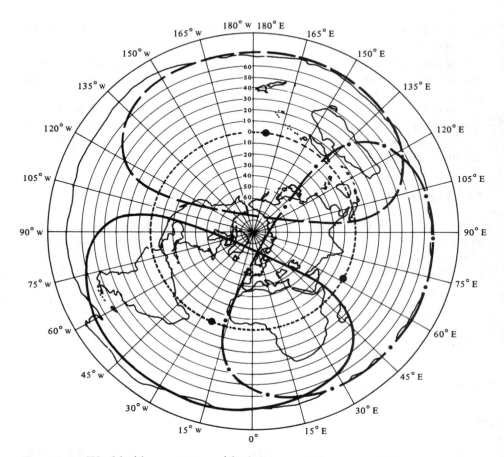

Fig. 13.6 Worldwide coverage with three geostationary satellites.

- a *high gain* (on the order of 60 dB) so as to increase the *equivalent isotropic radiated power EIRP* (sect. 12.4.1) at emission and the power collected at reception (effective area A_e);
- the *directivity* (related to the gain), i.e., the extreme fineness of the principal lobe of the beam (0.1° at emission at 6 GHz), necessary to avoid interference with other satellites in orbit. Furthermore, the secondary lobes must be minimal to reduce the interference with ground microwave links and the deleterious influence of the ground, which has a high noise temperature.

The antenna must be pointed toward the satellite continually and with a greater precision the more directive it is (order of magnitude: some minutes

Table 13.7 Examples of satellites.

	INTELSAT IVA	INTELSAT V
Year	1975	1981
Dimensions	6.8 m × 2.4 m ⌀	16 m × 6.4 m
Mass	1516 kg	1870 kg
Band	6/4 GHz	6/4 GHz and 14/11 GHz
Capacity	6,000 tph. circuits + 2 TV	12,000 tph. circuits + 2 TV

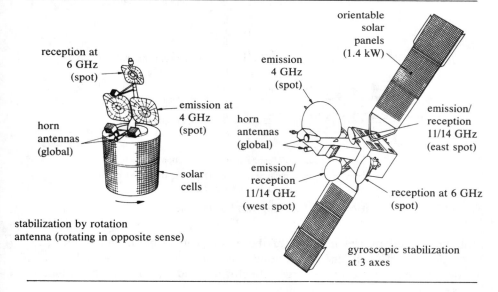

of arc), and this despite its considerable mass (several hundred tons), and even in the presence of a violent wind. An automatic tracking servomechanism is controlled by beacons emitted by the satellite.

On the other hand, inclement weather (rain, snow, frost) disturbs the antenna performance by increasing the noise. The originally recommended radome has proved to be disadvantageous. It is preferred to defrost the antenna, if necessary, by electric heating.

The most commonly used ground antennas are of the "Cassegrain" type (by analogy with the double mirror telescope of the same name). The large parabolic reflector is irradiated indirectly by the intermediary of a secondary hyperbolic reflector, which is confocal with the first (fig. 13.8). The mechanical

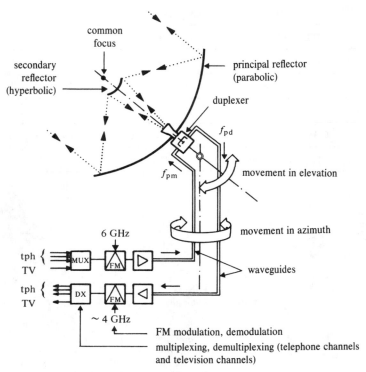

Fig. 13.8 Earth station with "Cassegrain" antenna.

problems posed by the mobility of the antenna in azimuth and in elevation have given rise to different solutions regarding its construction. It is also possible to avoid moving the large reflector by feeding it from a fixed emitter by use of a mobile secondary reflector.

13.4.3 Figure of merit of the receiver

Equation (13.5), applied to the descending path, can be written in the following way, taking into account (3.110):

$$10 \lg \xi'_{\text{lid}} = L_{E\sigma} - A_{\text{iso}} + G_{E\sigma} + G_{R\tau} - 10 \lg \frac{kT_N B_1}{1 \text{ mW}}$$

$$-20 \lg \delta_{\text{inom}} - 3 \text{ dB} \qquad (13.10)$$

where the index σ designates the quantities relating to the satellite and τ those relating to the earth station.

SATELLITE LINKS   487

In equation (13.10), the characteristics of the earth station at reception are represented by two quantities: the gain G_R of its antenna and the noise temperature T_N from (13.4). They are often grouped into a *figure of merit M*, wrongly designated by G/T and expressed, also incorrectly, in dB/K

$$M = G_{R_T} - 10 \lg T_N \qquad \text{dB} \qquad (13.11)$$

In operating a satellite telecommunications network, it is necessary to specify a value of M for all the receiving stations, so as to guarantee a uniform quality. This value still leaves a certain margin of freedom for the technological choice of G_{R_T} and T_N. For example, the Intelsat network requires a value of $M \geq 40.7$ dB at 4 GHz from its stations (of type A, antenna of approximately 29 m in diameter).

13.5 MULTIPLE ACCESS

13.5.1 Interest of multiple access

Each repeater, with its bandwidth of 36 MHz and its capacity of 972 channels (Intelsat IV A), can, in principle, be used for the unidirectional link between one specified earth station and another. Because, on the one hand, the number of pairs of earth stations to be interconnected is much larger than the number of repeaters of a satellite and, on the other hand, the capacity of a repeater is often exaggerated with respect to the traffic to be routed between two stations, it is indispensable to be able to freely distribute this capcity between several stations and to proceed to *multiple access* from the earth toward the *same* repeater of the satellite. It is a problem of multiplexing, similar to that encountered in terrestrial systems (sect. 4.4), and to which a frequency-division solution (FDM) or time-division solution (TDM) can be applied.

As with the analog repeaters or digital regenerators on lines, the *satellite ignores the structure of the multiplex* which crosses it. Besides, this structure can be modified after the satellite has been launched.

In contrast to classical multiplexes, multiple access does not necessarily concern isolated channels, but more often groups of channels, already multiplexed among themselves, and belonging to the same trunk group (same destination).

13.5.2 Frequency-division multiple access

The principle (fig. 13.9) of *frequency-division multiple access FDMA* consists of

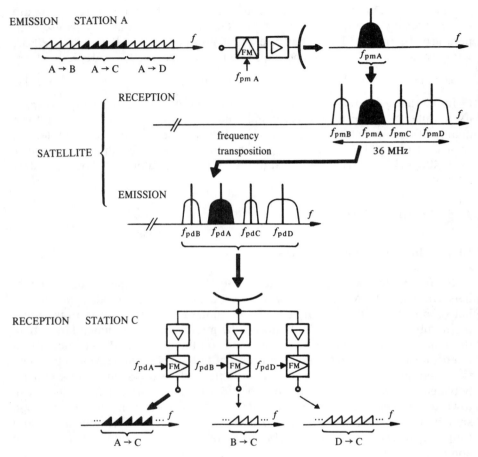

Fig. 13.9 Principle of frequency-division multiple access FDMA (example with four stations: A, B, C, D)

- placing n carriers (each frequency modulated by a frequency-division multiplex of z_n channels) within the band of a satellite repeater;
- assigning each of these carriers to **one** earth station for its emissions toward other stations;
- re-emitting **all** the n carriers from the satellite (by translating them in frequency) toward **all** the earth stations;
- receiving in each station **all** the carriers, demodulating them and extracting by frequency division the channels which are destined for it from other stations.

Each earth station thus contains one FM modulator, one transmitter (on its carrier), but several receivers and FM demodulators (as many as there are trunk groups coming from other stations).

13.5.3 Capacity reduction with FDMA

Frequency-division multiple access with n carriers has three consequences concerning the signal-to-noise ratio ξ'_{li} in a channel i according to (13.5):

- the emission power of the satellite repeater is shared between the n carriers. For one carrier, the emission level to be introduced into (13.5) is thus $L_E - 10 \lg n$, where L_E is the total emission level (unchanged) of the repeater;
- the nominal modulation index $\delta_{i\,\text{nom}}$ for a channel must be increased in such a way as to guarantee the same ratio ξ'_{li}, despite the reduction in level. Ideally, $\delta_{i\,\text{nom}}$ should be multiplied by $\sqrt{n}$;
- the available bandwidth for a carrier is reduced by a **factor** n with respect to the total bandwidth B of the repeater.

According to Carson's rule, (8.65), the bandwidth necessary for an FM modulated carrier is related to the modulation index $\delta_{i\,\text{nom}}$ and the maximum frequency $f_{i\,\text{max}}$ of the modulating signal which is here a group of z_n channels each of width B_1.

Assuming that

$$f_{i\max} \cong z_n B_1 \tag{13.12}$$

we can write

$$B \cong 2 z B_1 (\delta_{i\text{nom}} + 1) \qquad \text{for } n = 1 \tag{13.13}$$

$$\frac{B}{n} \cong 2 z_n B_1 (\sqrt{n}\, \delta_{i\text{nom}} + 1) \qquad \text{for any } n \tag{13.14}$$

In order for relations (13.13) and (13.14) to be simultaneously satisfied, it is necessary (but not sufficient) that

$$z_n < z / n \tag{13.15}$$

In conclusion, the distribution of power and bandwidth between n carriers, necessitated by multiple access FDMA, leads to a reduction of the total number nz_n of channels crossing the receiver, which is more important when n is large.

Table 13.10 gives a relative example of the satellite Intelsat IV.

Furthermore, with FDMA, the intermodulation products are particularly annoying. To avoid them, it is necesary to further reduce the total emission level L_E by some decibels.

Table 13.10 Capacity of one of the Intelsat IV repeaters with FDMA.

Number of carriers n	Numbers of channels per carrier z_n	Total capacity $n\,z_n$	Bandwidth per carrier B/n
1	972	972	35 MHz
7	60	420	5 MHz
14	24	336	2.5 MHz

13.5.4 Single channel per carrier FDMA system: SPADE system

An extreme case of FDMA consists of assigning only $z_n = 1$ channel per carrier. In this way a very flexible system is obtained, called a *single channel per carrier system*, whose carriers can be assigned by switching, from case to case to communications between two ground stations with low mutual traffic (demand assignment, sect. 15.2.7).

In FM modulation such a system would be impractical, because it would allow only a very restricted number n of carriers (and thus of channels) per repeater. However, PCM modulation, applied to each channel individually (no time-division multiplexing!) and followed by a PSK modulation, offers the possibility of placing a large number n of single channel carriers in the band of a satellite repeater (fig. 13.11), because it needs much lower power.

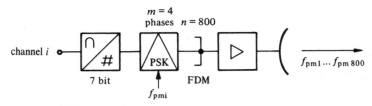

Fig. 13.11 Principle of emission with single channel carriers with PCM/PSK.

The SPADE system (Single channel per carrier PCM multiple Access Demand assignment Equipment) implements this idea with the following characteristics:

- $n = 800$ single channel carriers in frequency-division multiplex FDM;
- Seven-bit PCM modulation (for historical reasons, instead of the usual 8 bits), i.e., a bit rate of 56 kbit/s per telephone channel;
- modulation of the transmitter in PSK with $m=4$ phases, which reduces the symbol rate from (1.10) to $\dot{M} = \dot{D}/2 = 28$ kband.

SATELLITE LINKS 491

The bandwidth occupied by a PSK modulated carrier (sect. 8.11.5 and table 8.84) is around 1.6 $\dot{M}$, i.e., 45 kHz. In this way there is room for 800 carriers in the band of 36 MHz of a satellite receiver. The emission power is divided by 800, but the advantage of a digital transmission is precisely of maintaining a satisfactory quality (error probability and signal-to-noise ratio after demodulation) despite a very low signal-to-noise ratio at reception.

13.5.5 Time-division multiple access

Time-division multiple access TDMA safeguards efficiency, at the price, however, of extra synchronization difficulties (fig. 13.12). It is applied to digital transmission and consists of

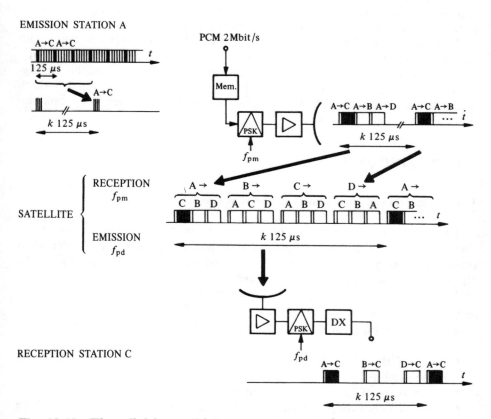

Fig. 13.12 Time-division multiple access (example with four stations: A, B, C, D).

- emitting synchronous periodic *bursts of pulses* from ground stations (PSK modulation of a carrier at frequency f_{pm} common to all the stations);
- organizing the emissions in such a way that these bursts arrive at the satellite exactly interleaved in time, without ever becoming confused. This problem is delicate and requires very strict time organization to take into account the different propagation time between the different ground stations and the satellite;
- supplementing each burst by a *preamble* which identifies its origin and destination;
- emitting all the pulse bursts from the satellite at the frequency f_{pd} toward all the earth stations.

The time-division multiplex avoids problems of intermodulation and leaves all the emission power of the satellite at the disposal of each emission. A slight loss of efficiency is due to the necessity of leaving a time margin between the packets to absorb fluctuations and synchronization imprecisions.

Time-division multiple access TDMA can be advantageously combined with *digital speech interpolation techniques* which consist of detecting pauses in a telephone communication in order to interleave into them the samples belonging to other communications. By this dynamic assignment of channels, which necessitates reliable detection and rapid switching, the capacity of a link can be doubled.

13.6 PERSPECTIVES

13.6.1 Increase of the capacity of the space sector

There is a risk of congestion of the geostationary orbit (celestial equator). The satellites must be placed there at a sufficient angular distance from each other, in relation to the directivity of ground antennas (vol. XIII, fig. 8.23). On the other hand, the available frequency range is also limited.

To satisfy the increasing demand (approximately 20% per year) in international and even national circuits, it is necessary to plan for

- satellites with larger capacity (Intelsat VI will offer 37,000 channels in 1986);
- reuse of the same frequencies on different spot beam antennas of the same satellite;
- use of two orthogonal polarizations to double the capacity of each carrier;
- exploitation of the 14/11 GHz and 30/20 GHz bands in the regions where the meteorological conditions permit it, thus offering the advantage of diminishing the dimensions of the antennas for the same gain.

13.6.2 Domestic satellites

Because of the larger market due to economic growth, satellites are needed to handle the national telecommunication traffic, especially in countries or regions whose topography is unsuitable for ground links.

It would be preferable to reduce the dimensions (and cost) of ground antennas (5 to 7 m in diameter) and to simplify the tracking apparatus. Very narrow beam antennas on the satellite allow a gain of 10–20 dB over the EIRP of the descending path and corresponding reduction of the gain of the ground antenna. Further, the passage to higher frequency bands also proceeds in the same direction.

13.6.3 Broadcasting satellites

Satellite projects destined for direct broadcasting of national audio or television programs are based on the 12 GHz band. Individual or collective domestic reception should be possible with antennas of 0.9–2 m in diameter.

Legal and political problems become superimposed on technical questions. Each country desires to have available a beam which covers at least its geographical surface, but does not necessarily wish to receive other emissions. It would be necessary to make use of antenna arrays, which would be highly directive and configured in such a way as to irradiate the desired zone of the earth.

Chapter 14

Optical Fiber Links

14.1 PRINCIPLES AND STRUCTURE OF OPTICAL FIBER LINKS

14.1.1 Optical information transmission

Light (defined in the broad sense, starting with the narrow range of visible light to include the much larger one of infrared), in principle belongs to the family of electromagnetic waves. In this role, it can serve as a transmission vector for information-carrying signals. It is, however, distinguished from classical electrical carriers by the following characteristics:

- its very high frequency range (100–1000 THz) requires a *very specific technology*;
- the light produced by the usable sources is generally incoherent and has a *random character* which makes it resemble a "noise" with a more or less limited spectrum (depending on the type of source), rather than a sinusoidal signal;
- consequently, the only currently modulable parameter of light is its "intensity" (optical power);
- in contrast to electrical radiocommunications (e.g., microwave links), in which the free space attenuation (assuming a lossless medium) predominates, the attenuation of an optical link, guided or not, is determined by *absorption* phenomena (losses) and *diffusion* phenomena. Furthermore, it is necessary to take into account the efficiency of optoelectronic transducers and their adaptation to the medium used.

The atmosphere lends itself only to optical transmission at short distances. However, the guiding of light in a medium whose dimensions and physical properties are optimizible, allows us to implement transmission of large quantities of information, at considerable distances, under quasi-ideal conditions. This is the objective of *optical fiber links*, to which this chapter is devoted. The intrinsic properties of fibers as a transmission medium are the object of section 3.7.

14.1.2 Structure of an optical fiber transmission system

The following principal elements appear in an optical fiber link (fig. 14.1):
- electro-optical *transducers* at emission, and optoelectronic transducers at

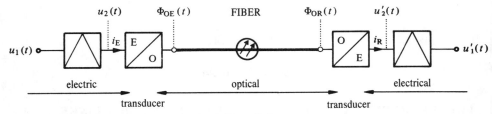

Fig. 14.1 Structure of an optical fiber link.

reception. They transform the electrical signal into a light signal and vice versa. Furthermore, they are completed by an optical adapter to the fiber. These transducers can be considered to be *amplitude modulators* (of the optical flux), respectively as an (optical) *envelope detector*;
- an *electric modulator* to place the information to be transmitted into an appropriate form (e.g., PCM, PPM, PFM, FM) *before* presenting it to the electro-optical transducer;
- *electrical demodulator* to carry out the reverse operation at reception and to restore the transmitted information;
- if an intermediate amplification is necessary (a repeater), it is done at the electrical level, because the direct amplification of optical signals is not (yet) possible. In the case of digital transmission, this amplification can be accompanied by a regeneration (in the sense of section 5.2).

14.2 OPTICAL TRANSDUCERS

14.2.1 Preliminary remarks

The domain of optoelectronic and electro-optical transducers, fundamental for optical fiber links, is characterized by recent and as yet unstabilized technological evolution. A detailed quantitative analysis of their performance is outside the scope of this work and would be very rapidly outdated. However, to a greater degree than in the case of electrical transmission by radio waves or lines, the properties of the emitters and receivers enter in a predominant manner into optical link planning. This is why a presentation, at least qualitative and comparative, of their characteristics appears to be indispensable in this context.

14.2.2 Characteristics of optical emitters

The electro-optical transducer plays the role of optical emitter; it carries out conversion between

OPTICAL FIBER LINKS

- an *electrical current* i_E at the input
- and a light flux or *optical power* Φ_{OE} at the output

In contrast to the usage in classical photometry (visible light) in which the light flux is expressed in lumens (taking into account the sensitivity of the human eye), here *optical power* Φ_O will be spoken of, in the electromagnetic sense (flux of the Poynting vector, vol. III, sect. 5.1.1), expressed in watts.

In the ideal case, the emitted optical power Φ_{OE} would be proportional to the electrical current, the proportionality coefficient defining the *optical yield* η_O (in W/A)

$$\Phi_{OE} = \eta_O \, i_E \qquad \text{W} \qquad (14.1)$$

Because the electrical power at the input of the transducer is proportional to i_E^2 it is found that

- the *optical power* emitted varies with the *square root* of the *electrical power* provided.

Consequently, in an optoelectronic system, it is essential to clearly distinguish these two types of power by different symbols Φ_O and P and to specify in each case which one is being referred to. It is also necessary to note that a variation of 1 dB of the *optical attenuation* between Φ_{OE} and Φ_{OR} leads to a variation of 2 dB of the *electrical attenuation* between $i_E(t)$ and $i_R(t)$, or between $u_2(t)$ and $u_2'(t)$.

The other interesting characteristics of an optical transmitter are

- the *linearity* of the law relating Φ_{OE} to i_E;
- the central wavelength λ_O (for $\Phi_{OE\,max}$) of its emission *spectrum* which must correspond to one of the optimal "windows" on the fiber (fig. 3.33);
- the *width* $\Delta\lambda$ of this spectrum (e.g., at $\Phi_{OE\,max}/2$) which determines the chromatic dispersion in the fiber (sect. 3.7.6);
- the *radiation diagram* which, as in the case of a radioelectric antenna, describes the spatial distribution of the optical power. In particular, the *numerical aperture* N of the emitter is defined, in the same way as in (3.77), by the sine of the cone opening angle containing half the radiated optical power. It should, as much as possible, correspond to the numerical aperture of the fiber;
- the *rise time* t_m and the *decay time* t_d of the optical step response;
- the *reliability*, in particular the average duration during which the characteristics will hopefully be satisfactory;
- the influence of the *temperature* on the characteristics.

The principal electro-optical tranducers, currently available as transmitters for optical fibers, are the *light emitting diode LED* and the *laser diode LD*. The orders of magnitude of their characteristics are compared in table 14.2.

Table 14.2 Comparison of the typical characteristics of electro-optical transducers.

	Light emitting diode LED	Laser diode LD
Spectrum	$\Delta\lambda \cong 50$ nm, centered at λ_0, with $\frac{1}{2}\Phi_{OE\,max}$ width	$\Delta\lambda \cong 1$ nm, centered at λ_0, with $\frac{1}{2}\Phi_{OE\,max}$ width
Conversion characteristic	Φ_{OE} vs i_E, curves for 0°C, 25°C, 50°C (approximately linear)	Φ_{OE} vs i_E, curves for 0°, 25°, 50°C with $i_{threshold}$
Nominal optical power	1 mW	5 mW
Linearity	médiocre	good for $i > i_{threshold}$
Rise time	10 ns	1 ns

14.2.3 Characteristics of optical receivers

At the end of the fiber, the optical power received Φ_{OR} is reconverted into an electrical current i_R by an optoelectronic transducer with a *responsivity* σ_O in A/W, such that

$$i_R = \sigma_O \, \Phi_{OR} \qquad \text{A} \qquad (14.2)$$

The restored *electrical power* (proportional to i_R^2) varies **with the square** of the received *optical power*. The behavior of the receiver is thus reciprocal to that of the transmitter.

Aside from the value of σ_O, characteristics analogous to those of an optical transmitter (linearity, spectral range of functioning, directivity diagram, numerical aperture, reaction time, reliability, influence of temperature) determine the quality of an optical receiver.

Currently, the optoelectronic transducers most frequently used for optical fiber transmission are the *PIN photodiode* (p-intrinsic-n-diode) and the *avalanche photodiode APD*, which makes use of the avalanche effect to increase considerably the output current i_R. Their characteristics are compared roughly in table 14.3.

Table 14.3 Characteristics of optical detectors

	PIN Photodiode (Si)	Avalanche photodiode APD (Si)
Spectral range	$\Delta\lambda \cong 500$ nm (400–1000 nm)	$\sigma_{O\,max}$; $\frac{1}{2}\sigma_{O\,max}$
Response coefficient $\sigma_{O\,max}$	0.5 A/W	5 A/W
Rise time and fall time	~ 1 ns	

14.2.4 Influence of noise on reception

Without going into the details of the physical process (presented in vol. VII, chap. 9), it must be recalled that the analysis of optoelectronic conversion is tied to the quantum theory of light, i.e., to its decomposition into *photons* with an elementary energy $h\nu$ where h is Planck's constant and ν designates the optical frequency (to distinguish it from the frequency f of the electrical signals which modulate the light wave).

Two consequences result for optical transmissions:

- the power spectral density of *thermal noise* can no longer be considered uniform. Relation (2.40), giving the thermal noise power dP_{Nth} in a frequency band df (here: $d\nu$) must be corrected (vol. VI, sect. 6.2.2) in

$$dP_{Nth} = \frac{h\nu}{\exp\left(\frac{h\nu}{kT}\right) - 1} d\nu \qquad W \qquad (14.3)$$

At optical frequencies, $h\nu$ ($\cong 2.10^{-19}$ J) is not negligible before kT ($\cong 4.10^{-21}$ J). It results in a lowering of the spectral power density at these frequencies (approximately -200 dB to 300 THz) with respect to the value kT, usual even in the microwave range;

- on the other hand, the random character of the arrival of photons is responsible for another noise, called the *quantum noise*, with effects which largely predominate over those of thermal noise.

If it is assumed that photons arrive in a purely random manner (Poissonian phenomenon), the probability of receiving x photons during a time interval T is given by **Poisson's law**.

$$\text{Prob}(x,T) = \frac{(\bar{n})^x}{x!} \exp(-\bar{n}) \tag{14.4}$$

where $\bar{n}$ represents the *average* number of photons received during T, corresponding to the optical energy received

$$\bar{n} = \frac{1}{h\nu} \int_t^{t+T} \Phi_{OR}(t)\, dt \tag{14.5}$$

The random fluctuations of x around $\bar{n}$ have repercussions on the output current i_R and on binary decisions which it may be necessary to take at reception.

14.2.5 Effect of quantum noise on an optical transmission with OOK

Optical transmission with OOK (sect. 8.11.3) is characterized by switching on and off light pulses, corresponding to an electrical signal modulated in time by analog information (PPM or PFM) or carrying coded digital infomation (for example, PCM or data).

In this case, the receiver must decide at each instant between two possibilities:

- presence of light (and, therefore, of photons) at the output of the fiber: binary condition 1;
- absence of light (no photon): binary condition 0;

In assuming (in the ideal case) that there is no other source of noise than the quantum noise, the decision threshold can be effectively fixed between zero and one photon during a time interval T.

The probability ϵ_0 of simulating the reception of a photon, while the binary state 0 is emitted, is practically zero

$$\epsilon_0 = \text{Prob}(1|0) \cong 0 \tag{14.6}$$

However, the random appearance of photons during an intentional light emission (state 1 emitted) can result in the fact that, despite everthing, no photon is detected during T (state 0 received by mistake). The probability ϵ_1, of interpreting a transmitted 1 as a 0 is given by the probability, according to

OPTICAL FIBER LINKS

Poisson's law (14.4), of receiving exceptionally $x=0$ photons during T, while on the average one would receive $\bar{n}$ according to (14.5).

$$\epsilon_1 = \text{Prob}(0|1) = \text{Prob}(x=0,T) = \exp(-\bar{n}) \tag{14.7}$$

In the case of a an OOK optical transmission of digital information, at a fixed rate $\dot{D}$, the time $T = 1/\dot{D}$ represents the duration of one bit.

If the binary states emitted are equally probable, the error probability ϵ per bit at reception is

$$\epsilon = \text{Prob}(0)\, \epsilon_0 + \text{Prob}(1)\, \epsilon_1 \cong \tfrac{1}{2}\epsilon_1 = \tfrac{1}{2}\exp(-\bar{n}) \tag{14.8}$$

We can draw the conclusion that, to guarantee for example $\epsilon \leq 10^{-9}$, at least $\bar{n} \cong 20$ photons per bit are necessary.

The optical energy W_0 necessary for the reception of one bit in state 1 proceeds from (14.8)

$$W_O = \bar{n}\, h\nu = h\nu \ln\left(\frac{1}{2\epsilon}\right) \tag{14.9}$$

The corresponding minimum optical power $\Phi_{\text{OR min}}$ is

$$\Phi_{\text{ORmin}} = \frac{W_O}{T} = h\nu\, \dot{D} \ln\left(\frac{1}{2\epsilon}\right) \tag{14.10}$$

It is important to note that

- to guarantee the same error probability ϵ, the *optical power received must increase proportionally to the bit rate* $\dot{D}$;
- relation (14.10) gives the optical power which must reach the receiver during the light emission (binary state 1) and not the average optical power;
- relation (14.10) represents the *lower theoretical limit* of the optical power at reception, taking into account only the quantum noise;
- in reality, other sources of noise intervene and oblige us to insert a higher optical power, called the *sensitivity* of the receiver, to guarantee the same error probability.

14.3 TRANSMISSION MODES ON OPTICAL FIBERS

14.3.1 Optical modulation

The variation of the optical power Φ_O according to the electrical input signal can be made in two ways:

- *continuous*, proportionally to the electrical signal, as with classical AM modulation. In optics, this procedure is more often called *intensity mod-*

ulation IM to emphasize the incoherent (non-sinusoidal) character of the optical carrier signal;
- *discrete*, generally binary, according to a procedure here called *optical on-off keying OOK*, by analogy with OOK modulation with a sinusoidal carrier (sect. 8.11.3). The electrical signal is thus a train of pulses of constant amplitude that carries analog or digital information in time.

14.3.2 Analog transmission

Direct IM modulation (fig. 14.4(a)) by the electrical baseband signal is possible, but meets two obstacles:
- *the linearity* required of such a transmission to avoid intermodulation products is hardly compatible with the scarcely linear behavior of optoelectronic transducers;
- *the noise effect* at reception has direct repercussions for the restored analog information.

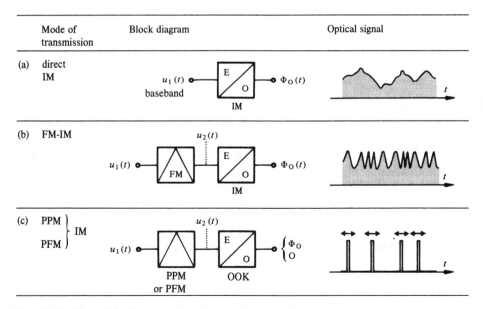

Fig. 14.4 Possible forms of analog optical modulation.

The combination of an electrical FM modulation, with, furthermore, an IM optical modulation by the signal $u_2(t)$ at a constant amplitude, but variable frequency (FM-IM modulation, fig. 14.4(b)), avoids part of the linearity and noise problems, but enlarges the band necessary for transmission. Furthermore, the chromatic dispersion, (sect. 3.7.6) is a source of phase distortions which provoke nonlinear distortions after demodulation (sect. 8.6.12).

The ideal solution for an optical analog transmission is rather a time-based electrical modulation of pulses: PPM or PFM (fig. 14.4((c)). The signal $u_2(t)$ thus takes only two amplitude values(U_p and 0), which permits *optical on-off keying* (OOK). This method of operation does not require linearity. It is particularly favorable to the functioning of optical transmitters. The effect of noise depends on the slope of the pulse edges, i.e., on the occupied bandwidth (sect. 8.10.8).

For analog transmission of a single television channel by optical fiber, the frequency pulse modulation PFM is recommended. PPM modulation, which is more complex, however allows the implementation of a time-division multiplex of several channels.

14.3.3 Digital transmission

The most favored use of optical links is digital transmission. The very broad bandwidths available open the way to high bit rates, with minimal error probabilities.

However, the *transmission mode* must be adapted to the particular conditions of optical links, namely:

- the electrical signal $u_2(t)$ must be *unipolar* (no negative luminous intensity!);
- a *binary mode* ($m = 2$) is preferable to a mode with $m \geq 2$, in such a way as to modulate the light on and off (OOK modulation);
- the *dc component* of $u_2(t)$ does not present difficulties (in contrast to line transmission); on the contrary, it is even inevitable (as a consequence of the first condition);
- however, the *average* optical power over a long random binary sequence must also be as *constant and independent* of the binary statistics as possible; this results in requiring a very low power spectral density at low frequencies (despite the presence of a line at $f = 0$);
- a high *clock rate content* is indispensable for synchronizing the receiver to $\dot{D}$;
- the economy of the *bandwidth* is less critical than on lines, but remains an important concern at very high bit rates;
- the *mode conversion* between the electrical input interface and the signal $u_2(t)$ applied to the optical modulator must be as simple as possible;

- the *error detection* or at least the monitoring of the error rate must be possible in operation.

The *unipolar binary NRZ mode* (fig. 14.5(a)) has the disadvantage of no longer containing the synchronization information in the case of long trains of identical bits (impossible to exclude, especially in data transmission). By means of a *scrambling* operation, i.e., by logical combination with a pseudo-random sequence at emission, one artificially increases the number of transitions, and thus the clock rate content. An inverse operation of *descrambling* at reception, restores the original binary information by combination with the *same* pseudo-random sequence.

By astute transcoding of x elements at the input into $y > x$ binary elements offered to the optical transducer (code of type xByM, sect. 9.6.6), we can at the same time guarantee a sufficient clock rate content and monitor the error rate (due to the introduction of a slight redundancy). With this model, the modes 5B6B, 7B8B, 17B18B, *et cetera* have been proposed.

Coded mark inversion CMI, presented in figure 9.23 and proposed for the interface of digital systems at 140 MBit/s is also a binary mode favorable for optical transmission (fig. 14.5(b)). It is very poor in low frequencies (high constancy of the average optical power) and its spectrum presents a line at D. Its only inconvenience is that it occupies a broader bandwidth than the simple binary modes.

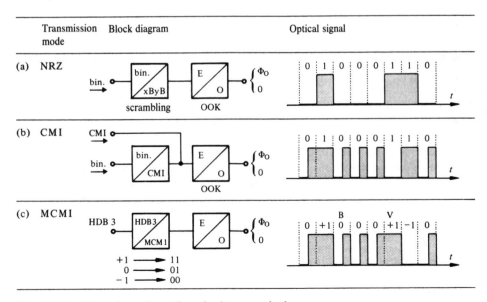

Fig. 14.5 Digital modes of optical transmission.

The ***modified coded mark inversion MCMI*** allows us to adapt directly the electrical HDB3 mode (sect. 9.6.5) to binary optical emission (fig. 14.5(c)) for transmission at 8.448 Mbit/s.

14.4 PLANNING OF A DIGITAL OPTICAL FIBER LINK

14.4.1 Parameters

Because of the importance of this form for optical fiber links, only digital transmission will be considered for planning in the outline of this chapter.

From the point of view of the user, the following parameters are determining factors:

- bit rate $\dot{D}$ to be transmitted;
- length l of the link;
- acceptable error probability ϵ per bit.

For a given type of fiber, the characteristic parameters are

- the ***attenuation*** A_O of the envelope of the optical signal after the distance l (sect. 3.7.9);
- the rms duration $d_{h\,\text{eff}}$ of the optical impulse response at the end of the link (sect. 3.7.8), which expresses the modal and chromatic ***dispersion***.

To these are added less obvious but non-negligible parameters in the case of optical links, such as:

- the supplementary attenuation due to the splices and connectors;
- the attenuation due to imperfect optical coupling between transducers and the fiber;
- the response time of the transducers.

14.4.2 Methodology

As with any digital transmission, the error probability essentially depends on two factors whose combined effect is illustrated overall by the ***eye pattern*** (sect. 5.6):

- the influence of the ***noise*** which surrounds the characteristic binary values of the optical signal with a zone of uncertainty. In the case of an optical link, the ***quantum noise*** must be considered as well as the thermal noise;
- the appearance of ***intersymbol interference*** due to the stretching out of pulses resulting from dispersion in the course of transmission in the fiber.

The planning of an optical digital link considers these two effects separately, (as presented in a general manner in chapter 5):
- the *energy budget* allows us to assure that the optical signal-to-noise ratio at reception is sufficient to respect the tolerable error probability ϵ;
- the evaluation of the *dispersion* (or of the bandwidth, which is related to it) gives an idea of the encroachment of pulses on each other, and thus of the intersymbol interference, and of the increase of the error probability which results from this increase in the presence of noise.

14.4.3 Energy budget

The ratio between the optical power Φ_{OE} produced by the transmitter and the minimal optical power $\Phi_{OR\,min}$ necessary for satisfactory reception, or the difference of corresponding absolute optical power levels L_O, gives the maximum tolerable attenuation for the optical link

$$A_{Omax} = 10 \lg \frac{\Phi_{OE}}{\Phi_{ORmin}} = L_{OE} - L_{ORmin} \qquad (14.11)$$

The energy budget consists of evaluating **all the optical attenuations** between transmitter and receiver (fiber, connectors, splices, *et cetera*), The sum must remain lower than $A_{O\,max}$, including a safety margin.

The theoretical limit of $\Phi_{OR\,min}$ is imposed by the quantum noise according to (14.10). It varies linearly with the bit rate D and logarithmically with $1/\epsilon$. The real optoelectronic transducers require a higher minimum optical power (sensitivity), specified by the manufacturer, according to the type of photodiode and as a function of ϵ.

Figure 14.6 gives the orders of magnitude of L_{OE} and of $L_{OR\,min}$ as a function of the bit rate, for present transducers and for an error probability $\epsilon = 10^{-9}$.

The energy budget can serve to estimate the maximum rate possible on a link in which all the optical parameters are known. Because this calculation does not take dispersion into account, it is also necessary to be sure (sect. 14.4.4) that the intersymbol interference remains small; if not, the level of optical power at reception must be increased.

If we wish the energy budget to give an indication of the maximum possible length of the fiber (regenerator spacing) with given transducers, for a specific rate and error probability, it is necessary to consider the following points:
- due to mode coupling (sect. 3.7.7), the optical attenuation does not increase linearly with the length l;
- the number of splices increases with l, in relation to the (limited!) manufacturing length of fiber;

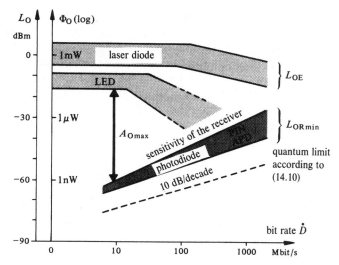

Fig. 14.6 Maximum tolerable optical attenuation for $\epsilon = 10^{-9}$.

- depending on the optical conditions, dispersion plays a more important role than attenuation in the determination of the regenerator spacing (fig. 14.10).

14.4.4 Effect of dispersion

The effects of chromatic dispersion and of the response times of the transducers are added to the rms duration of the optical impulse response d_{heff} of the fiber, evaluated in section 3.7.8 on the basis of *modal dispersion only*. Very roughly, a quadratic addition of these effects can be assumed (square root of the sum of the squares).

$$d_{hefftot} = \sqrt{\sum_i d_{heffi}^2} \qquad (14.12)$$

For multimode fiber links, the modal dispersion nevertheless remains the most important factor, due to the fact that d_{heff} increases with the length of the fiber. This increase is only linear when the fiber length is less than a *critical length* l_{crit} which depends greatly on the physical state of the fiber (curves, irregularities, imperfections of the splices) and can even vary for the same fiber between the laboratory measurements and those in the field. Above this length, mode coupling increases d_{heff} proportional to $\sqrt{l}$ (fig. 3.37).

The uncertainty concerning the increase law of the intermodal dispersion has implications for the planning results based on the intersymbol interference. In particular, there is the possibility that the regenerator spacing may be unnecessarily short if it is based on the hypothesis (pessimistic for multimode fibers) of a linear increase. On the contrary, the hypothesis of an increase in $\sqrt{l}$ could be too optimistic and lead to regenerators which are too far apart.

Single mode fibers, exempt from modal dispersion, are not subject to this amgibuity ($d_{\text{h eff}} \sim l$).

14.4.5 Eye pattern

The knowledge of the (quasi-Gaussian) optical impulse response allows us to trace the *eye pattern* of the optical signal at reception and to thus evaluate the intersymbol interference, while they succeed each other at a rate $\dot{D}$ (fig. 14.7).

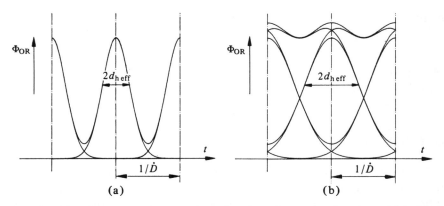

Fig. 14.7 Eye pattern at reception, in the absence of noise: (a) without intersymbol interference $1/\dot{D} > 2d_{\text{heff}}$; (b) with intersymbol interference $1/\dot{D} \cong 2\,d_{\text{heff}}$.

At a higher rate, the rms duration of the impulse response must be reduced so as not to increase the interference, which, for a given fiber, means reducing the regenerator spacing.

The noise effect on reception is superimposed on the eye pattern. For the *binary state 0*, if it is assumed that no light is emitted, the principal contribution is that of the *thermal noise* of the receiver. However, at *binary state 1*, when the optical transmitter is active, the *quantum noise* predominates. Its power is proportional to the number of photons transmitted, and thus to the instantaneous optical power. At the output of the optoelectronic transducer, the electrical signal $u_2'(t)$ is thus disturbed by a noise whose rms value is proportional to $u_2'(t)$. The result is a dissymetry of the noise effect on the eye pattern, (fig. 14.8). Consequently, it is preferable to place the decision threshold for the regeneration of the binary states, **under** the non-disturbed half of the eye openings.

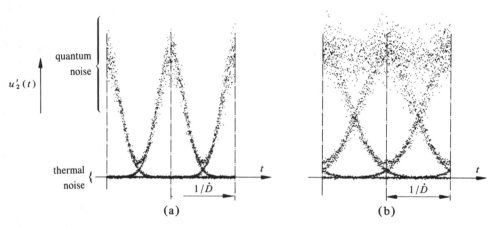

Fig. 14.8 Eye pattern at reception, disturbed by noise: (a) without intersymbol interference; (b) with intersymbol interference.

14.4.6 Optical bandwidth

Although this concept is very ambiguous, the bandwidth of an optical fiber is often mentioned in the specialized literature. Conforming to the conventional definition (sect. 2.4.8), it is considered to be the frequency of the electrical signal $u_2(t)$ for which **the attenuation of the optical signal envelope increases by 3 dB** with respect to its value at $f = 0$ (continuous optical emission, constant envelope). The *electrical* attenuation between $u_2(t)$ and $u_2'(t)$ increases thus by 6 dB, according to the remark made in section 14.2.2.

For an optical transfer function $H_O(f)$ of Gaussian type, according to relation (3.88), the optical bandwidth B_O, which corresponds to a decrease of optical power by a factor 2 (3 dB), becomes, in setting $H_O(B) = H_O(0)/2$:

$$B_O = \frac{\sqrt{\ln 2}}{\sqrt{2}\pi} \frac{1}{d_{\text{heff}}} \cong \frac{0.19}{d_{\text{heff}}} \quad \text{Hz} \qquad (14.13)$$

It is therefore inversely proportional to the rms duration of the impulse response d_{heff} which depends on the length l of the fiber.

The *specific optical bandwidth* b_O of an optical link (including transducers) of a length of 1 km is often mentioned. For a multimode fiber, the extrapolation law of this value b_O to the bandwidth B_O of a link of any length l reflects the same uncertainty as the function which relates d_{heff} to l, depending on the critical length l_{crit} (fig. 14.9)

$$B_O \sim \frac{b_O}{l} \quad \text{for } l < l_{crit} \tag{14.14}$$

$$B_O \sim \frac{b_O}{\sqrt{l}} \quad \text{for } l > l_{crit} \tag{14.15}$$

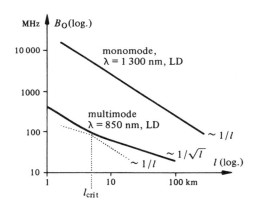

Fig. 14.9 Typical evolution of the bandwidth as a function of the length.

Let us recall that, in the case of real channels, which certainly includes optical links, Nyquist's theorem (4.7) is approached by the empirical relation (4.9), i.e., in the case of a binary optical transmission:

$$\dot{D}_{max} = 1.25 \, B_O \tag{14.16}$$

The maximum possible bit rate, taking dispersion into account, thus depends on the length l according to the same law as B_O (fig. 14.9).

14.4.7 Regenerator spacing

The maximum distance l_{max} which can be covered without intermediary regeneration is one of the most important parameters for the operational cost

OPTICAL FIBER LINKS

of optical transmission systems. This maximum regenerator spacing is determined by the two approaches presented earlier:

- **at low and medium rates**, the energy budget gives the tolerable *attenuation* $A_{O\max}$ for the fiber (including the connectors and splices). Figure 14.6 shows that $A_{O\max}$ decreases by 10 dB when the rate is increased tenfold, which requires a corresponding decrease of the regenerator spacing;
- **at high rates**, the *dispersion* becomes the determining factor due to the intersymbol interference that it introduces. Provided that $l > l_{\text{crit}}$ (multimode fiber), the regenerator spacing l must therefore be proportional to $1/B_O^2$, and thus proportional to $1/\dot{D}_O^2$, according to (14.15) and (14.16).

Figure 14.10 qualitatively illustrates these two laws. The real values of the spacing l envisaged for optical fiber links greatly depend on the fiber and transducer technology. Table 14.11 gives some orders of magnitude. As we move from the first "window" (fig. 3.33) at $\lambda \cong 0.85$ μm, to the second window

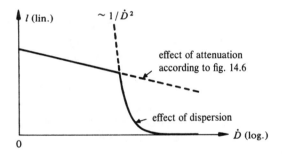

Fig. 14.10 Regenerator spacing as a function of the bit rate (multimode fiber).

Table 14.11 Typical values of regenerator spacing.

Bit rate $\dot{D}$ (Mbit/s)	Fiber type	λ (nm)	Spacing l (km)
8	multimode, gradient	850	10—12
34	multimode, gradient	850	8—10
		1300	30—60
140	multimode, gradient	850	7—9
		1300	20—25
	single mode	1300	30—70
565	single mode	850	4—5
		1300	20—45

at $\lambda \cong 1.3$ μm in which the chromatic dispersion becomes negligible and the attenuation approximately 4 times weaker, the regenerator spacing can be multipled 3 to 6 times, depending on the type of transducer. On the other hand, the single mode fiber is free of modal dispersion, which further increases the regenerator spacing.

These figures must be compared with those of table 9.31 which gives the regenerator spacing for the same digital systems on coaxial cable.

14.5 TRANSMISSION SYSTEMS ON OPTICAL FIBERS

14.5.1 Long-distance systems

The advantages of optical fibers, mentioned in section 3.7.10, are particularly attractive for the long-distance network for the following reasons:
- the progressive introduction of digital transmission techniques favors implementation in optical form;
- the transmission needs (number of telephone channels or equivalent channels) increase and will require systems with larger and larger bandwidths;
- the optical regenerator spacing is notably higher at the same rate than that of regenerators on a coaxial cable;
- the transmission quality (error probability ϵ) is excellent.

Consequently, only *digital systems* enter into consideration, principally those of higher stages of the CCITT hierarchy (sect. 9.4.1), with rates of 34 Mbit/s (480 channels), 140 Mbit/s (1920 channels), or even 565 Mbit/s (7680 channels), or multiples of these rates.

These are "four-wire" systems, with each transmission direction carried by a separate fiber.

The intermediate optical repeaters must be remotely supplied with dc power, which require special metallic conductors as well as fibers, in the optical cable. The interest of the single mode fibers, with their very long regenerator spacing (table 14.11), would be precisely to avoid this problem by freeing the distance between two urban centers of any regenerators.

Submarine cables with (single mode) optical fibers are also envisaged for transoceanic links, in the second (or perhaps third) optical "window" (1.3 μm or 1.55 μm). Thus important reliability problems arise for the submerged regenerators, as well as those concerning the water-tightness of the cable and its resistance to traction during placement. Spacing of 25 to 50 km between regenerators at rates of 140 or 280 Mbit/s would contribute to making these systems economically feasible, even vis-a-vis satellite links. The optical transatlantic cable TAT8, planned for 1988, will have a capacity of 8000 telephone circuits on two pairs of single mode fibers.

14.5.2 Optical transmissions in the local network

The penetration of optical fibers into the local network is less evident, especially for economic reasons. However, with the hypothesis of a *local broadband network* (sect. 15.8.5), the optical fibers can open attractive perspectives for

- telephony in digital form, e.g., in PCM;
- duplex data transmission;
- transmission of one or several television programs, according to the desires of the subscriber (selection by return channel);
- audio program broadcasting.

The form in which these different transmissions would take place (analog, digital, mixed) is not yet clear.

The remote power supply of the terminal equipment, at least for minimal function (telephony), poses a serious problem. It is not impossible that it could be solved by the optical power itself (optical power feeding).

14.5.3 Optical multiplexing

When it is a matter of transmitting signals corresponding to several telephone channels (long-distance systems) or to different services (local network) on the same fiber, the classical form of multiplexing (FDM and especially TDM) can be applied *at the electrical level*. The electrical multiplex signal $u_2(t)$ thus modulates the light emission, which takes place with a single wavelength.

Optical transmission nevertheless offers another possibility: *wavelength-division multiplexing WDM* or optical multiplexing, which consists of simultaneously emitting light waves of different wavelengths on the same fiber. The number of channels thus multiplexed would remain modest (2–8), taking into account the favorable spectral domain (three "windows" between 800 and 1600 nm) and the difficulty of separating these channels at the ends. To this aim, optical interference filters have been developed, with a bandwidth of approximately 50 nm. The supplementary attenuation (non-negligible) due to these filters must be included in the energy budget of the link and further shortens the regenerator spacing.

On this basis, it is possible to conceive of a *bidirectional link* using the same fiber with different wavelengths in the two directions (fig. 14.12). One thus realizes an optical "frequency-division pseudo-four-wire" system (sect. 4.6.4).

Despite the technological problems it poses, optical multiplexing is of economic interest, in particular in the cases in which no regenerator is necessary.

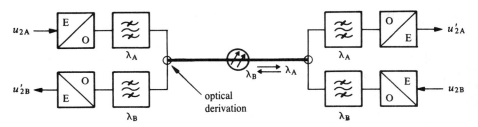

Fig. 14.12 Bidirectional transmission with optical multiplex.

14.5.4 Maintenance problems

The long-term stability (aging) of the parameters of the fiber, of splices, and of optical transducers is still poorly known. This is why the operation of an optical fiber transmission system requires permanent monitoring of the error rate, considered to be representative of the transmission quality. This measurement can be made in service by the statistical observation of violations of a coding law with redundancy (5B6B, CMI, *et cetera*).

If intervention is necessary on an installed fiber, leading to supplementary splices, then the fact that the attenuation due to a splice (0.5 to 1 dB) is equivalent to that of several hundreds, or even thousands of meters of fiber must be taken into account. A sufficient margin for such operations must be designed into the system from the start when we plan the link.

Chapter 15

Networks and Switching

15.1 TYPES AND STRUCTURE OF TELECOMMUNICATIONS NETWORKS

15.1.1 Concept of a network: definitions

The concept of a *network* appears when *several* sources or *several* (permanent or potential) receivers share the same service. The network is composed of a group of similar transmission channels, as well as the means necessary to link them and assign them to users.

Depending on the type of service, (sect. 1.1.2), the assignment is *fixed* (static) or *variable* (dynamic). In the latter case, a *switching* operation is necessary.

This follows the classification of table 15.1.

Table 15.1 Types of networks.

Number of sources	1	$n > 1$	$n \geqslant 1$
Number of receivers	$n > 1$	1	$n \geqslant 1$
Type of assignment	fixed	fixed	variable
Type of network	DIFFUSION	COLLECTION	SWITCHED

⊙ = source   ⊗ = receiver

In the diffusion and collection network, transmission is, by nature, *unidirectional*. In a switched network, it can be *bidirectional*, if each user plays both roles of source and receiver.

15.1.2 Shared switched network

For economic reasons, the switched network is most often implemented in *shared* form (fig. 1.2), i.e., commonly used by numerous customers. The network itself includes:
- groups of standardized transmission channels, which are equivalent from the viewpoint of traffic routing (same origin, same destination, similar transmission quality), called *trunk groups*;
- switching centers, called *exchanges* or *central offices*, capable of assigning one channel to each communication according to the temporary needs of the users, as a function of the final goal, but independently of the origin of the request;
- *control and management units* the services of which are requested by any user and assigned to each from case to case.

In this way, sharing consists of placing a *limited number of units*, which are functionally specialized, (channels, switching nodes, *et cetera*), but generally accessible, at the disposal of a *large number of users*, who are not predetermined. Based on statistical considerations of the number and duration of communications, i.e., the traffic offered A according to (2.59), this principle implies a *risk of congestion* when the network is not quantitatively capable of responding to the users' requests (sect. 2.5).

15.1.3 Configurations of switched networks

The interconnection of several points by a switched network, in principle, can be realized by different topologies, of which the principal ones are (fig. 15.2):
- the *mesh network* in which the points are linked two-by-two by a direct line;
- the *star network* in which a switching unit, called the *transit exchange*, interconnects the two trunk groups corresponding to each of the points;
- the *linear network*, or *bus*, in which a single trunk group is directly accessible at each point;
- the *ring network*, which is a closed linear network;

The criteria for choosing among these structures and their numerous variants are essentially economic and introduce the following aspects:
- *transmission*: good use of the available bandwidth (analog) or bit rate (digital), path length, and adaptation of quality to distance;
- *switching*: number and types of exchanges, complexity of the routing

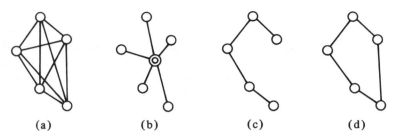

Fig. 15.2 Network configurations: (a) mesh; (b) star; (c) linear; (d) ring.

(signaling), problems and conflicts of access, and dimensioning of trunk groups;
- *reliability*: vulnerability of the entire network and its component lines, monitoring and maintenance, and possibility of redundancy.

15.1.4 Mesh network

The number N of trunk groups necessary to completely interconnect n points is

$$N = \tfrac{1}{2} n (n - 1) \tag{15.1}$$

Thus, it increases proportionally to the square of the number n of points, which quickly makes the meshed network prohibitive in terms of cost. On the other hand, the direct links are poorly utilized if the traffic between their ends is low. However, the transmission quality of each link can be optimized as a function of length.

Consequently, this network is rarely completely meshed. Direct links only exist, if they are justified by an important cross traffic. The other links are established by a detour across one or several auxiliary nodes, functioning as a transit switch, as in a star network (fig. 15.3).

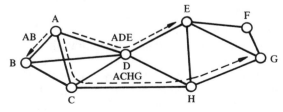

Fig. 15.3 Partially meshed network.

The reliability of the mesh network is better the more complete the interconnection. The breakdown of one link affects only the direct traffic between the two centers it joins. If the exchanges and signaling permit it (which is not always the case), the traffic thus interrupted can even be routed through another center of the network, which reduces the switching quality (sect. 2.5), but allows us to avoid the interruption of traffic.

15.1.5 Star network

In the star configuration, the number of trunk groups is minimal. However, each link can become a part of a transmission chain that extends to any distance. Therefore, it must strictly respect quality requirements in proportion to its own length.

The star structure is typical of the following networks:

- *local networks* for connecting customers with low traffic volume to a *terminal exchange* (or local exchange), which gives them access to a shared network;
- *rural networks* that link several terminal exchanges to a *nodal exchange*, which interconnects them and provides access to the long distance network.

The reliability of the star network is threatened by the vulnerability of the links (since there is no possible detour) and of the transit exchange.

15.1.6 Linear network

In its most general form, the linear network represents the common placement of a certain total bandwidth B_{tot} (analog network), or of a certain bit rate $\dot{D}_{tot}$ (digital network), between several users for their communications, each of whom momentarily uses only a part of B_{tot} or $\dot{D}_{tot}$.

The analog linear network formed by frequency-division multiplexing of communications is difficult to implement due to the problems which variable assignment of different frequency bands from case to case would pose. Nevertheless, it has an application in certain radiocommunications networks.

However, in the digital case, the flexibility of time-division multiplexing lends itself well to variable assignment of any part of the bit rate $\dot{D}_{tot}$ to users, according to their momentary needs. The partial rates thus assigned can even be different, provided that their sum remains lower than, or equal to, $\dot{D}_{tot}$.

Ultimately, if the users produce their information in a discontinuous manner, it is possible to interlace some between others at the rate of their appearance, and thus to compose a *dynamic multiplex* with variable structure in time. This is notably the case with packet switching (sect. 15.5.4).

The assignment of transmission capacity to a user and the access of the user to this capacity present coordination problems, which must be dealt with in order to avoid conflicts. Strict discipline is necessary; the access, exchange, and procedures are regulated by *protocols*, which are agreed upon between users.

15.1.7 Ring network

The closed ring network has the same properties as the linear network, with the additional advantage of requiring only a single support for the two transmission directions, the forward and return paths between two points representing a complete turn in the ring.

In principle, the ring configuration lends itself to an entirely *decentralized* solution to switching operations and even control operations. The bit rate $\hat{D}_{tot}$ can be completely shared and placed at the disposal of any pair of users for transmission or information exchange from case to case according to their needs. The blocks of information sent on the ring are supported by the *address* of the receiver, and each user extracts the blocks addressed to him from the ring. However, regulation of access to the ring, surveillance of its functioning, and especially its reliability, present delicate problems [54].

Due to the extreme flexibility in assigning the ring's capacity to users, this structure can be envisioned for *domestic or local information networks*, in which the bit rates and the traffic are highly varied and often irregular. Decentralized switching leads to proportionality between cost and number of users, even if this number is low, and allows easy extension of the network to new terminals.

15.2 TELEPHONE NETWORK

15.2.1 General structure

The classical telephone network is typically a *shared network* to which the users, called *customers* or *subscribers*, have access by individual lines, which constitute the *local network*.

Its structure is hierarchical at several stages. Conceived as a star network in the lower stages, it becomes more or less completely meshed toward the summit of the hierarchy. The detailed structure varies from one country to the next, according to its geographic dimensions and its telephone density. Figure 15.4 gives a typical example.

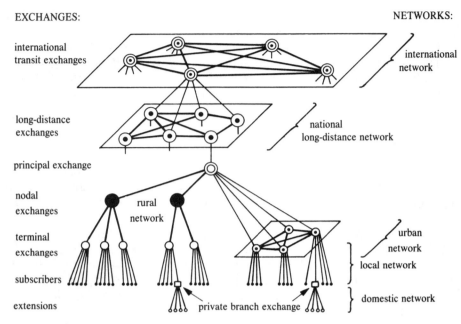

Fig. 15.4 Example of the structure of a telephone network.

15.2.2 Distribution of investments

Four principal categories of network equipment can be distinguished. We can compare the proportion of the total investment that they represent (Fig. 15.5):

- *terminals*: customer telephones and installations, including the private branch exchanges;
- *customer lines*: balanced pairs assigned individually and exclusively to each customer;
- *exchanges*: terminal and transit switching equipment to all levels of the worldwide network, except the private branch exchanges;
- *transmission equipment*: lines, cables, carrier systems, digital systems, microwave links, satellites, *et cetera*.

The proportion of approximately 45% taken by individual equipment (terminals and customer lines) is striking. It has the tendency to increase because of the increasing optimization of the shared network. An important portion of this investment, not included in the proportions given in figure 15.5, is devoted to the acquisition of *land* and the construction of *buildings*. It also includes the

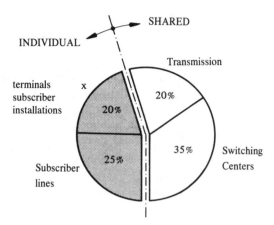

Fig. 15.5 Relative investment in an average national network.

costly air conditioning systems required by modern electronic and computer systems.

15.2.3 Private branch exchanges

When a subscriber (private, office, enterprise) wishes to have several interconnectable telephones available (internal traffic) or connectable with the public network (external traffic), a *private automatic branch exchange PABX* takes charge of the switching operations at the subscriber's location. The size of such an exchange can range from two telephones to several thousand extensions.

The specifics of internal and external traffic, the presence of one or several manual operators, and an ever-increasing range of *special services* (inquiry call, call transfer, abbreviated dialing, restriction of access to the public network, telematics, *et cetera*) particularly complicate the control of the PABXs, and make it one of the privileged domains of the application of real time data processing to telecommunications (stored program control SPC).

A new generation of private branch exchanges, entirely digital, unites in-house data transmission and voice communication with digital modulation in the same network (integration of services, sect. 15.8.3).

15.2.4 Local network

Composed almost exclusively of balanced pair lines, open wires, or cables, its geographic coverage greatly depends on the telephone density D of users (sect. 2.5.4), on the demographic density of the region, and on the size of the local exchange. Average lengths of 1.5 to 2 km are usual, while longer subscriber lines can exceed 10 km.

The local network generally does not contain amplifiers. It is composed of "two-wire" circuits, which are loaded if necessary. The maximum length of the lines as a function of the gauge of the conductors is limited by:

- the *resistance* with direct current, in relation to the voltage of the battery (generally 48 V), which supplies the subscriber lines from the central exchange with the minimum current necessary for good functioning of the microphone;
- the *attenuation* at 800 Hz (overall loss according to sect. 2.4.1), according to the requirements of the national transmission plan (sect. 15.4).

While indispensable for ensuring the connection of dispersed subscribers, the local network is, however, very poorly utilized, given the very low traffic of an average subscriber (sect. 2.5.3). When economic considerations justify it, the rate of utilization of local lines can be improved, at the cost of a decrease of switching quality (loss, delay), depending on the case, by the following means:

- *party lines*: connection of two subscribers on the same line with *exclusive assignment* of the line to one or the other of the subscribers, by way of switching, for each communication (fig. 15.6(a));
- *multiplexing* (pair gain system) of two or several channels on the same physical support (pair or quad), assigned in a *fixed* manner to each of the subscribers (fig. 15.6 (b)). This solution does not at all degrade the switching quality;
- *concentration of traffic* by a concentrator located near the subscribers, which proceeds to a *variable assignment* of a reduced number z of channels to the subscribers, at the moment when they need them (fig. 15.6(c)). This concentrator is, in fact, a switching stage of the local exchange, shifted in the direction of the subscribers. It can be combined with a multiplexer, which regroups the z channels on the same physical support.

15.2.5 National shared network

This network consists of the following means of transmission:

- *balanced-pair lines* (sometimes open wires, more often cables) used at short and medium distances, either in baseband (one communication per pair, with terminal loading or amplification, if necessary), in frequency-division multiplexing (low capacity carrier systems), or with time-division multiplexing (digital systems at 2 Mbit/s or 8 Mbit/s);

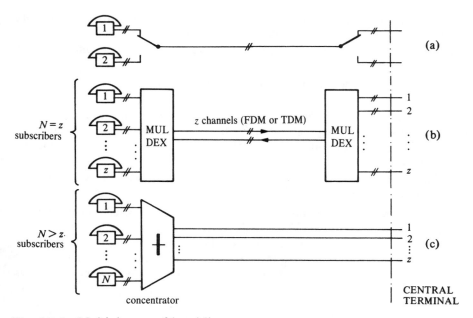

Fig. 15.6 Multiple use of local lines.

- *coaxial cables* for carrier systems at medium or long distances, increasingly replaced by the installation of digital systems at 8 Mbit/s, 34 Mbit/s, or 140 Mbit/s;
- *microwave links* when geographical conditions permit (mountainous terrain, towers) or require it (jungle, desert, bodies of water, *et cetera*);
- optical fibers (planned or being introduced).

The entire long-distance network and part of the rural network are "four-wire" because of the requirements of multiplexing and amplification.

15.2.6 International network

The international network uses the same means as national long-distance networks. Routing is variable, with priority for direct links, if such exist, otherwise through a detour across a transit exchange of a higher hierarchical level.

For intercontinental communications, three means coexist:
- *short-wave* links (10 m wavelength) for radiotelegraphy and, occasionally, for radiotelephony (notably for ships at sea). Their precarious and unpredictable quality limits their use to very specific needs;

- *transoceanic coaxial cables*, in limited number, but of increasing capacity (table 10.9), and soon optical cables;
- geostationary *satellite links* with rapidly increasing numbers and capacity.

15.2.7 Insertion of satellite links into the network

Satellite capacity of z telephone channels (limited by the available bandwidth) allows us to design $z/2$ forward and return circuits with frequency-division (FDMA) or time-division (TDMA), "pseudo four-wire," multiple access (sect. 13.5).

The assignment of these circuits to communications can be made according to two different principles:

- *fixed assignment*, or preassignment, which consists of assigning a specified group of channels (links) exclusively to a pair of earth stations. This solution is interesting for constructing large links between countries with high mutual traffic. Despite the fact that the satellite appears to be at the center of a star, this structure is actually equivalent to a *mesh network* (fig. 15.7(a)). The satellite uniformly processes (by amplification and frequency transposition) all the channels without distinction, independently of their assignment;

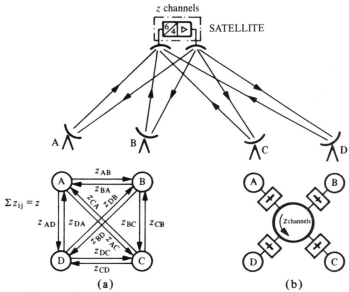

Fig. 15.7 Modes of channel assignment for a satellite and equivalent network topology. (a) fixed assignment; (b) demand assignment.

- *demand assignment*, whereby z satellite channels (all or part of its capacity) are shared among several earth stations and assigned from case to case, in pairs, to any communication. From the viewpoint of traffic flow, we thus have the equivalent of a *ring network*, which links all the stations with z channels (fig. 15.7(b)). This procedure implies a *switching* operation in the earth stations (assignment of one channel in each direction) and *signaling* between them to regulate the access and exchange procedures. However, the satellite ignores everything of the channel assignments, and does not intervene in this operation. Demand assignment offers great flexibility for low volume traffic flow between small countries. On the other hand, differences in local time allow us to make the satellite traffic more uniform by successively covering the "busy hour" of several countries located at different longitudes.

For future generations of satellites, it is foreseen that direct satellite-to-satellite links will become available, and that switching will be carried out aboard the satellite, by means of an embarked transit exchange, becoming thus the center of a *star network*.

15.2.8 Economics of choosing a transmission system

The parameters that determine the choice of a transmission system are so numerous and diverse that it is not possible to indicate with certainty which system is the most suitable for each type of application. The principal criteria of choice are economic in nature, particularly the *unit cost of a transmission channel*. It depends on, among other things, the *capacity* (number of channels z), on the medium used (cables with several balanced or coaxial pairs, microwave links) and on the *transmission distance l*.

Figure 15.8 illustrates qualitatively the influence of the number of channels z on the annual cost of transmission equipment (cables, repeaters, antennas) for one channel and a unit distance. This figure shows that the cost per kilometer of a channel decreases when the capacity of the system increases, and it is generally better to obtain the total number of channels required, with a few systems of large capacity, rather than with a larger number of systems comprising each fewer channels.

If we assume, from relation (2.80), that the total cost of the system is an exponential function of the number of channels, its exponent m, which proceeds very roughly from figure 15.8, is approximately equal to 0.6.

For purposes of illustration and comparison, figure 15.9 shows how the costs of a channel (terminal equipment and transmission equipment) varies as a function of distance in three possible types of use of existing *balanced pairs* in the rural network. Because its initial cost ($l = 0$, only terminal equipment)

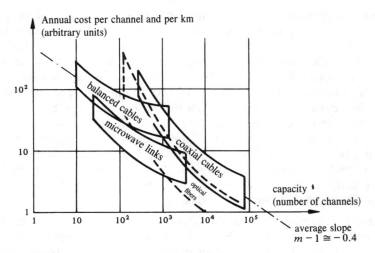

Fig. 15.8 Effect of size on the unit cost of a channel.

is lower than low capacity carrier systems, the digital system proves to be of economic interest for medium distances.

Figure 15.10 compares the influence of distance on the cost of a multiplexed channel in the case of a coaxial cable (carrier systems of different capacities), a microwave link (with hops of approximately 50 km on the average), and a satellite link (the cost is very high, but independent of distance).

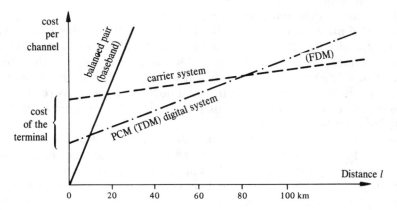

Fig. 15.9 Comparative cost of a channel on balanced pairs, with and without multiplexing.

NETWORKS AND SWITCHING

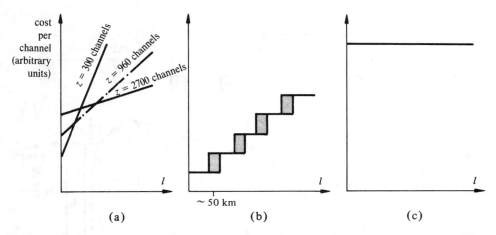

Fig. 15.10 Influence of distance on the unit cost of a channel: (a) coaxial cable; (b) microwave link; (c) satellite link.

It is evident that in order to become decision elements in the choice of a system these diagrams must be specifically stated with quantitative data, which greatly depend on local conditions, technology, and the market.

15.3 ECHOES AND STABILITY

15.3.1 Origin of echoes

Subscriber lines are, for clear economic reasons, constructed as "two-wire", whereas the optimal use of the shared network by multiplexing, as well as the amplification necessary to compensate for the attenuation of long-distance transmissions, imposes a "four-wire" solution, at least in part of this network (sect. 4.6, specifically fig. 4.31).

If we assume that the sum of the repeater gains distributed in the "four-wire" network exceeds by a value G_4 (residual gain) the sum of the attenuations of the line sections which separate them, including the crossing attenuation A_t of each hybrid, then the level diagram of a complete "two-wire" to "two-wire" link in one transmission direction from A to B has the form given in figure 5.11.

The impossibility of perfectly balancing the hybrids makes a reflected signal appear in B (fig. 4.33). This signal returns to the "four-wire" network, propagates there in the reverse direction, undergoes the same residual gain G_4 and reappears on the "two-wire" side after having recrossed hybrid A. This para-

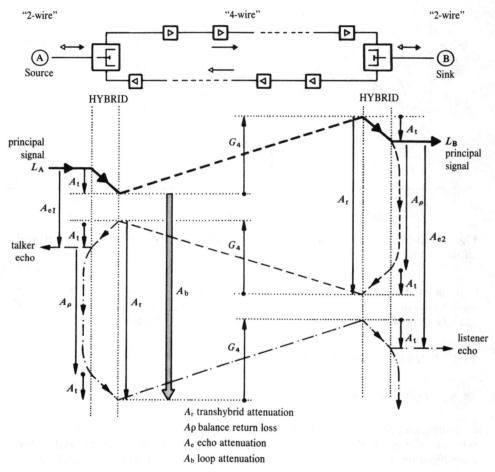

Fig. 15.11 Level diagram of a "2-wire" to "4-wire" to "2-wire" link.

sitic signal, resulting from a reflection at the far end B of the "four-wire" network, is called the *talker echo*. It is itself at the origin of a reflected signal which leaves from A toward B and reappears at B in the form of a *listener echo*, and so on.

15.3.2 Importance of echoes

The effect of an echo is more undesirable the lower its attenuation with respect to the principal signal and the longer its delay.

The attenuation A_{e1} of the talker echo comes from figure 15.11

$$A_{e1} = A_t - G_4 + A_r - G_4 + A_t =$$
$$= A_r + 2A_t - 2G_4 \qquad \text{dB} \qquad (15.2)$$

where A_r represents the transhybrid attenuation, defined according to (4.23).

It is delayed, with respect to the signal emitted by the source, by double the propagation time across the "four-wire" network.

The listener echo undergoes the same delay with respect to the principal signal received. However, its attenuation A_{e2} is generally higher, which makes it less perceptible.

$$A_{e2} = A_p + A_t - G_4 + A_r - G_4 + A_t$$
$$= 2A_r - 2G_4$$
$$= A_{e1} + A_p \qquad \text{dB} \qquad (15.3)$$

In telephony, the echoes are particularly noticeable and annoying when their delay exceeds approximately 50 ms (or approximately twice 5000 km of coaxial cable). This case occurs in very long distance links, particularly by satellite. *Echo suppressors* are therefore necessary. In their simplest form, they interrupt or strongly attenuate one of the transmission directions when a level detector establishes the presence of a telephone signal in the other direction (fig. 15.12). They thus create an automatic half-duplex operating mode and therefore do not allow full duplex transmission, which can be troublesome in data transmission.

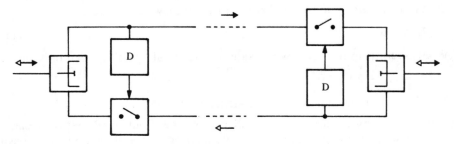

Fig. 15.12 Principle of echo suppressors: D, level detector.

The return of the reflected signal by hybrid B (fig. 15.11) and the second reflection which it undergoes in hybrid A constitute a *reaction loop* which can become unstable:

- if the signal which is reinjected after two reflections is of a higher level than that of the principal signal;
- and if the phase is in opposition to that of the principal signal.

Steady oscillations appear in the "four-wire" link at a frequency such that the phase shift of the reaction loop is exactly π. Their amplitude is limited only by the saturation of the repeaters. If such an instability takes place in any of the channels of a carrier system, the common repeater becomes saturated and all the other channels are affected by intermodulation.

This phenomenon, called *singing*, is extremely dangerous and must be avoided at all costs.

15.3.4 Stability condition

To avoid it, the attenuation A_b in the reaction loop must absolutely remain *positive*. By referring to figure 15.11, the stability condition is thus

$$A_b = 2A_r - 2G_4 > 0 \text{ dB} \tag{15.4}$$

In reality, an extreme case occurs on the "2-wire" side when the link is not yet established in the following exchange. The "two-wire" output is thus practically open, and we thus have $Z_2 = \infty$ and, according to (4.21), a reflection coefficient $\rho = 1$. The transhybrid attenuation thus takes a minimal value from (4.23)

$$A_{rmin} = 2A_t = 7 \text{ dB} \tag{15.5}$$

In order to respect condition (15.4) even in this transitory situation, it is necessary to limit the residual gain in the "four-wire" network:

$$G_4 < 7 \text{ dB} \tag{15.6}$$

When seen from the "two-wire" side, the residual "gain" G_2 can only be

$$G_2 = G_4 - 2A_t < 0 \text{ dB} \tag{15.7}$$

This is thus an attenuation! The conclusion is deceptive:

- to guarantee stability, the total gain of the "four-wire" network can, at most, just compensate for the attenuation of the lines and the crossing attenuation of the two hybrids!

It is, therefore, unrealistic to hope that the benefits from the amplification in the "four-wire" network will compensate for the attenuation of the "two-wire" network.

The *singing point* defined by condition (15.4) must be avoided at all costs, taking into account all the tolerances and variations of attenuation possible in the network. This is why the limit imposed by (15.6) on the residual gain is again lowered by a safety margin of several dB.

15.4 TRANSMISSION PLAN

15.4.1 Objectives of the national transmission plan

One of the principal quality criteria of a telephone link is the intensity of sound perceived by the receiver. Any link within the world network must respect condition (2.30) imposed on the reference equivalent. Let us recall that the limit of 36 dB includes the electroacoustic efficiency of the transducers (sect. 2.4.4). Each country is free to distribute the national part of this reference equivalent between the transducers (microphone and receiver) and the attenuation (overall loss, according to sect. 2.4.1) of its network. This distribution constitutes the *national transmission plan*.

The values of overall loss A_{eq} which it contains are defined by convention at 800 Hz. The transmission plan allows:

- dimensioning of the non-amplified "two-wire" part of the network, particularly:
- determination of the gauge (diameter of the conductors) necessary for subscriber lines and "two-wire" links between exchanges;
- determining the need for loading.

15.4.2 Factors influencing the transmission plan

The establishment of a national transmission plan is an important and delicate operation when we are striving for a compromise between cost and quality. It must particularly take into account

- the type of telephone set used (reference equivalents at emission and reception);
- the demographic and telephone density of the country;
- the structure of the network;
- the routing mode (fixed or variable) of communications across this structure;
- the stability condition of the "four-wire" network (sect. 15.3.4);
- the crossing attenuation through the exchanges;
- the transition period between the old plan and the new plan (particularly, coexistence of terminal units of different types);
- the progressive introduction of digital transmission and switching systems.

15.4.3 Example of a national transmission plan

Figure 15.13 gives a very simplified example of a national transmission plan based on a star structure with two "two-wire" stages and on a subscriber telephone set with a reference equivalent of 8.7 dB at emission and 1.7 dB at reception. The corresponding level diagram is presented in figure 5.14.
This example brings up the following remarks:

- the residual gain in the "four-wire" network has been fixed in this case at $G_4 = 0.5$ dB, with a margin of 6.5 dB with respect to the stability condition (15.6);
- between the extreme cases of subscribers who are very close to the inputs into the "four-wire" network and the tolerable limit of 7.8 dB for each "two-wire" section, the difference of overall loss from end to end of the national network is 13.6 dB;
- this example is typical of a small country and high telephone density.

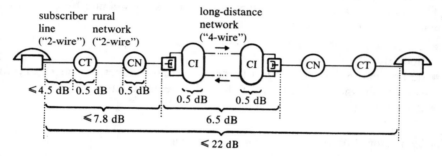

Fig. 15.13 National transmission plan (simplified example). Equivalent distribution at 800 Hz.

15.5 SWITCHING PRINCIPLES

15.5.1 General switching functions

In a shared network, switching has two fundamental objectives:

- *to concentrate the traffic* coming from sources with low activity toward common transmission means, by assigning part of their capacity to each active source, from case to case;
- *to route* information from one source toward the corresponding sink, according to a fixed or variable itinerary across the network, from one exchange to the next.

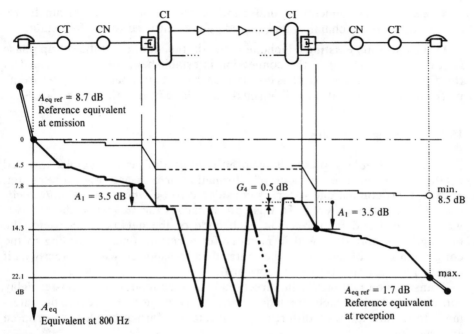

Fig. 15.14 Level diagram of a national network (example according to the transmission plan of fig. 15.13).

15.5.2 Circuit switching

The principle of *circuit switching*, typical of telephony, consists of establishing by interconnection a link of several channels end to end *before* the communication proper. The link is held as long as the users desire it, independently of their effective activity rate during the communication. It is generally bidirectional, with the same channel for the two directions ("two-wire" switching) or two distinct channels forming a *circuit* ("four-wire" switching). Each communication passes through three successive phases:

- *setting up* the link: an active switching phase to detect the request for service, receive and interpret the identity of the terminal requested (dialing), search for and occupy a route leading to it, inform the required terminal of the arrival of a call (ringing), recognize its response and interconnect the two terminals;
- *holding* the link during the entire duration of the communication with supervision of the two terminals, and charging, generally according to the duration and distance of calls;

- *releasing* the connections under the control of one of the terminals, returning all the channels and shared equipment used to the free state.

In the currently existing telephone network, the duration of the set-up phase is several seconds. Once the connection is established, the circuit placed at the disposal of the terminals has constant characteristics, particularly regarding its bandwidth or its bit rate. Transmission is done in real time (sect. 4.1.8).

15.5.3 Message switching

Message switching proceeds from an entirely different principle, inspired by the mail service: each block of information or message (e.g., a telegram, page of text, commercial letter) is considered to be whole and is ***individually*** routed across the network. To this end, it carries the ***address*** of the receiver which is read and interpreted at each node of the network. En route, the messages are sorted, stored in memory for a certain time, according to the congestion state of the rest of the route, then transported whole, exactly as if it were physical mail (store-and-forward).

Transmission is typically unidirectional (simplex mode). The transfer delay can be very long, which precludes a true conversation. However, the intermediate storage allows a difference in bit rate and format between the source and the sink.

15.5.4 Packet switching

In principle, *packet switching* is related to message switching with, however, the following differences:
- packets are parts of messages, with normalized formats and including control bits designed to protect the packet from transmission errors;
- segmentation of the message in packets is done by the network and not by the user;
- storage time of the packets in the waiting queues at the network nodes is a few milliseconds. It gives users the impression of a bidirectional transmission in real time.

As seen from the terminal, packet switching more closely resembles circuit switching because of its very short set-up time. A "virtual circuit" is established, but it has no physical form. It allows the automatic addressing of the packets relative to a given communication. Liberation consists of erasing this software link between the identity of the caller and the address of the called receiver.

As seen from the network, packet switching is in fact a mode of operation rather than true switching. It allows optimal use of the channel bit rate, by juxtaposing packets from any origin or destination in time, without discontinuity (dynamic multiplexing). Unlike circuit switching, the line capacity is only occupied as long as the terminals are effectively active.

Although it is well-adapted to telematics, especially in conversational mode (in which the terminal activity is low), packet switching would pose serious problems if it were applied to digital telephony, because the packets would be thus constituted from coded samples (PCM octets). In effect, the packets reach the receiver with variable and unpredictable delays. "Packetizing" of speech is, however, the subject of study and experimentation.

15.5.5 Comparison of the three types of switching

Table 15.15 summarizes and compares these types of switching. It must be noted that message transfering is also possible in a circuit-switched or packet switching network, provided that there is sufficient memory. However, the complete "transparency" of the channel, in time and regarding the code, is an

Table 15.15 Comparison of the three types of switching.

	Circuit switching	Packet switching	Message switching
Delay	short and constant (except satellites) 0 to ms	variable and medium 10 ms to s	long and variable s to min
Degree of line use	low to medium	high	medium
Protection against errors	by the user	by the network	depending on the case
Bidirectional?	yes	yes	no
Possible change of format? of bit rate?	no no	yes yes	yes yes
Memory requirements en route	zero	low (rapid)	high (slow)
Application	telephony classical telex telematics	telematics	modern telex

exclusive property of circuit switching. Packet switching makes use of a judicious compromise between transparency and degree of line use.

15.5.6 Signaling

While the transmission function respects the integrity of the information entrusted to it by users as scrupulously as possible, the switching function requires, consumes, and produces information. This auxiliary information, called *signaling information*, is transmitted

- from the user toward the network (request for service, dialing or selection, end of communication, *et cetera*);
- from the network toward the user (calling signal or ringing, tones, caller's response, charging, *et cetera*);
- from one exchange to another within the network (seizing a line, dialing, communication management, maintenance, *et cetera*).

The signaling information is digital in nature (dialed digits, charging units, commands, and acknowledge signals).

In telematics, aside from the above three categories of signaling, a complex exchange of auxiliary information takes place from end to end across the network with the goal of enabling monitoring (transmission error detection) and managing the proper transmission of data. This signaling in the broad sense includes all the *protocols*, i.e., all the rules of dialogue and access to services.

In *circuit switching*, each channel usually has a certain transmission capacity for its own signaling, in the form of

- an appropriate frequency band in analog carrier systems (sect. 10.2.5);
- an appropriate bit rate in digital systems (sect. 9.2.3).

In the case of modern processor-controlled exchanges, the signaling relative to all the channels of the same trunk group is advantageously grouped on a *common channel* (common channel signaling), which carries all the signaling messages in the form of direct data transmission, with dynamic multiplexing, from one processor to the other. Each message has a *label* indicating to which channel it pertains.

In *packet switching*, signaling makes use, on the one hand, of a *header* which accompanies each packet and contains the address of its destination as well as control information, and on the other hand, of special packets, used especially to establish a virtual circuit between the two terminals.

15.6 DATA NETWORKS

15.6.1 Private networks

Aside from the particular cases in which private users make use of their own lines (domestic or local area data networks), the private data networks are composed of

- *transmission* means borrowed from the *public* telecommunication network (leased circuits, sect. 11.4.1);
- *switching* centers (circuit, message, or packet switching) which belong to them and which are adapted to the particular needs of data switching.

These networks thus represent an alternative to the use of switched circuits across the public network (telephone or telex). The leasing costs of the circuits are not justified unless there is substantial traffic. The occupation rate of the leased circuits can be improved by a fixed multiplexing or, even better, dynamic multiplexing (traffic concentration).

15.6.2 Public network

By nature, private networks are limited to a closed group of users. Furthermore, they are generally incompatible with each other, due to differing protocols and software.

A public data network, analogous to the existing telephone network, attempts

- to cover on a national and international basis the telematics needs of users of different categories (private, professional, the general public);
- to offer effective, flexible, and economical shared transmission, switching, and information storage services;
- to guarantee a certain universality in the rates, formats and modes of operation of the terminals connected to the network;
- to permit a qualitative and quantitative evolution of needs.

A major effort toward standardization has been made by the ISO (International Standard Organization) and by the CCITT (Recommendations of series X) with the aim of preventing incompatibility and unfavorable options.

Message switching does not enter into consideration for a public network, because from the start it excludes the conversational mode.

Packet switching is particularly well adapted for this application due to the large variety of existing bit rates and traffic structures available. Figure 15.16 illustrates the different ways of connecting terminals to such a network. Packet

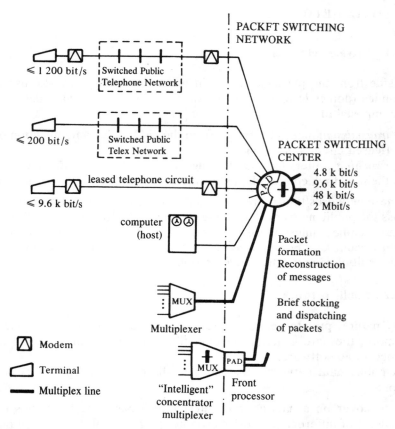

Fig. 15.16 Access to a packet switching public network.

assembly and disassembly PAD can take place in switching centers or in *front end processors*, which are more peripheral, and which manage the dynamic multiplexing of information in a concentrator-multiplexer.

A public network based on *circuit switching* is also possible, but with less flexibility and a less favorable line efficiency. It has, however, the advantage of being more directly compatible with the digital telephone network (services integration, sect. 15.8).

15.6.3 Videotex service

Videotex is typically a telematics service designed for the general public. Its principal aim is, originally, to permit long-distance consultation of data

banks containing information of general interest. It is characterized by the following points:
- use of the public switched telephone network as a vehicle of digital information to be transmitted bilaterally between the subscriber and the data bank;
- conversational mode (interactive);
- questioning by means of a special keyboard, transmission at 75 bit/s;
- response displayed on a domestic television screen (alphanumeric characters and graphic elements, color); transmission at 1200 bit/s, with modem, in the telephone channel;
- 128 alphanumeric characters in the form of matrices of 7 x 10 dots, 24 lines of 40 characters per "page" (full screen). Possibility of a graphic complement by means of other matrices (alphamosaic system) or by geometrical elements (alphageometric system).

Modification of the data bank contents by the information providers, as well as access across the network to a remote data bank, can be done advantageously by packet switching, once the communication has been established between the terminal and the videotex service (virtual circuit).

Aside from the consultation of data banks, numerous other possible applications of videotex can be envisioned: interpersonal messages (electronic mail), financial transactions, reservations and orders, interconnection of personal computers, *et cetera*.

Videotex is perhaps the embryo of a **new form of communication**, which is more oriented toward text, pictures, and data than toward sound.

15.7 INTEGRATED DIGITAL NETWORK (IDN)

15.7.1 Advantages of the digital form for switching

The two principal aspects of digital systems, (sect. 9.1.1) namely
- *binary* representation of analog information, for example by PCM modulation;
- *time division* multiplex,

whose advantages for transmission have already been emphasized (sect. 9.12), are also of interest for switching:
- the *binary* form allows the easy interconnection of any input and output of an exchange, by a simple operation of logical conjunction (AND function, vol. V., sect. 1.3.3) between the input and the connection command. The linearity problems are eliminated and the crosstalk problems are considerably reduced;

- the *time-division multiplex* structure allows the implementation of several different links across the same connecting point by activating it at different (periodic) moments, which leads to a considerable reduction of topological complexity and of physical size of the switching equipment;
- furthermore, the *binary* form easily lends itself to momentary storage which gives the possibility of changing the time slot between the input and the output of the switch.

Finally, an *economic motivation*, powerful and decisive, in the development of microelectronics and microcomputers, has led to the current introduction of large-scale and high-speed digital transmission techniques in telecommunication sustems. Since digital transmission is economically justified (fig. 15.9), the practical realization of high-speed *digital* transmission is *advantageous* in comparison with cable, but is dependent upon analog techniques, requiring analog-digital conversion at network input (fig. 15.17(c)).

15.7.2 Integration of switching and transmission

The convergence of technical and economic arguments in favor of digital transmission and switching, considered independently, leads to their envisioned unification, not on the level of their respective functions (which will always remain distinct), but at the level of a common digital form, in an *integrated digital network IDN*. Figure 15.17 shows the path which leads to this idea.

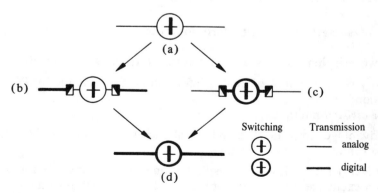

Fig. 15.17 Integration of transmission and switching: (a) classical analog network; (b) digital transmission and analog switching; c) digital switching and analog transmission; (d) integrated digital switching and transmission.

The extra advantages of this integration are:

- *economy* of analog-digital converters (PCM "codecs"). In the ideal case of an entirely digital network, "codecs" are only necessary at the periphery of the network;
- *improvement of the transmission quality* of analog information by avoiding the accumulation of quantizing noise in successive "codecs";
- possibility of *digital transmission from end to end*, with code and bit rate transparency (circuit switching).

15.7.3 Evolution toward an integrated digital network

Even if the final objective of perfect integration into a completely digital network appears attractive, it would be entirely unrealistic to forget that its progressive installation will take decades and that, during this long transitional phase, the two forms, analog and digital, must coexist harmoniously. Depending on the case, the digital conversion of the network can follow two ways:

- by *substitution*, for the replacement of old, defective, or saturated transmission or switching equipment or for network extensions (new lines, new exchanges). This approach can be motivated by economic opportunism, without the deliberate intention of reaching an integrated digital network (fig. 15.18 (a));
- by *overlay*, i.e., by constructing a digital network designed in integrated form from the beginning, in parallel with the existing analog network. The economic evidence for this approach is uncertain. However, it is justified by the argument of early service integration (sect. 15.8), in a region and for certain subscribers (fig. 15.18 (b)).

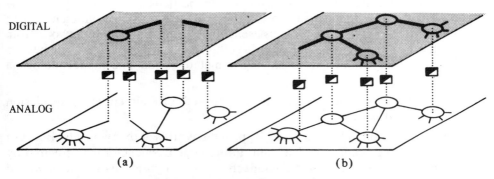

Fig. 15.18 Progressive introduction of a digital network: (a) substitution; (b) overlay.

In the two cases, the problem of respecting the national transmission plan occurs with even more severity because digital systems are, by nature, "four-wire" systems, whereas the analog links they are replacing or supplementing can be "two-wire".

15.7.4 Penetration of digital technology into the local network

The ultimate stage of the digital conversion of the telecommunications network will consist of placing the analog-digital conversion as close as possible to the subscriber, i.e., in the subscriber's telephone set. Only in this way will the subscriber have direct access to the digital channel, with a bit rate which is imposed by digital speech transmission (e.g., 64 kbit/s in PCM), but which could also serve for other uses (sect. 15.8).

A number of technical and economic problems are involved with this objective, in particular, we have:
- if we use the present local star network, with balanced copper lines, a bidirectional digital transmission on only two wires must be implemented;
- if we had a newly designed local network, what structure (star, bus, ring) and what medium (balanced line, coaxial pair, optical fiber) should be chosen?
- the local signaling (currently dc current or in-band) must be completely revised and integrated into the bit rate on the line;
- the terminal unit must be synchronized and framed on the binary flux it receives;
- the reliability of the local digital network must be equivalent to the present situation, despite the large number of active electronic circuits which the digital terminal unit must contain. In particular, functioning must be assured even in the case of breakdowns in the power distribution network;
- the local network is the most sensitive part of the public network from the economic viewpoint. On the other hand, it also lends itself best to mass production. This results in a stimulating economic and technological challenge.

Three types of solutions have been proposed for bidirectional transmission on a balanced pair:
- transmission of the *"two-wire"* type with hybrids at the two ends. Because the balance cannot be sufficient in the large frequency range covered by pulses to be transmitted, the reflections in the hybrids disturb the received signals in the other direction in an unacceptable manner. It is indispensable to proceed to *echo cancellation* which consists of anticipating the

reflected signal from the emitted signal and of subtracting it from the received signal before extracting information from it;
- *"time-division pseudo-four-wire"* transmission (burst mode) according to the principle presented in section 4.6.5. The arrangement is relatively simple. The instantaneous rate must be more than doubled to take the propagation time on the line into account, which, in turn, limits the range of the network;
- *"frequency division pseudo-four-wire"* transmission (sect. 4.64), i.e., the allocation of two distinct frequency bands for the two directions, with appropriate shaping of the transmitted pulses to limit their spectrum.

15.8 INTEGRATED SERVICES DIGITAL NETWORK (ISDN)

15.8.1 Service integration

Service integration of services is understood to mean the use of one part or all of the telecommuniations nework to route *in common* the different informations relating to services of a different nature, such as, for example

- telephony
- telematics, in the broad sense
- telex and teletex
- facsimile
- visiophony
- broadcasting of audio or video programs
- teleaction, telemetry, remote alarms, *et cetera*

These services may be switched or not, some of them allowing a bilateral exchange (dialogue), others a purely unilateral transmission (monologue). They deal with analog information (speech, music, pictures) or digital information (data, texts, *et cetera*.). The bandwidths or the bit rates required are very different (table 1.10).

The number and nature of the services integrated in the same network can give rise to multiple variations.

15.8.2 Present situation in an analog nework

The present situation of public telecommunications is characterized by a still pronounced predominance of analog systems, by the progressive introduction of digital systems and overall, by the three types of networks, represented very schematically in figure 15.19, with the four same categories of equipment as in figure 15.5:

544 TELECOMMUNICATION SYSTEMS

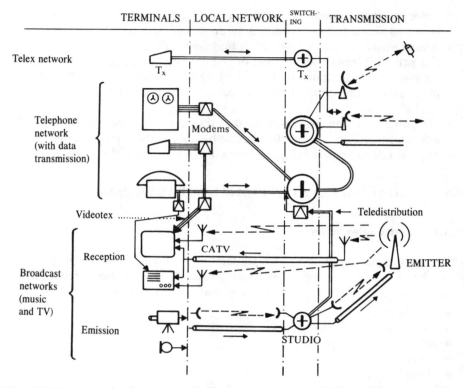

Fig. 15.19 Actual telecommunication networks and their applications.

- the switched *telephone* network, originally conceived (and still used by most customers) for speech transmission;
- the switched *telex* network, designed for the transport of digital information (text) at low speed;
- the *broadcast* networks of voice, audio or video information ("radio" and television), by radio waves or by cable.

In reality, these networks partially overlap one another. It is found, for example, that

- the telex network uses telephone channels for its transmission, except in the local network, where it has its own lines; however, it has its own exchanges;
- radio and television programs are transmitted from one point to another (studio, transmitter, re-transmitter) by using microwave links and cables similar or even common to multichannel telephone systems;

- telephone channels are used for data transmission;
- videotex is based on the telephone network and on the domestic television receiver;
- the subscriber line can carry, as well as telephone signals or data, alarm signals, charging pulses, or even, in some countries, six broadcasted audio programs (wire broadcasting by AM modulation between 160 and 355 kHz),
- the teledistribution networks, which are general private, (cable television, CATV) are frequently used for remote monitoring operations (remote sampling of domestic energy counters).

In a certain sense, the current analog networks thus offer a partial integration of some services. The situation of telematics, which is often considered a guest in the present telephone network, is described in figure 15.20. It is found that data (digital by nature) are adapted to the (analog) telephone channel by a discrete analog modulation in a modem, but that this channel may traverse part of the route in digital form in a PCM system (a situation which is more and more likely). Such an accumulation of conversions in a cascade arrangement is deleterious to the quality of transmission.

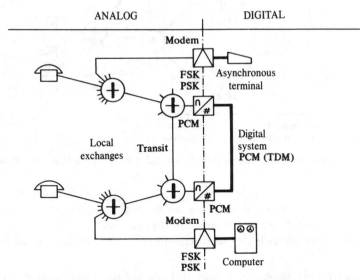

Fig. 15.20 Integration of data transmission in an analog network, or in digital form for part of the transmission.

15.8.3 Digital network with service integration

Economic and qualitative objectives on one hand, and the growth of needs on the other, stimulate two tendencies in the telecommunications network:
- integration of telephone transmission and switching in digital form (integrated digital network, sect. 15.7.2);
- creation of public data networks, especially for the routing of professional data traffic (sect. 15.6.2).

The conjunction of these parallel tendencies (figure 15.21) brings up the possibility of a *common* digital network for telephony (in PCM) and data transmission, called an *integrated services digital network ISDN*.

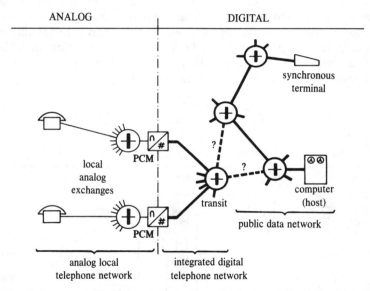

Fig. 15.21 Integrated digital network for telephony and the public data network.

The perfect integration of services would not however be attained unless there is a long-term digital local network (figure 15.22).
The perspective of such an integration nevertheless brings up the following remarks:
- telephony very probably would still remain the predominant service for a long time. Predictors evaluate the proportion of subscribers interested in other services in the long run at 1%;

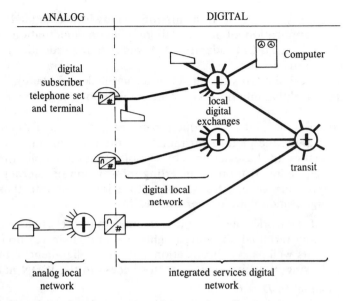

Fig. 15.22 Hypothetical network, with services integration.

- the offer of new services must not be a pretext for increasing the price of the basic telephone service;
- the impatience of data users conflicts with the techno-economic inertia of the existing telephone network. It is, however, necessary to make sure that transitory solutions (public data network) do not compromise the possibility of future integration;
- It is not certain whether the very specific needs of professional users can be covered in an integrated service public network (with circuit switching due to the telephony); a special data network, with packet switching, will precede the ISDN and very probably coexist with it.

15.8.4 Services compatible with the digital telephone channel

Imposed by the transmission quality expected from a PCM telephone communication (sect. 7.5.12), the bit rate

$$\dot{D} = 64 \text{ kbit/s} \tag{15.8}$$

is considered to be the *unit base rate* for an integrated network. Aside from ordinary telephony, this rate alternatively allows

- higher quality telephone transmission (bandwidth extended beyond 3.4 kHz) by application of more elaborate digital modulation and source encoding (e.g., DPCM, adaptive ΔM, redundancy reduction);
- fast data transmission;
- efficient facsimile (10 s for an A4 page, in black and white);
- transfer of alphanumeric texts from one memory to another (electronic mail);
- transmission of fixed or infrequent pictures (slow scan TV) at the rate of approximately one picture per minute (without redundancy reduction);
- a combination of these services by exploiting, for example, the pauses of a telephone conversation by inserting information of another nature, or by compressing speech transmission to a lower bit rate than 64 kbit/s with a digital modulation other than PCM.

In the local network, the 64 kbit/s channel must be enhanced by extra capacity to allow terminal signaling (highly elaborated, in relation to certain new services), as well as synchronization and frame alignment of the terminal unit. The following rates are envisioned for access to the ISDN network:

- base channel b: $\dot{D}$ = 64 kbit/s;
- auxiliary channel Δ: $\dot{D}$ = 16 kbit/s for "out-of-octet" signaling and data transmission;
- besides the simple access b + Δ with $\dot{D}$ = 80 kbit/s, enlarged access of the type nb + Δ (e.g., at 144 kbit/s) are also possible for special services.

15.8.5 Broadband multiservice network

Beyond services compatible with the digital "telephone" rate of 64 kbit/s mentioned in section 15.8.4, a significant rate increase must be made to meet the needs of other services, such as, for example:

- high fidelity audio broadcasting, stereophony: $\dot{D}$ = (4–6) x 64 kbit/s
- visiophony (bidirectional transmission and switching of moving pictures in real time, in a modest format, in parallel with telephone communication): $\dot{D}$ = 80–140 Mbit/s
- television broadcasting, or video switching on demand: $\dot{D}$ = 80–140 Mbit/s

By the introduction of the *switching function* into broadband (audio, video) services which, up till now, could only be unilaterally *broadcasted*, these services functionally approach the classical switched services such as telephony or telematics. The idea of uniting all these services, those of the telephone type ($\dot{D} \leq$ 64 kbit/s) and those said to be "large band" ($\dot{D} \gg$ 64 kbit/s), into the same digital network is therefore not absurd. However, the implementation of this idea, especially in the local network in which it would be more attractive, poses serious technical, economic, and human problems:

- sufficient interaction must take place between the *offer and demand*: the cost and profitability of such a network strongly depend on the traffic flow, which itself increases if the network is efficient and if the service is reasonably priced;
- the investments to be made are very substantial, whether the network be based on the application of *optical fibers* or whether it is proposed to use *domestic satellites*. It is in any case clear that the present local network with balanced pairs, despite its still unexploited possibilities for digital transmission, cannot serve as a basis for a broadband multiservice network;
- the public's *acceptance of new services* is never certain. Numerous historical examples show that even services such as the telephone have found a different application than that which their inventor originally imagined (Bell wanted to use it as a means of musical broadcasting) while others, such as visiophony, have been rejected by the public;
- the complexity of certain new services necessitates *user guidance* by a particularly well studied man-machine dialogue, so that the lay user will not be confused.

15.9 EPILOGUE

15.9.1 The spiral

In the introduction to this volume, the approach adopted for the presentation of telecommunication systems was described as a *spiral path*. Geometrically, this path is characterized by the fact that it has a center, but no end. Therefore, there cannot be a conclusion to this book. The epilogue which takes its place gives only a glimpse of the meaning and the trend of this spiral.

At the center of telecommunications is man and his existential need for relationship. At infinity, toward which the spiral extends, is located an ideal of service, transparency, and communion. Meanwhile, the evolution of systems and networks is driven by creativity and imagination, but also limited by economic and physical constraints which have been discussed often in this book. It is nevertheless essential that this evolution neglects neither its center, nor its aim.

Faced with an unknown future, different attitudes are possible, from resigned pessimism to impatient exaltation. In the particular case of telecommunications, we can hold four of these attitudes:

- *complacency*: Concerning the quantititative and qualitative performance of existing networks, we might think that it is useless to wish to do better or more than at present. This satisfied and passive attitude forgets, however, firstly that the density and quality of service currently offered are very unequally distributed (90% of telephones and television receivers are found in 25 of the 157 member countries of the ITU!) and, on the other hand, that the evolution of society will probably bring with it new needs and demands;
- *imagination*: By its search for new and original ideas, human creativity can offer a vision of the future in which it is difficult, at first, to distinguish fiction from possible reality. In any case, the great inventions are not always those recognized as such at the moment; they are rare and unpredictable by nature;
- *extrapolation*: In connection with a slow and continuous evolution, this attitude is particularly comprehensible in the case of telecommunications due to the considerable techno-economic inertia of the system. It consists of postulating a continuous quantitative and qualitative development and denotes a certain fatalism;
- *projection*: As opposed to extrapolation, this approach initially defines an objective (technical, human, or social) and then elaborates the means of attaining it. The development is thus coherent and planned as a function of the defined goal.

To imagine as clearly as possible what telecommunications could be at the dawn of the 21st century, we must simultaneously proceed to realistic extrapolation and dare to make idealistic projections.

15.9.2 A set of specifications for future telecommunications

The projection of desirable objectives allows us to draft something of a set of specifications for the telecommunications of tomorrow, with a view toward constructing the future, instead of having to undergo it.

At the risk of disappointing futurologists, this set of specifications contains nothing sensational. It tends rather to concentrate on the needs of human beings and society, the principal need in this context appearing to be that of bilateral communication, i.e., of information *exchange*.

Some suggestions in this direction could be:
- favoring the *interactive conversational modes*, starting with the telephone, prolongment of live dialogue, by completing it through exchange of text, pictures, and data;
- contributing to the solution of certain "*civilization problems*" rather than spreading information pollution. In particular, the replacement of certain

movements of persons by telecommunications means could contribute to reducing traffic congestion, the waste of energy, and the monstrous concentration of business districts. To a certain degree, telecommunications can also compensate for social isolation and economic disparity, by-products of our civilization, by facilitating contacts and exchanges;
- decreasing *response time* so as to keep pace with accelerating economic and social processes;
- safeguarding in all humility an *ideal of service* and playing, modestly but well, this role of conveyor by respecting the interests of the users, their anonymity, and the privacy of transmitted messages;
- contributing to elevating *information quality*, in the sense presented below.

15.9.3 Information quality

In the energy domain, it is well known that it is not sufficient to know the quantity of energy available. It is still necessary to specify what quality level this energy is located at. The same goes for information: the *information quantity* (sect. 1.5.3) is an intrinsic and objective measure, but does not at all consider the subjective effects that it brings up in the receiver. It appears to be necessary, in a human context, to complete this measurable quantity with a notion which is much more difficult to define, that of *information quality* which expresses the *extent to which the information fits the needs of the person who receives it*.

In matters of information and communication, there are three types of needs to which telecommunications attempts to respond:

- the need to be *informed*, in the sense of receiving news about what happens around us and in the world. The quality expected is expressed in terms of objectivity, actuality, and completeness, but also implies the possibility of a conscious and free choice;
- the need to *know*, i.e., to consult and retrieve existing information, stored in "memories" (books, tables, encyclopedias, files, timetables, lists, catalogues, and directories, including that of the telephone). Quality here is synonymous with speed and ease of searching, selectivity of access, precision, and being up-to-date;
- the need for *relationship*, in the form of interpersonal contacts and exchanges, on the private or professional level. The quality of human communication completely escapes quantitative criteria; it cannot be measured in decibels

When it comes to *information broadcasting*, the role of telecommunications is not as neutral as is believed. In fact, by favoring the avalanche of information which continually unfurls before us, telecommunication has suppressed the filter that sifts out information as set by the time lag between the event and its announcement, between the thought and its expression. We now confuse the importance of a piece of information with its immediacy, and true knowledge with the quantity of information received in real time. In contrast to systems dealt with in information theory, and whose disorder (entropy) diminishes when information is provided, human beings react to a plethora of information with a feeling of interior disorder and perplexity, or by an indifference which generates a "mortal lukewarmness" (K. Lorenz [59]).

It is' perhaps in the domain of *interactive information retrieval* and *information processing* that the increasing interpenetration of telecommunications and informatics will permit, let us hope, an improvement of information quality. The collective and massive diffusion of low level information could thus give way to more individualized bilateral communication, allowing a selective access to information sources, according to the differentiated and momentary needs of the users. Videotex is a step in this direction.

From its beginnings, telecommunication has sought to satisfy its users' need for *relationship*. The rapid and enduring success of systems such as the telegraph, and later telex and telephone, have demonstrated their usefulness. However, these systems do not themselves guarantee the quality of the human communication that their links allow. Of what use is it really to be linked to 500 million other telephone subscribers, if we have nothing to say to our neighbor? Installing telephones in the Tower of Babel would not make it more livable!

Suspended between the infinity of an extremely perfected communication technology and the zero of a difficult, disputed, or even refused human communication, we must use the quantity of the first to improve the quality of the second.

How much knowledge
  have we lost in information?
How much wisdom
  have we lost in knowledge?
              T.S. Eliot

Chapter 16

Appendices

16.1 CONVENTION FOR THE REPRESENTATION OF SPECTRA

16.1.1 Unilateral and bilateral spectrum

From Euler's equivalence

$$\cos(2\pi ft) = \frac{1}{2}[\exp(j2\pi ft) + \exp(-j2\pi ft)] \qquad (16.1)$$

follow two possible expressions of the Fourier series for a periodic signal, of period $T=1/f$. The result is two possible representations (spectra) of a periodic signal in the frequency domain (vol. VI., sect. 4.5):

- **unilateral** spectrum, corresponding to a sum of *sinusoidal waves* of frequencies nf, all positive, and of a dc component ($n \geq 0$). The amplitudes $U_{nI} f$) and the phases $\phi_{nI}(f)$ of these sinusoidal waves define the amplitude spectrum and phase spectrum;

bilateral spectrum, corresponding to a sum of *complex conjugate exponential pairs* with arguments $j(2\pi nf + \phi_{nII})$ and with modulus U_{nII}. The variable nf, by extension called the "frequency", is thus positive, zero or negative.

Because of (16.1), the following relations exist between these two representations, which are strictly equivalent:

bilateral spectrum			unilateral spectrum
for $\|n\| \geq 1$	$U_{-nII} = U_{nII}$	$= \frac{1}{2}U_{nI}$	for $n \geq 1$
	U_{0II}	$= U_{0I}$	$n = 0$
for $\|n\| \geq 1$	$-\phi_{-nII} = \phi_{nII}$	$= \phi_{nI}$	for $n \geq 1$

The bilateral amplitude spectrum can be converted to the unilateral spectrum by folding the demi-axis of the "negative frequencies" over that of the positive frequencies, followed by an addition, which leads to doubling all the components except that at $f = 0$. This rule is equally valid in the case of non-periodic signals (continuous spectra).

16.1.2 Adopted convention

The *bilateral spectrum* is a very useful representation for theoretical arguments, not only for periodic and determined signals, but also for random signals (power spectral density), because it is the immediate result of the Fourier transform (chapter IV. 7 and chapter VI. 4). It lends itself well to the operations of convolution, which are often found in modulation procedures.

However, in the practical applications and notably in the *measurement* of spectra, only the positive frequencies make sense and it is naturally the *unilateral spectrum* which interests us.

Faced with the dilemma of the elegance of the mathematical tool offered by the Fourier transform, on the one hand, and the concern for remaining close to a physical reality accessible to measurement, on the other hand, this book has systematically adopted the following convention:

- in *equations and formulas*, the Fourier transforms are used according to their usual definition (VI. 4.1.2) and emphasized by capital letters:

$$u(t) \circ \xrightarrow{\text{F}} \bullet \; U(f) \tag{16.2}$$

   The range of $U(f)$ extends from $f = -\infty$ to $f = +\infty$;

- the *graphic presentation* of spectra is, however, *always unilateral*. This fact is emphasized in each spectrum by a special black and white arrow on the vertical axis at $f = 0$, to emphasize the fact that only the positive frequencies are represented and that all the spectral components with non-zero frequencies are doubled with respect to the bilateral spectrum.

16.2 GAUSSIAN FUNCTION

16.2.1 Expression and significance

The Gaussian function $g(x)$ (fig. 16.1) gives the probability density function $p(x)$ of a random variable x with a *normal distribution* having the following characteristic parameters:

- mean value (mathematical expectancy) equal to zero $\mu_x + 0$;
- standard deviation $\sigma_x = 1$.

It is also called the *reduced normal law*. It is expressed

$$p(x) = g(x) = \frac{1}{\sqrt{2\pi}} \exp\left(-\frac{1}{2} x^2\right) \tag{16.3}$$

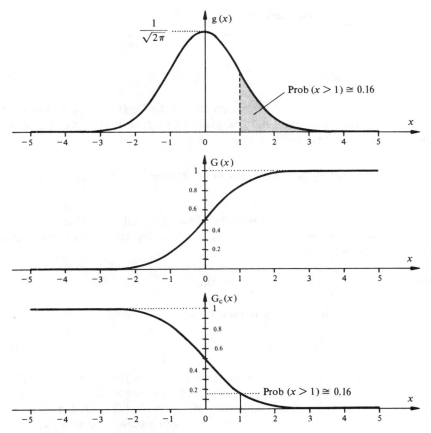

Fig. 16.1 Gaussian function g(x), its integral form G(x), and complementary integral form $G_c(x)$.

The probability density function p(y) of a random variable y with normal distribution, with *any* mean μ_y and *any* standard deviation σ_y, is derived from the reduction relation

$$p(y) = \frac{1}{\sigma_y} g\left(\frac{y - \mu_y}{\sigma_y}\right) \tag{16.4}$$

In the case (very common in telecommunications) of random Gaussian signals u(t) with a *zero mean value*, the standard deviation is equal to the r.m.s. value U_{eff} of the signal. Its probability density function is thus

$$p(u) = \frac{1}{U_{\text{eff}}} g\left(\frac{u}{U_{\text{eff}}}\right) \tag{16.5}$$

16.2.2 Integral Gaussian function

The function

$$G(x) = \int_{-\infty}^{x} g(x') dx' \tag{16.6}$$

is called the *integral Gaussian function* (figure 16.1). It expresses the cumulative distribution function, i.e., the probability that a reduced variable x' would have a value between $-\infty$ and x.

16.2.3 Complementary integral Gaussian function

It is often more useful to know the *probability of exceeding a given value* x $\text{Prob}(x' > x)$, i.e, the probability that the reduced Gaussian variable x' exceeds a threshold located at x. It is reduced by the probability density function $p(x')$ by an integration

$$\text{Prob}(x' > x) = \int_{x}^{\infty} p(x') dx' = \int_{x}^{\infty} g(x') dx' \tag{16.7}$$

In comparing this with (16.6), it is noted that

$$\int_{x}^{\infty} g(x') dx' = 1 - G(x) = G_c(x) \tag{16.8}$$

$G_c(x)$ is called the *complementary integral Gaussian function* (fig. 16.1). Its values, by successives decades, are presented in figure 16.2.

In particular, it allows the evaluation of the probability that a signal $u(t)$ with Gaussian distribution and zero mean would exceed a certain value U_0. With (16.5), we have, in effect:

$$\text{Prob}[u(t) > U_0] = \frac{1}{U_{\text{eff}}} \int_{U_0}^{\infty} g\left(\frac{u}{U_{\text{eff}}}\right) du = G_c(x) \tag{16.9}$$

by setting

$$x = U_0 / U_{\text{eff}} \tag{16.10}$$

For example, the r.m.s. value ($x = 1$) is exceeded (unilaterally) with a probability $G_c \cong 0.16$, while $U_0 = 5 U_{\text{eff}}$ is only exceeded with a probability of 3×10^{-7}.

16.2.4 Approximations

For large values of x, the following asymptotic formula can be useful:

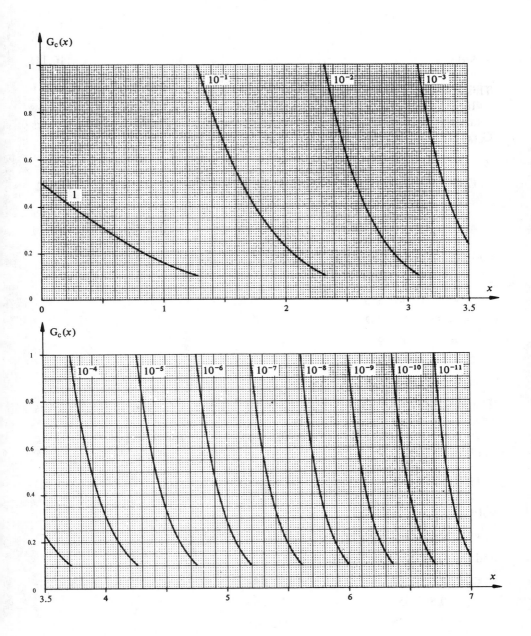

Fig. 16.2

$$G_c(x) \cong \frac{1}{\sqrt{2\pi}} \frac{\exp(-\frac{1}{2} x^2)}{x} \left(1 - \frac{1}{x^2}\right) \tag{16.11}$$

This leads to a relative error lower than 1% for $x \geq 4$

If tables are unavailable and for precise numerical calculation, the following relation, given in 183, lends itself well to computer programming and provides $G_c(x)$ with an excellent approximation, for $x \geq 0$:

$$G_c(x) \cong \frac{1}{\sqrt{2\pi}} (c_1 z + c_2 z^2 + c_3 z^3 + c_4 z^4 + c_5 z^5) \exp\left(-\frac{x^2}{2}\right) \tag{16.12}$$

$$z = \frac{1}{1 + c_0|x|} \tag{16.13}$$

and the following coefficients:

$C_0 = 0.231\ 641\ 900$
$C_1 = 0.319\ 381\ 530$
$C_2 = 0.356\ 563\ 782$
$C_3 = 0.781\ 477\ 937$
$C_4 = 1.821\ 255\ 978$
$C_5 = 1.330\ 274\ 429$

The absolute error is lower than 7.5×10^{-8}. For $x < 0$, relation (16.12) gives $G(x) = 1 - G_c(x)$ instead of $G_c(x)$.

In the same way, the *inverse* function $G^{-1}_c(y)$ for $0 < y \leq 0.5$ can be approached by

$$G_c^{-1}(y) \cong w - \frac{C_0 + C_1 w + C_2 w^2}{1 + K_1 w + K_2 w^2 + K_3 w^3} \tag{16.14}$$

with

$$w = \sqrt{-2 \ln y} \tag{16.15}$$

and the following coefficients:

$C_0 = 2.515\ 517$ $K_1 = 1.432\ 788$
$C_1 = 0.802\ 853$ $K_2 = 0.189\ 269$
$C_2 = 0.010\ 328$ $K_3 = 0.001\ 308$

The absolute error is lower than 4.5×10^{-4}. For $0.5 < y < 1$, we use the symmetry relation

$$G_c^{-1}(y) = -G_c^{-1}(1-y) \tag{16.16}$$

and calculate $G_c^{-1}(1-y)$ by (16.14).

16.2.5 Relationship with the error function

In the English literature, an *error function* erf(x), is used, and defined in the following way:

$$\text{erf}(x) = \frac{2}{\sqrt{\pi}} \int_0^x \exp(-y^2) dy \tag{16.17}$$

In the same way a *complementary error function* erfc(x) is defined

$$\text{erfc}(x) = 1 - \text{erf}(x) \tag{16.18}$$

These two functions are not entirely identical to the integral Gaussian function $G_c(x)$, and the complementary integral Gaussian function $G_c(x)$, respectively. Between them exist the following relations:

$$G(x) = \frac{1}{2}[1 + \text{erf}(x/\sqrt{2})] \tag{16.19}$$

$$G_c(x) = \frac{1}{2}\text{erfc}(x/\sqrt{2}) \tag{16.20}$$

$$\text{erf}(x) = 2G(x\sqrt{2}) - 1 \tag{16.21}$$

$$\text{erfc}(x) = 2G_c(x\sqrt{2}) \tag{16.22}$$

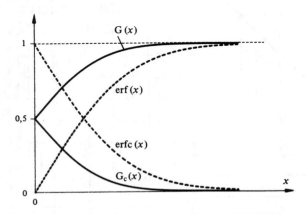

Fig. 16.3 Comparison of the error function with the integral Gaussian function.

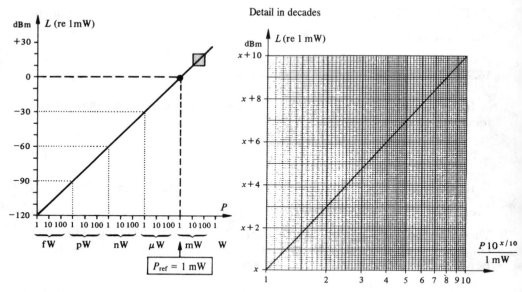

Fig. 16.4 Absolute power level.

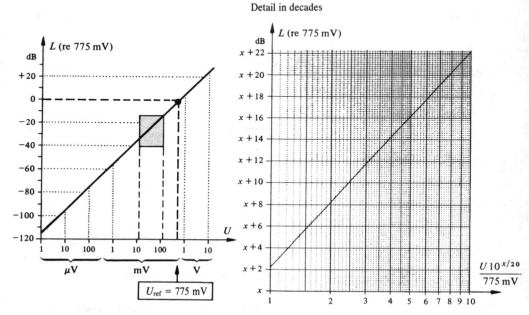

Fig. 16.5 Absolute voltage level.

16.3 ABSOLUTE POWER AND VOLTAGE LEVELS

In relation to the definitions given in section 2.3.4, figures 16.4 and 16.5 illustrate the equivalence between absolute levels and power (reference: 1 mW), on the one hand, and voltage (reference: 775 mV) on the other. They allow rapid logarithmic representation of known quantities and vice versa.

For the same voltage level, the power level increases by 9.03 dBm if the impedance passes from 600 Ω to 75 Ω and by 10.8 dBm if it passes from 600 Ω to 50 Ω.

Bibliography

[1] J. DE ROSNAY, *Le Macroscope, vers une vision globale*, Seuil, Paris, 1975.
[2] M. T. HILLS, B. G. EVANS, *Transmission Systems*, G. Allen & Unwin, London, 1973.
[3] H. TAUB, D. SCHILLING, *Principles of Communication Systems*, McGraw-Hill Kogakusha, Tokyo, 1971.
[4] P. Z. PEEBLES, JR., *Communication System Principles*, Addison-Wesley, Reading, 1976.
[5] R. L. FREEMAN, *Telecommunication Transmission Handbook*, (2nd Ed.), Wiley, New York, 1981.
[6] R. L. FREEMAN, *Telecommunication System Engineering. Analog and Digital Network Design*, Wiley, New York, 1980.
[7] ITT, *Reference Data for Radio Engineers*, (6th Ed.), Howard W. Sams, Indianapolis, 1975.
[8] M. ABRAMOWITZ AND I. STEGUN, *Handbook of Mathematical Functions*, Dover Publications, New York, 1972.
[9] S. J. ARIES, *Dictionary of Telecommunications*, Butterworths, London, 1981.
[10] H. POOCH et al., *Fachwörterbuch des Nachrichtenwesens*, Schiele & Schön, Berlin, 1976.
[11] CCITT, *Yellow Book*, VIIth Plenary Assembly, Geneva, November 1980, ITU, Geneva, 1981.
[12] R. E. MATICK, *Transmission Lines for Digital and Communication Networks*, McGraw-Hill, New York, 1969.
[13] R. CROZE, L. SIMON, J.-P. CAIRE, *Transmission téléphonique. Théorie des lignes*, (4th Ed.), Eyrolles, Paris, 1968.
[14] ITU, *Graphique en couleurs indiquant l'attribution des bandes de fréquences entre 10 kHz et 275 GHz*, ITU, Geneva, 1973.
[15] H. SCHMID, *Theorie und Technik der Nachrichtenkabel*, Hüthig, Heidelberg, 1976.
[16] D. GLODGE, "Impulse Response of Clad Optical Multimode Fibers," *Bell System. Techn. J.*, vol. 52, no. 6, July / Aug. 1973, pp. 801-816.
[17] J. E. MIDWINTER, *Optical Fibers for Transmission*, Wiley, New York, 1979.
[18] L. THOUREL, *Les antennes*, Dunod, Paris, 1971.
[19] H. NYQUIST, "Certain Topics in Telegraph Transmission Theory," *Trans. AIEE*, vol. 47, no. 2, Feb. 1928, pp. 617-644.

[20] J. R. PIERCE, *Symboles, signaux et bruit. Introduction à la théorie de l'information*, Masson & Sofradel, Paris, 1966.
[21] J. DUPRAZ, *Théorie de la communication. Signaux, bruits et modulations*, Eyrolles, Paris, 1973.
[22] W. R. BENNETT, J. R. DAVEY, *Data Transmission*, McGraw-Hill, New York, 1965.
[23] W. R. BENNETT, "Spectrum of Quantized Signals," *Bell Syst. Techn. J.*, vol. 27, no. 7, July 1948, pp. 446-472.
[24] K. W. CATTERMOLE, *Principles of Pulse Code Modulation*, Iliffe, London, 1969.
[25] R. STEELE, *Delta Modulation Systems*, Pentech Press, London, 1975.
[26] P. BYLANSKI, D. G. W. INGRAM, *Digital Transmission Systems*, Peregrinus, Stevenage, 1976.
[27] D. E. NORGAARD, "The Phase-shift Method of Single Sideband Signal Generation," *proc. IRE*, vol. 44, no. 12, Dec. 1956, pp. 1718-35.
[28] D. K. WEAVER, "A third Method of Generation and Detection of Single Sideband Signals," *Proc. IRE*, vol. 44, no. 12, Dec. 1956, pp. 1703-1705.
[29] J. H. ROBERTS, *Angle Modulation: The Theory of System Assessment*, Peregrinus, Stevenage, 1977.
[30] A. ANGOT, *Compléments de mathématiques à l'usage des ingénieurs de l'électrotechnique et des télécommunications*, Masson, Paris, 1972.
[31] E. HOLZLER, H. HOLZWARTH, *Pulstechnik*. Vol. I: *Grundlagern*; Vol. II: *Anwendungen und Systeme*, Springer, Berlin, 1975.
[32] J. A. BETTS, *Signal Processing, Modulation and Noise*, English Universities Press Ltd., London, 1970.
[33] H. STARK, R. B. TUTEUR, *Modern Electrical Communications*, Prentice-Hall, London, 1979.
[34] M. SCHWARTZ, *Information Transmission, Modulation and Noise*, McGraw-Hill, New York, 1970.
[35] R. BEST, *Theorie und Anwendungen des Phase-Locked Loops*, AT-Verlag, Aarau, 1981.
[36] F. M. GARDNER, *Phaselock Techniques*, Wiley, New York, 1966.
[37] H.-J. HILDEBRANDT, *Trägerfrequenztechnik*, Oldenbourg, München, 1975.
[38] M. MATHIEU, *Télécommunications par faisceau hertzien*, Dunod, Paris, 1979.
[39] H. POOCH (ed.), *Richtfunktechnik*, Schiele & Schön, Berlin, 1970.
[40] H. BRODHAGE, W. HORMUTH, *Planung und Berechnung von Richtfunkverbindungen, Planning and Engineering of Radio Relay Links*, Siemens, Erlangen, 1977.
[41] J. PARES, V. TOSCER, *Les systèmes de télécommunication par satellites*, Masson, Paris, 1977.

[42] J. J. Spilker, *Digital Communications by Satellite*, Prentice-Hall, London, 1977.
[43] V. K. Bhargava, D. Haccoun, R. Matyas, P. Nuspl, *Digital Communications by Satellite: Modulation, Multiple Access and Coding*, Wiley, New York, 1973.
[44] S. D. Personick, *Optical Fiber Transmission Systems*, Plenum Press, New York, 1981.
[45] A. Cozannet, J. Fleuret, H. Maître, M. Rousseau, *Optique et télécommunications*, Eyrolles, Paris, 1981.
[46] R. M. Gagliardi, *Introduction to Communications Engineering*, Wiley, New York, 1978.
[47] D. L. Richards, *Telecommunication by Speech*, J. Wiley, New York, 1973.
[48] J. Lehnert, *Initiation à la télégraphie*, Siemens, Berlin 1969.
[49] NCC, *Handbook of Data Communications*, The National Computer Centre, Oxford, 1982.
[50] D. W. Davies, D. L. A. Barber, *Communication Networks for Computers*, Wiley, New York, 1973.
[51] J. Clavier, G. Coffinet, M. Niquil, F. Behr, *Théorie et technique de la transmission de données*, Masson, Paris, 1977.
[52] C. Macchi, J-F. Guilbert, *Téléinformatique*, Dunod, Paris, 1979.
[53] D. R. Doll, *Data Communications. Facilities, Networks and Systems Design*, Wiley, New York, 1978.
[54] E. R. Hafner, *Digital Communication Loops—A Survey*, International Zurich Seminar on Digital Communication, 1974, Paper D 1.
[55] H. Inose, *An Introduction to Digital Integrated Communications Systems*, Peregrinus, Stevenage, 1971.
[56] "Réseau intégrant téléphone et données, (numéro particulier)," *Echo des Recherches*, no. 111, first trimester 1983.
[57] "Réseau numériques avec intégration des services," *Revue des Télécommunications* (ITT), vol. 56, no. 1, 1981.
[58] A. Glowinski, *Télécommunications, objectif 2000*, Dunod, Paris, 1980.
[59] K. Lorenz, *Les huit péchés capitaux de notre civilisation*, Flammarion, Paris, 1973.
[60] S. C. Littlechild, *Elements of Telecommunications Economics*, Peregrinus, Stevenage, 1979.
[61] International Telecommunication Union, *From Semaphore to Satellite*, ITU, Geneva, 1965.
[62] S. von Weiher, *Tagebuch der Nachrichtentechnik von 1600 bis zur Gegenwart*, VDE-Verlag, Berlin, 1980.
[63] N. Jéquier, *Les télécommunications et l'Europe*, Centre de recherches européennes, Lausanne, 1976.

Select Bibliography

The Traité d'Électricité, listed below by volume number, is published by the Presses Polytechniques Romandes (Lausanne, Switzerland) in collaboration with the École Polytechnique Fédérale de Lausanne. The title of each volume is given with the year of publication in parenthesis. English translations by Artech House are denoted by an asterisk with the year of publication in parenthesis.

Vol.	Author	Title
I	Frédéric de Coulon & Marcel Jufer	Introduction à l'électrotechnique (1981).
II	Philippe Robert	Materiaux de l'électrotechnique (1979).
III	Fred Gardiol	Electromagnétisme (1979).
IV	René Boite & Jacques Neirynck	Theorie des reseaux de Kirchhoff (1983).
V	Daniel Mange	Analyse et synthèse des systèmes logiques (1979). *Analysis and Synthesis of Logic Systems (1986).
VI	Frédéric de Coulon	Theorie et traitement des signaux (1984). *Signal Theory and Processing (1986).
VII	Jean-Daniel Chatelain	Dispositifs à semiconducteur (1979).
VIII	Jean-Daniel Chatelain & Roger Dessoulavy	Electronique (1982).
IX	Marcel Jufer	Transducteurs électromécaniques (1979).
X	Jean Chatelain	Machines électriques (1983).
XI	Jacques Zahnd	Machines séquentielles (1980).
XII	Michel Aguet & Jean-Jacques Morf	Energie électrique (1981).
XIII	Fred Gardiol	Hyperfréquences (1981). *Introduction to Microwaves (1984).
XIV	Jean-Daniel Nicoud	Calculatrices (1983).
XV	Hansruedi Bühler	Electronique de puissance (1981).
XVI	Hansruedi Bühler	Electronique de réglage et de commande (1979).
XVII	Philippe Robert	Mesures (1985).
XVIII	Pierre-Gérard Fontolliet	Systèmes de télécommunications (1983). *Telecommunication systems (1986).
XIX	Martin Hasler & Jacques Neirynck	Filtres électriques (1981). *Electric Filters (1986).
XX	Murat Kunt	Traitement numérique des signaux (1980). *Digital Signal Processing (1986).
XXI	Mario Rossi	Electroacoustique (1984).
XXII	Michel Aguet & Mircea Ianovici	Haute tension (1982).

List of Symbols

PRINCIPAL GRAPHIC SYMBOLS

input output   Direction of transmission
→   (for all the symbols below)

* Converter, transducer (general)

* Generator of sinusoidal signals

* Pulse generator

   Noise generator

* Amplifier, repeater

* Low-pass filter

* Band-pass filter

* Band-stop filter

* Modulator (general)

* Analog/digital converter

* Digital/analog converter

   Interrupter, sampler

   Sampler with hold

---

*Graphic symbol conforms to IEC standards.

LIST OF SYMBOLS

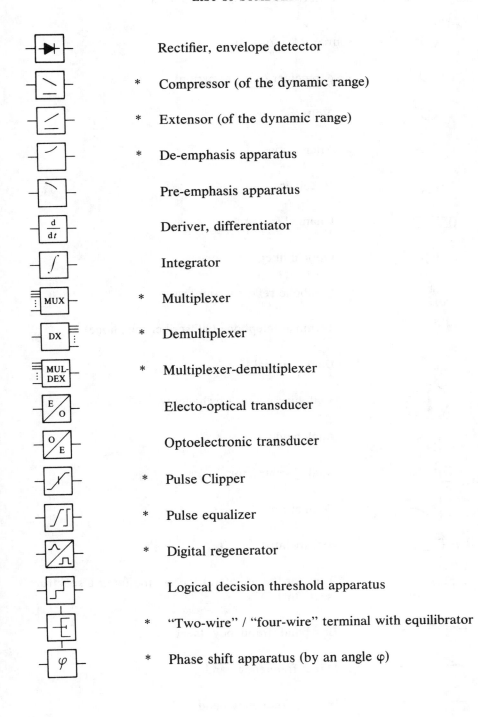

▶|		Rectifier, envelope detector
	*	Compressor (of the dynamic range)
	*	Extensor (of the dynamic range)
	*	De-emphasis apparatus
		Pre-emphasis apparatus
$\frac{d}{dt}$		Deriver, differentiator
$\int$		Integrator
MUX	*	Multiplexer
DX	*	Demultiplexer
MUL-DEX	*	Multiplexer-demultiplexer
E/O		Electo-optical transducer
O/E		Optoelectronic transducer
	*	Pulse Clipper
	*	Pulse equalizer
	*	Digital regenerator
		Logical decision threshold apparatus
	*	"Two-wire" / "four-wire" terminal with equilibrator
φ	*	Phase shift apparatus (by an angle φ)

TELECOMMUNICATION SYSTEMS

⊙–	Source (of information)
–⊗	Sink, receiver (of information)
	* Voltage source
	* Microphone
	* Receiver
	Channel (general)
	Optical fiber
	Parabolic reflector antenna
	* Terminal telephone (subscriber telephone)
	Data terminal
	Switching center (exchange)
	Analog adder
	Analog subtractor
	Comparator
	Extrapolator
	Axes for the unilateral representation of a spectrum (sect. 16.1.2)
	Baseband (frequency) (sect. 4.2.2)
	* Direct frequency band
	* Inverse frequency band

List of Abbreviations and Acronyms

Abbreviation	Description	Section
AMI	Alternate mark inversion (pseudoternary mode)	9.6.4
AM	Amplitude modulation	8.2
APD	Avalanche photodiode	14.2.3
ARQ	Automatic request	11.2.4
ASK	Amplitude shift keying (discrete amplitude modulation)	8.11.2
AΔM	Adaptive delta modulation	7.9.2
BER	Bit error rate	9.10.1
CATV	Cable television	15.8.2
CCIR	Comite consultatif international des radiocommunications	2.2.2
CCITT	Comite consultatif international telegraphique et telephonique	2.2.2
CEPT	Conference europeenne des postes et telecommunications	9.3.1
CMI	Coded mark inversion	9.6.6
CODEC	Encoder-decoder	9.3.4
DCTE	Data-circuit terminating equipment	11.4.6
DPCM	Differential pulse code modulation	7.8.6
DSBSC	Double sideband suppressed carrier	8.3
DTE	Data terminal equipment	11.4.6
EIRP	Equivalent isotropic radiated power	12.4.1
FDMA	Frequency-division multiple access	13.5.2
FDM	Frequency-division multiplex	4.4.2
FEXT	Far-end crosstalk	3.4.2
FM	Frequency modulation	8.6
FSK	Frequency shift keying	8.11.4
HDB	High density bipolar	9.6.5
HF	High frequency	
HRC	Hypothetical reference circuit	9.7.4
IDN	Integrated digital network	15.7
IFRB	International Frequency Registration Board	2.2.2
IF	Intermediate frequency	10.7.2
IM	Optical intensity modulation	14.3.1

INTELSAT	International Telecommunications Satellite Organization	13.3.6
ISDN	Integrated services digital network	15.8
ISI	Intersymbol interference	5.1.8
ISO	International Standard Organization	
ITU	International Telecommunication Union	2.2.2
LD	Laser diode	14.2.2
LED	Light emitting diode	14.2.2
LF	Low frequency	
MCMI	Modified coded mark inversion	14.3.3
MFC	Multifrequency code	10.2.5
MODEM	Modulator-demodulator	11.4.6
MSK	Minimum shift keying	8.11.9
MTBF	Mean time between failures	2.6.6
MTTR	Mean time to repair	2.6.6
MULDEX	Multiplexer-demultiplexer	
NEXT	Near-end crosstalk	3.4.2
NOSFER	*Nouveau systeme fondamental pour la determination des equivalents de reference*	2.4.2
NRZ	Non-return to zero	5.1.6
NTSC	National Television Systems Committee	10.7.7
OC	Short wave	3.8.3
OL	Long wave	3.8.3
PABX	Private automatic branch exchange	15.2.3
PAD	Packet assembly and disassembly	15.6.2
PAL	Phase alternation line	10.7.7.
PAM	Pulse amplitude modulation	8.10.1
PCM	Pulse code modulation	7.5
PDM	Pulse duration modulation	8.10.2
PFM	Pulse frequency modulation	8.10.4
PLL	Phase locked loop	8.6.10
PPM	Pulse position modulation	8.10.3
PSK	Phase shift keying	8.11.5
QAM	Quadrature amplitude modulation	11.4.6
RZ	Return to zero	5.1.6
SECAM	Sequential coder and memory	10.7.7
SNR	Signal-to-noise ratio	2.4.15
SPADE	Single channel per carrier PCM multiple access demand assignment equipment	13.5.4
SSB	Single sideband modulation	8.4
SSB	Single sideband	8.4
TASI	Time assignment speech interpolation	1.7.3

LIST OF ABBREVIATIONS AND ACRONYMS

TDMA	Time-division multiple access	13.5.5
TDM	Time-division multiplex	4.4.3
TV	Television	
UHF	Ultra-high frequency	3.8.2
VHF	Very high frequency	3.8.2
VHF, UHF	Ultra-short wave	3.8.3
VSB	Vestigial sideband modulation	8.5
WDM	Wavelength division multiplex (optical multiplex)	14.5.3
ΔM	Delta modulation	7.8.3
$\Delta\Sigma M$	Delta-sigma modulation	7.8.7
ΦM	Phase modulation	8.7

Glossary

Symbol	Unit	Description
a	1	Attenuation factor (linear)
A	dB (Np)	Attenuation (logarithmic)
A	E	Offered traffic
A	1	Availability
A	1	A-law constant (quantization)
A_e	m^2	Effective surface of an antenna
A_r	dB	Reflection attenuation
A_t	dB	Attenuation of traversing (the terminal)
A_p	dB	Equilibration attenuation
b	rad	Phase shift (line)
b	1	Number of bits per PCM word
b_0	Hz · m	Specific optical bandwidth
B	Hz	Bandwidth
c_m	s^{-1}	Mean occupation frequency
c_0	m · s^{-1}	Speed of light in a vacuum
C	bit · s^{-1}	Transmission channel capacity
C	1	Compression ratio (in quantization)
C'	F · m^{-1}	Linear capacity of a line
d	1	Distortion rate
d_h	s	Duration of the impulse response of a fiber
D	bit	Decision rate
D	1	User density in a network
$\dot{D}$	bit · s^{-1}	Bit rate
$e(t)$	–	Step response
$e(t)$	1	Sampling function
$\hat{e}$	1	Clipping probability
$\underline{E}(f)$	1	Transfer function at the equalizer
f	Hz	Frequency
$f(t)$	Hz	Instantaneous frequency
f_{IF}	Hz	Intermediate frequency
F	–	Fourier transform
F	dB	Noise factor
g	1	Number of levels
g	1	Gain (linear)
$g(x)$	1	Gaussian function

$G(x)$	1		Integral Gaussian function
$G_c(x)$	1		Complementary integral Gaussian function
G	dB (Np)		Gain (logarithmic)
G'	$\Omega^{-1} \cdot m^{-1}$		Linear loss of a line
h	m		Height, altitude
h	$J \cdot s$		Planck's constant
$h(t)$	–		Impulse response
H	Sh		Information content, entropy
$\dot{H}$	$Sh \cdot s^{-1}$		Information rate
$\underline{H}(f)$	1		Transfer function of a channel
$\underline{i}(t)$	A		Instantaneous current
j	1		Imaginary unit $\sqrt{-1}$
$J_k(x)$	1		Bessel function of order k
k	$J \cdot K^{-1}$		Boltzmann constant
k'	$F \cdot m^{-1}$		Mutual linear capacity between lines
l	m		Length, distance
lb			Binary logarithm (base 2)
lg			Decimal logarithm (base 10)
ln			Natural logarithm (base e)
L	dB		Level
L'	$H \cdot m^{-1}$		Linear inductance of a line
L_p	H		Repeater inductance
m	1		Number of values per symbol
m	1		Modulation rate (AM)
m'	$H \cdot m^{-1}$		Mutual linear inductance between lines
$\dot{M}$	$Bd = s^{-1}$		Symbol rate
n	1		Refractive index of a medium
N	1		Numerical aperture
N	1		Number of repeaters or regenerators
$p(x)$	1		Probability density of the continuous variable x
p	$W \cdot m^{-2}$		Surface power
P	W		Power
Prob $(x > x_0)$	1		Probability that $x > x_0$ (continuous variable)
Prob (x_0)	1		Probability that $x = x_0$ (discrete variable
q	1		Number of quantization levels
r	m^{-1}		Image resolution
r	1		Yield factor of an antenna
R	Ω		Resistance
R	bit		Redundancy
R	m		Radius of the earth
R'	m		Hypothetical radius of the earth
R'	$\Omega \cdot m^{-1}$		Linear resistance of a line

Symbol	Unit	Description
t	s	Time
t_m	s	Rise time of a pulse
T	K	Absolute temperature
T_a	K	Absolute ambient temperature
T_0	K	Conventional absolute temperature (290 K)
T	s	Period, duration of a bit or a moment
$u(t)$	V	Instantaneous voltage
$u_{BE}(t)$	V	Emitted elementary base signal
$u_{BR}(t)$	V	Received elementary base signal
U	V	Amplitude of $u(t)$
$U(f)$	V	Fourier transform of $u(t)$
v_φ	m·s^{-1}	Phase velocity
y	E	Intensity of routed traffic
z	1	Number of channels of a multiplex
Z	Ω	Impedance
$\overline{Z}_c$	Ω	Characteristic impedance of a line
α	dB (Np)·m^{-1}	Linear attenuation of a line
α	1	Compression factor
β	rad·m^{-1}	Linear phase shift of a line
γ	m^{-1}	Linear exponent of propagation of a line
$\overline{\Gamma}$	1	Exponent of propagation of a line
δ	rad	Angle of loss of a dielectric
δ	1	Deterioration of the signal-to-noise ratio
δ	1	Index of modulation (FM, ΦM)
Δ	V	Interval (or step) of quantization
ΔU	V	Amplitude deviation (AM)
Δf	Hz	Frequency deviation (FM)
$\Delta\varphi$	rad	Phase deviation ΦM, phase shift
ϵ	1	Error probability of regeneration per moment
ϵ	1	Relative permittivity
ϵ_0	A·s·V.$^{-1}$·m^{-1}	Permittivity of a vacuum
ζ_c	rad	Argument of the characteristic impedance of a line
η	1	Mode efficiency
η_O	W·A^{-1}	Yield of an electro-optic transducer
ϑ	m	Depth of penetration
λ	m	Wavelength
μ	V·s·A^{-1}·m^{-1}	Permeability
μ	1	Number of moments per character
μ	1	Constant of the μ-law (quantization)
ν	Hz	Optical frequency
ξ	1	Signal-to-noise ratio

GLOSSARY

ξ_0	1	Reference signal-to-noise ratio
ρ	$\Omega \cdot m$	Resistivity
$\underline{\rho}$	1	Reflection factor
$\sigma(x)$	–	Standard deviation of the variable x
σ_O	$A \cdot W^{-1}$	Coefficient of the optoelectronic response
τ	s	Delay, propagation time
τ_φ	$s \cdot m^{-1}$	Linear phase propagation time
τ_g	$s \cdot m^{-1}$	Linear group propagation time
τ_0	s	Mean time to repair (MTTR)
τ_1	s	Mean time between failures (MTBF)
T	s	Duration of a pulse
φ	rad	Phase
$\varphi(t)$	rad	Instantaneous phase
$\varphi_x(t)$	–	Autocorrelation function of the variable x
$\Phi(f)$	$W \cdot Hz^{-1}$	Power spectral density
Φ_O	W	Optical power
ω	s^{-1}	Angular frequency
Ω	sr	Solid angle

GENERAL INDICES

cp	Composite
crit	Critical
C	After compression
e	Sampling
eff	Effective (rms)
eq	Equivalent
E	At emission
i	Relative to channel i
inf	Lower, inferior
iso	Isotropic
k	Relative to symbol k
max	Maximum
min	Minimum
M	Relative to a moment
nom	Nominal
N	Of noise
O	Optical
p	Of the carrier
q	Of quantization
r	Relative

ref	Reference
res	Resultant
R	At reception
sup	Superior, higher
S	Of the signal
th	Thermal
tph	Telephone
x	At the input of an amplifier
xp	Near-end crosstalk
xt	Far-end crosstalk
y	At the output of an amplifier
0	Set to zero (level)
1	Primary
2	Secondary
′	After transmission/after equalization
*	After quantization

INDEX CONVENTIONS

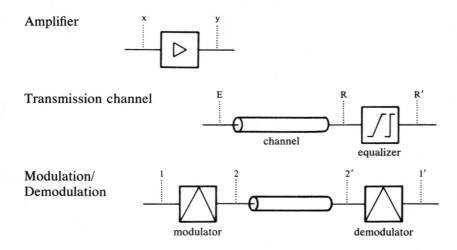

INDEX

Adaptive delta modulation (AΔM), 276
Aliasing, 156
Alternate mark inversion (AMI), 386–387
Amortization (of costs), 72–73
Amplification, 211
Amplitude modulation (AM), 282–283
Amplitude shift keying (ASK), 352
Analog
    information, 11
    signal, 15
Anisochronous, 377
Antenna, 130
    for ground stations, 483
    isotropic, 130
    large surface, 132–133
    parabolic reflector, 458–459
    satellite, 480–481
    wire, 133
Armoring, 100
Armstrong modulator, 320–321
Assignment, 515, 522, 524–525
    demand (satellites), 525
    fixed, 515, 522, 524
    variable, 515, 522
Asynchronous, 378–379
Atmosphere, 124
Attenuation, 35, 416, 530
    balance return loss, 171
    coefficient, 82, 84, 99–100, 110
    composite, 35–36
    crosstalk, 93, 99, 104
    image, 36, 82
    intrinsic (optical fiber), 116, 122
    transhybrid, 171, 530
    wave, 133
Audio broadcasting, 426
Automatic request, 438
Availability, 64

Balance return loss, 171
Balanced pair, 98–100, 102, 104–105, 194, 414
Balancing impedance, 169
Bandwidth, 46–47, 140–141, 466–467
    optical, 509–510
Baseband, 147, 444
Basic group, 408–409
Beacon signals, 482
Bell, A. G., 9, 549
Bessel function, 304–305, 309
Bit, 13
Bit error rate (BER), 401
Bit interleaved, 368
Bit rate, 14
Boltzmann's constant, 52
Breakdown, 62
    software failure, 68
Broadcasting network, 137–138
Broadcasting system, 426
Burst mode, 543
Busy hour, 58

Cables, 77
    balanced pair, 98–100
    coaxial pair, 109
Capacitive, 98
Capacity (of a channel), 139, 144
Carrier, 154, 281, 406–407
    carrier spacing, 406–407
Carrier system, 405–406, 410, 414
Carson's rule, 308–309, 314
Channel, 14, 144–146, 160
    common channel signaling, 370, 536
    real, 15–16
Character, 11, 13, 436
Characteristic impedance, 82, 84, 86, 110

Chrominance, 23, 433–434
Circuit, 533–534
  dedicated, 443
  hypothetical reference circuit, 421, 469
  hypothetical reference circuit, digital, 394
  switched, 443
Circuit switching, 533–534
Cladding (optical fiber), 115
Clipping, 235, 239, 419–420
  probability of clipping, 419–420
Clock, 174
Clocked
  mode, 175
  regenerator, 179–180
  transmission, 174, 441
Coaxial pair, 109–114, 400, 414–415
Code (*see also* Coding), 255–256, 437
  AMI, 175, 386–387, 444
  ASCII, 438
  CCITT, 435, 438
  CCITT No. 5, 438
  CMI, 390, 504
  diphase (Manchester), 445–446
  folded, 256
  4B3T, 389
  Gray's, 256
  high density (HDB), 388–389, 444
  MCMI, 505
  Morse, 437
  polar, 175, 200
  simple binary, 255
  telex No. 2, 435, 438
  telex No. 3, 438
Code immunity, 436
Codec, 373, 375–376, 403
Coded mark inversion (CMI), 390
Coding, 14, 253–255
  channel, 16
  nonuniform, 257
  source, 16
  transition, 451
Communication, 14
Companding, 226, 246, 257, 275
  adaptive, 275

instantaneous, 229–230, 247
syllabic, 230–231
Companding improvement, 257–258
Compandor, 226
Compatibility, 33, 76
  mono-stereo, 429–430
  black and white/color, 433
Complementary error function, 559
Conception (of a system), 27
Conciseness (of a code), 436
Conformal (transmission), 45–46
Congestion, 58, 516
Core (optical fiber), 114–116
Cost, 71, 525–527
Couplings, 92–93, 96
Crosstalk, 51, 299, 395, 422
  balanced pair, 102–104
  coaxial pair, 113
  far-end crosstalk (FEXT), 93–98
  intelligible, 290
  line, 92–93
  near-end crosstalk (NEXT), 93–98
  time, 347, 349–350
Cross modulation, 290

Data, 24, 435, 443, 537
Data transmission, 443
  in analog telephone channel, 445–446
  in baseband, 444
  in digital telephone channel, 445
Decibel, 35
Decision content, 13
Decision rate, 14, 144
Decoder, 15
Delay, 44–45
  envelope delay, 288
  group delay, 289
Delay system, 58
Delta modulation (ΔM), 267–268, 270–271
  adaptive, 275–276
Delta sigma modulation ($\Delta\Sigma M$), 273–275
Demand assignment, 525
Demodulation,
  by envelope detection, 285–286, 314
  coherent, 293, 315, 356

FM, 314
  incoherent, 356
  isochronous, 293, 356, 358
  SSB, 296–297
  VSB, 301–302
Design, 29–30
Deviation (variation), 288, 302–304
  amplitude, 288
  frequency, 302–303
  instantaneous, 302–303
  phase, 304
De-emphasis, 417
de Forest, L., 9
Diagram
  eye pattern, 187, 202, 392, 508–509
  radiation (phasor), 306, 308
  reliability, 65
Diffraction, 455–456
Digital
  information, 11
  numerical aperture, 117–118, 497
  signal, 15
  system, 365–367
Diode
  laser, 497
  light emitting, 497
Diphase Code, 445–446
Discrete (source), 14
Dispersion, 119–120, 507–508
  chromatic, 119
  guide, 119
  modal, 119–120, 507–508
Distortion, 46
  AM, 289–290
  attenuation, 46, 48, 122, 158, 238, 445
  factor, 210, 243
  FM, 317–318
  linear, 46–47
  nonlinear, 48–49
  phase, 46, 112, 114, 445
  quantizing, 236–237
  SSB, 298
Diversity, 129–130, 457, 471
Documentation, 32
Double sideband suppressed carrier (DSBSC), 290–294

Drop and insert, 410
Duplex
  half-duplex, 165
  full-duplex, 165
Duration
  rms duration, 121
Dynamic compression, 226–227
Dynamic range (of a signal), 226

Earth station, 483
Echo, 527–529
  cancellation, 542–543
  suppressor, 529
Economics
  aspects, 71
  choices, 525–527
Efficiency (of a mode), 386–387
Elementary sound, 18
Encoder, 254–255
Energy budget (optical), 506–507
Entropy, 12–13
Envelope, 286, 312, 450
  detection, 285–286
  format, 450
Envelope delay, 288
Envelope detection, 285–286
Equalization, 184–185, 188, 192, 211, 471
Erlang, 57
Error
  effect on PCM, 262
  function, 559
  probability, 190–192, 194, 196–200,
    394–395, 443–444, 459
  rate, 401
  regeneration, 173, 189
  residual, 436
  transmission, 177
Euler's equivalence, 553
Exchange, 516, 518–521
  local, 518–519, 521–522
  main (central), 516, 520
  nodal, 518
  private automatic branch (PABX), 521
  trunk, 516
Extrapolation, 272, 550
Eye pattern, 186–189

Facsimile, 4, 22, 435, 543, 548
Fading, 135, 457–459, 462–463
Failure, 63
Filters, 156–158, 384, 398
Four-wire, 166
  frequency division, 166–167, 543
  time division, 167, 543
  network, 527
Frame, 162, 367, 370–372, 378, 381–383
  stuffing frame, 381–383
Frame alignment, 368–372
Framing, 368–372, 402
  framing bit, 450
Frequency
  division multiplex (FDM), 161–162
  modulation (FM(, 302–303
  shift keying (FSK), 352, 354–355, 357
Frequency ranges, 126
Fresnel zones (ellipsoids), 456–457
Front end processor, 538
Full duplex, 165

Gain, 35–36, 417, 530
  composite gain, 35–36
  of an antenna, 130
Gaussian,
  complementary integral function, 555–556
  Gaussian distribution, 192, 554
  Gaussian integral function, 555–556
Graded index fiber, 115, 119
Granular noise, 268
Gray, E., 9
  Gray's code, 255
Group
  basic group, 409
  mastergroup, 409
  supergroup, 409
  supermaster group, 409
Group delay, 83, 289
Group velocity (*see* Phase velocity)
Growth of the telephone, 9

Half duplex, 165
Hardware, 29
Harmonics, 48–50
Hearing, 20

Heterodyne receiver, 427
Hierarchy
  of analog systems, 408–411
  of digital systems, 376
High density bipolar code (HDB), 388–389, 444
Histogram (*see* Level diagram)
History of telecommunication, 8–10, 473
Hop, 460
Hybrid, 169–171
Hypothetical reference circuit
  analog, 421, 469
  digital, 394

Idle noise, 268
Image
  color, 23–24
  fixed (still pictures), 22
  frequency, 427
  mobile (motion pictures), 22–23
  parameters, 82
Impulse noise, 195
Impulse response, 89
Inavailability, 64, 66
Information, 11, 551
  analog, 11
  content, 11–12
  digital, 11
  quality, 551
Information broadcasting network, 2, 551
Instability, 530
Instantaneous angular frequency, 281
Integrated digital network (IDN), 539–540
Integrated services digital network (ISDN), 543, 546–547
Interface, 447
Interference, 51
  AM, 323, 326–327
  digital transmission, distortion of, 261
  discrete analog modulation, 352
  FM, 324, 328–329
  ΦM, 325, 329–330
  PPM, 350–351
  sinusoidal signal, 322, 352
  SSB, 324, 327–328

INDEX 581

Intermediate frequency (IF), 427
Intermodulation
 noise, 51, 212, 423, 425
 product, 49–50, 212
International Telecommunication Union (ITU), 33
Intersymbol interference, 177, 180–181
Investments, 71, 75–76
 per subscriber, 75–76
Ionosphere, 124–126
Isochronous (coherent)
 demodulation, 293, 315, 356
 signals, 41
 transmission, 174
Isotropic radiator, 130

Jitter, 205
 phase jitter, 205, 451
 waiting time jitter, 381

Krarup cable (continuous loading), 105

Laser diode, 497
Law
 A-law (nonuniform quantization), 250–252, 257–259
 logistic, 60
 μ-law (nonuniform quantization), 252–253
 normal (Gaussian) distribution, 554
 Poisson's, 194, 500
Leakage conductance, 79, 85
Leased circuit (see Circuit, dedicated)
Level, 37–38
 absolute, 38
 in a multichannel system, 417–418
 maximumm (clipping), 419–420
 meter, 39
 noise level set referred to zero, 55–56
 nominal, 55, 419–420
 real level, 55, 418
 relative, 39
 system level referred to zero, 55
 thermal noise, 52
Level diagram (histogram), 40–41
Light emitting diode, 497
Line, 77

balanced pair, 98, 100–102, 104, 108–109, 394, 415
 coaxial pair, 109–114, 399–401, 415–416
 equipment, 366, 414
 loaded, 105–106
 open wire, 98–99
 party, 522
Loading, 104–108
Local network, 519, 521–522
Logistic law, 60
Loss, 42
 overall, 42
Luminance, 23, 431, 433–434

Maintenance, 70, 514
Man, 4–5, 549
Manchester code, 445–446
Marconi, G., 9
Mastergroup, 409
Merit, figure of, 468, 487
Mesh network, 517
Message switching, 534
Microwave link, 128, 453
Minimum shift keying (MSK), 359–361
Mode (see also Code), 174–175, 195
 transmission, 175, 385–386, 503
Modified coded mark inversion (MCMI), 505
Modulation, 148–150
 adaptive, 275–276
 amplitude modulation (AM), 282, 337
 analog, 149, 281
 anglar modulation, 283, 463
 cross modulation, 290
 delta modulation (ΔM), 267, 270
 delta-sigma ($\Delta\Sigma M$), 273–275
 differential phase shift keying, 358–359
 differential pulse code modulation (DPCM), 235, 272–273
 digital, 149, 233, 235
 digital differential, 235, 265–267
 discrete analog, 352, 446
 double, 410
 double sideband suppressed carrier (DSBSC), 290–291, 337

frequency modulation (FM), 302–303, 336
frequency shift keying (FSK), 352
intensity modulation (optical) (IM), 501–502
inverse, 292
minimum shift keying (MSK), 359, 361
mixed, 163
negative (TV), 431
on-off keying (OOK), 352–355, 500–501, 503
optical, 501–502
phase (ΦM), 319–320, 334
phase shift keying (PSK), 353, 460
pulse, 154, 337
pulse code modulation (PCM), 235, 248–249
single sideband (SSB), 294–300, 327, 333, 405
stepwise modulatiion, 410
vestigial sideband (VSB), 300–302
Modulation factor, 284–285
Modulation index, 303–304, 320
Monitoring, 70, 402
Monomode fiber, 115, 119
Monopoly (telecommunications), 6
Morse, S. F. B., 8, 437
Morse code, 437
Multiframe, 371
Multimode fiber (single-mode), 115
Multiplexing, 148, 160–161, 365, 522
dynamic multiplexing, 519, 535
frequency division multiplex (FDM), 161, 405
time division multiplex (TDM), 162, 365
wavelength division multiplex (WDM), 513
Multiple access, 487
frequency division (FDMA), 487–490
time division (TDMA), 491–492
Multiplicative noise, 457–458
Music, 21

Neper, 35, 82
Network, 2, 515

broadcast, 137, 544
integrated digital network (IDN), 539–540
integrated services digital net work (ISDN), 10, 543, 546–457
international, 523
linear, 516
local, 519, 521, 542, 549
local broadband, 515, 549
mesh, 516–518, 524
national, 521–522
private data, 537
public data, 537–538
ring, 516–517, 519, 525
rural, 518, 523
shared, 3, 516, 519, 522
star, 516, 518, 525
switched, 2, 516
telephone, 519, 544
telex, 544
urban, 522
Noise
additive, 458
background, 51, 212, 216, 224, 423, 425
figure, 213–214
Gaussian, 52
granular, 268
idle, 268
impulse, 195
intermodulation, 50, 212, 423, 425–426
internal, 212–213
quantizing, 236–237, 240–241
quantum, 500
thermal, 52, 212, 499
white, 52, 194, 241
Noise-in-slot, 425
Noise-to-signal ratio, 54–55, 218
Noise distribution, 422
Noise budget, 216, 469, 478
line, 216
microwave link, 469
satellite, 478
Noise factor, 212–214, 422
Noise temperature, 213–214, 477
Nominal level, 55, 419–420
Nominal signal, 54, 219

INDEX

Numerical aperture, 117–118, 497
Nyquist
   expanded criterion, 183–184, 353
   first Nyquist criterion, 181
   Nyquist's theorem, 141–143, 510
   second Nyquist criterion, 181–183

Octet, 367, 449
On-off keying (OOK), 352–353
Open-wire line, 98–99
Operation
   of analog systems, 425
   of digital systems, 401
Optical
   power, 497
   power efficiency (yield), 497
   transmission system, 512–513
Optical fiber, 114–115
   optical fiber links, 495
Optical spacing, 510–511
Orbit (satellite), 473–475
Overall loss, 42–522
Overlay (network), 541

Packet
   assembly and disassembly, 537–538
   switching, 534–539
Pair
   balanced, 98–100, 102, 104–105, 194, 414
   coaxial, 109–114, 400, 414–415
Pair gain system, 522
Parallel system (structure, reliability), 65–67
PCM word, 367–368
Phantom circuit, 107–108
Phase
   delay, 83
   locked loop (PLl), 294, 315
   modulation (PM), 319–321
   shift keying (PSK), 353, 356
Phase change
   coefficient, 82, 85
   composite, 36
Phase velocity, 83
Phasor, 282
Phoneme, 18
Pilot, 412–413

Planck's constant, 499
Plan
   modulation plan, 414
   transmission plan, 44, 531–532
Planning, 27, 195
   of an analog system, 421
   of a digital system, 392
   of an optical link, 505
   of a satellite link, 477–478
Plesiochronous, 377, 391
Poisson distribution, 194–195, 500–501
Polar
   antipolar, 175, 200, 444
   unipolar, 175, 503
Present value calculation, 74
Pre-emphasis, 417, 429, 464–465
Primary line parameters, 78
Primary signal, 148–149, 281
Primary system, 370–374
Principal exchange, 516, 520
Principal group
   15-supergroup assembly, 409
Private automatic branch exchange (PABX), 521
Projection, 550
Propagation, 455
Proximity effect, 81
Pseudo-ternary code (AMI), 175, 386–389, 444
Psophometer, 52
Psophometric weighting, 52–53
Pulse amplitude modulation (PAM), 337, 343
Pulse code modulation (PCM), 235, 248–249
Pulse duration modulation (PDM), 344
Pulse frequency modulation (PFM), 345–346
Pulse position modulation (PPM), 345

Quad, 100–101
   DM-quad (twin quad) 100
   star-quad, 100
Quadrature amplitude modulation (QAM), 448, 460
Quality
   figure of merit (microwave link), 468

figure of merit (ground station), 487
information, 551
switching, 28, 57
service, 27
transmission, 27, 42
Quantizing, 16–17, 233, 235
distortion, 237, 239, 241
interval, 235
noise, 236–237, 240–241
nonuniform, 245–247
step, 236
uniform, 239
Quantum noise, 500

Radio broadcasting, 125, 286, 429
Rate
bit rate, 14
decision rate, 14, 144
information rate, 14, 139
symbol rate, 17–18, 139, 144
Rayleigh diffusion, 116–117
Real time, 139, 145
Redundancy (information), 13, 386, 436
Redundancy (reliability), 62, 66
Reeves, A., 9
Reference equivalent, 42–43
Refraction
atmospheric 455
refractive index, 114
Regeneration, 173, 189, 191
cumulative error, 203, 393
error, 173, 178, 189, 393
Regenerator, 178–179, 203, 205, 390–391, 400, 403, 506–507, 510–511
spacing, 398, 400, 506–507, 510–511
Reis, P., 9
Relay, 178
Reliability, 28, 60–61
satellite reliability, 482–483
Remote control, 4, 435, 481–482
Repeater, 220, 406, 415–416, 481
Repeater spacing, 423–425
Response
frequency, 88, 122–123
impulse, 89, 120–121
step, 89–92, 140–141

Ring network, 516, 519
Rise time, 141

Sampling, 150, 233, 237
ideal, 150
natural, 346–347
real, 150
sample-and-hold, 158
sampling theorem, 143, 154–155
uniform, 346
Satellite, 480
broadcast, 493
domestic, 493
geostationary, 475
link, 473, 524–525
Saturation, 216
Scrambling, 504
Secondary parameters, 82–83, 87–88
Secondary signal, 148, 282
Self-interference, 51, 236, 298
Sensitivity, 500
Series system (structure, reliability), 64–66
Service, 2–4, 435
integration, 543
Shannon, 12
Shared network, 3, 516, 519
Sheath, 100
Sideband, 284
Sidetone, 172
Signaling, 369–370, 413, 536
bit stealing, 369
common channel, 370, 536
in-band, 369, 413
out-of-band, 369, 413
Signal addition, 41
Signal, 15
analog, 15
digital signal, 15
multiplex, 417–419
nominal, 54, 219
Signal-to-crosstalk ratio, 94
Signal-to-disturbance ratio, 394–397
Signal-to-noise ratio, 54–55, 219–221, 263–264, 421, 463, 479

Signal-to-quantizing noise ratio, 241–244, 251, 259, 262
Simplex, 166
Singing point, 530
Single-mode fiber (*see* Monomode fibre)
Single sideband modulation (SSB), 294–300
Skin depth (depth of penetration), 79
Skin effect, 79–80, 91
Slips, 378
Slope overload, 269–272
Slow-scan TV, 548
Social effects, 6–7, 551
Software, 29
Software failures, 68, 70
Spacing
    carrier spacing, 406–407
    frequency spacing between broadcast channels, 429–430
    frequency spacing between TV channels, 432–433
    optical regenerator spacing, 510–511
    regenerator spacing, 398, 400, 506–507
    repeater spacing, 423–425
Specifications, 29–30
Spectrum, 553
Speech, 18–20
Stability, 527, 530
    condition, 530
Staircase signal, 160, 237
Standards, 32–33
Star network, 516, 518
Start-stop, 441
Step index fiber, 115
Step response, 89–92
Stereophonic broadcasting, 429–431
Store-and-forward, 534
Strowger, A., 9
Stuffing, 379, 451
Submarine cable systems, 414, 512
Subscriber, 519
    subscriber (customer) cable, 102–103, 520
Superheterodyne reception, 427–428
Supermaster group, 409

Surface
    effective (area), 131
Switched circuit, 443
Switching, 1, 532–536
    circuit, 533–534, 536
    message, 534
    packet, 535–536
Symbol, 17, 173
Symbol rate, 17–18
Synchronization
    stuffing, 379
System, definition, 7

Telecommunications, definition, 1–2
Telecomputing, 435
Telecontrol, 435
Telecopying, 4, 435
Teledistribution (CATV), 138, 545
Telegraphy, 4, 21, 435
    data under voice (infra-acoustic), 442
    voice-frequency, 442
Telematics, 4, 435, 543
Telemetry, 4, 435, 481–482, 543
Telephone band, 47–48
Telephone user density, 59
Telephonometry, 44
Teleprinter, 435
Teletex, 4, 21, 435, 543
Television, 4, 22–23, 300, 431–434, 15.8
    black-and-white TV, 22–23, 433–434
    cable TV, 138, 545
    color TV, 23–24, 433–434
Telex, 4, 21, 435, 441, 543
Temperature
    effect of temperature on cables, 81, 416
    noise temperature, 213, 477
Terminal
    data terminal, 446
    subscriber (customer) terminal, 520
Terminal equipment (analog), 406, 411
Terminal equipment (digital), 366, 372
Terminal exchange, 518
Text, 21
Threshold effect, 330, 468
Threshold FM, 330

Time
- differed time, 145–146
- group delay, 83, 289
- phase delay, 83
- real time, 139, 145
- rise time, 141

Time assignment speech interpolation (TASI), 21, 415
Time division multiplex (TDM), 162
Traffic concentration, 523
Traffic, 56–58
- intensity, 57
- offered, 58

Transducers, 15, 42, 44, 119, 168, 495, 504
Transfer function, 88, 122
Transhybrid attenuation, 171, 530
- reflection factor, 171

Translation, 410
- channel translation, 410
- frequency translation, 297, 427

Transmission, 1
- conformal, 45
- equipment, 366, 406, 520
- plan, 44, 531–532

Transmultiplexer, 383
Transponder, 481
Transposition, 99, 481

Tributary, 377
Troposcatter link, 129
Troposphere, 124
Trunk group, 57, 516
Twisted (stranded), 99
Two-wire, 166

User guidance, 549

Van Allen belt, 13.1
Vestigial sideband modulation (VSB), 300–302
Video signal, 431–432
Videotex, 4, 21, 539–540, 545
Visiophony, 4, 543, 548
Voice, 18
Voice-frequency telegraphy, 442
Voiced sounds, 18

Wavelength division multiplex (WDM), 513
Wire broadcasting, 4
Woodward's theorem, 312
Word interleaved, 368
Worldwide reference equivalent, 43–44

Yield factor, 132, 459

z-channels, 522–523